Transcriptomics in Health and Disease

Geraldo A. Passos
Editor

Transcriptomics in Health and Disease

 Springer

Editor
Geraldo A. Passos
Department of Genetics
University of São Paulo
Ribeirão Preto
São Paulo
Brazil

Videos to this book can be accessed at http://www.springerimages.com/
videos/978-3-319-11984-7

ISBN 978-3-319-11984-7   ISBN 978-3-319-11985-4 (eBook)
DOI 10.1007/978-3-319-11985-4

Library of Congress Control Number: 2014957482

Springer Cham Heidelberg New York Dordrecht London

Printed on acid-free paper

Springer is part of Springer Science+Business Media (www.springer.com)

This book is dedicated to the Department of Genetics, Ribeirão Preto Medical School, University of São Paulo, Brazil, which in 2015 celebrates its 50 years of research, teaching and clinical activities.

Preface

The completion of the human genome, with its more than 3 billion base pairs (bp) of sequenced DNA, has provided an unprecedented wealth of knowledge. With the additional investigation of single nucleotide polymorphisms (SNPs), we have also learned how little genetic variability there truly is in the human genome. Moreover, genome-wide association studies (GWAS) have revealed important genotype-phenotype correlations. Nevertheless, our understanding of the functionality of the genome is still lacking.

Dispersed among its 3 billion bp, the human genome features approximately 20–25,000 functional genes that encode various proteins and their isoforms. In recent years, however, scientists realized that the functionality of the genome is not restricted to only protein-encoding genes, which are transcribed into messenger RNAs, but also to the transcription of non-coding RNAs [e.g., microRNAs (miRNAs)], which play important roles in the posttranscriptional control of gene expression and, consequently, influence the resulting phenotypes.

Broadly speaking, it is at this point—from studies investigating where the functions of the genome first begin—that the science of transcriptomics emerged. For example, how are RNA molecules transcribed, what are the different species of RNA, what are the functions of each of these species and how are they differentially expressed among cells, tissues and organs?

Transcriptomics can therefore be thought of as the molecular biology of gene expression on a large scale. It is derived from functional genomics studies with a focus on transcription. Since its inception, transcriptomics has benefitted from and will continue to benefit from microarray technology. Sequencing is undoubtedly the ultimate tool when the objective is to delve into the differences at the sequence level or to confirm the specific RNA isoform involved. Even more so now, with the emergence of new technologies for high-throughput RNA sequencing (RNA-Seq), we can answer more questions about the structure of RNAs, such as those found in alternative splicing. However, the bottleneck remains in the data analysis because sequences are currently being obtained in quantities that have never been previously achieved.

However, as microarray bioinformatics has reached a very advanced stage (with more than 15 years to perfect the analysis pipeline) and as microarray slides themselves have become increasingly "large", currently encompassing sequences from the entire functional genome plus the complete set of known non-coding RNAs, researchers have not neglected the applications of this important technology.

Recent comparative analyses have indicated a strong concordance between exon microarrays and RNA-Seq data. Therefore, the goal is now to use these two complementary strategies for in-depth transcriptomics studies.

This book was organized on the basis on these assumptions. It includes 17 chapters and covers the fundamental concepts of transcriptomics, as well as the current analytical methods. We provide examples in high-level technical and scientific detail, using accessible language whenever possible, as each chapter is written by experienced and productive researchers in the field.

Over the first six chapters (Part I), we introduce the concept of the transcriptome, as well as how microarrays or RNA-Seq can be used to trace expression signatures, measure transcriptional expression levels and establish connections between genes based on their transcriptional activity in normal cells, differentiating cells and organs.

Chapters 7–17 (Part II) then provide examples of the state of the transcriptome associated with major human diseases, such as inflammatory diseases, autoimmune diseases, metabolic diseases (such as type 2 diabetes mellitus), genetic diseases (such as Down syndrome), cancer and infections caused by pathogenic microorganisms, such as tuberculosis mycobacteria, fungi and the protozoan *Trypanosoma cruzi*, which is the causative agent of Chagas disease.

Special attention is also given to Chap. 17, which was strategically placed at the end of this book. The author of this chapter, who was one of the original developers of microarray technology in the mid-1990s, discusses the medical potential of transcriptomics from an analytical point of view.

I hope this book will be useful to researchers who wish to gain a comprehensive view of transcriptomics in health and human disease. I would like to thank all of the authors for their dedication and time spent writing these chapters. Finally I thank Springer for providing this opportunity and for its continued support during the writing and organization of this work.

Internet Access to Video Clip

The owner of this text will be able to access these video clips through Springer with the following Internet link: http://www.springerimages.com/videos/ 978-3-319-11984-7.

Ribeirão Preto, Brazil   Geraldo A. Passos

Contents

Part I   Basic Principles of the Transcriptome and Its Analysis

1   **What Is the Transcriptome and How it is Evaluated?**............................ 3
Amanda F. Assis, Ernna H. Oliveira, Paula B. Donate,
Silvana Giuliatti, Catherine Nguyen and Geraldo A. Passos

2   **Transcriptome Analysis Throughout RNA-seq** 49
Tainá Raiol, Daniel Paiva Agustinho, Kelly Cristina Rodrigues Simi,
Calliandra Maria de Souza Silva, Maria Emilia Walter,
Ildinete Silva-Pereira and Marcelo Macedo Brígido

3   **Identification of Biomarkers and Expression Signatures** 69
Patricia Severino, Elisa Napolitano Ferreira and Dirce Maria Carraro

4   **Methods for Gene Coexpression Network Visualization**
and Analysis.. 79
Carlos Alberto Moreira-Filho, Silvia Yumi Bando, Fernanda Bernardi
Bertonha, Filipi Nascimento Silva and Luciano da Fontoura Costa

5   **Posttranscriptional Control During Stem Cells Differentiation**............ 95
Bruno Dallagiovanna, Fabiola Holetz and Patricia Shigunov

6   **Transcriptome Analysis During Normal Human**
Mesenchymal Stem Cell Differentiation 109
Karina F. Bombonato-Prado, Adalberto L. Rosa, Paulo T. Oliveira,
Janaína A. Dernowsek, Vanessa Fontana, Adriane F. Evangelista
and Geraldo A. Passos

Part II Transcriptome in Disease

7   **Thymus Gene Coexpression Networks: A Comparative
    Study in Children with and Without Down Syndrome** 123
    Carlos Alberto Moreira-Filho, Silvia Yumi Bando, Fernanda Bernardi
    Bertonha, Filipi Nascimento Silva, Luciano da Fontoura Costa
    and Magda Carneiro-Sampaio

8   **Transcriptome Profiling in Autoimmune Diseases**................................. 137
    Cristhianna V. A. Collares and Eduardo A. Donadi

9   **Expression of DNA Repair and Response to Oxidative Stress
    Genes in Diabetes Mellitus**... 161
    Paula Takahashi, Danilo J. Xavier and Elza T. Sakamoto-Hojo

10 **MicroRNAs in Cancer**.. 181
    Adriane F. Evangelista and Marcia M. C. Marques

11 **Transcriptome Profiling in Chronic Inflammatory Diseases
    of the Musculoskeletal System** 195
    Renê Donizeti Ribeiro de Oliveira and Paulo Louzada-Júnior

12 **Transcriptome Profiling in Experimental Inflammatory Arthritis**....... 211
    Olga Martinez Ibañez, José Ricardo Jensen and Marcelo De Franco

13 **Transcriptome in Human Mycoses**.. 227
    Nalu T. A. Peres, Gabriela F. Persinoti, Elza A. S. Lang,
    Antonio Rossi and Nilce M. Martinez-Rossi

14 **Transcriptomics of the Host–Pathogen Interaction in
    Paracoccidioidomycosis**.. 265
    Patrícia Albuquerque, Hugo Costa Paes, Aldo Henrique Tavares,
    Larissa Fernandes, Anamélia Lorenzetti Bocca, Ildinete Silva-Pereira,
    Maria Sueli Soares Felipe and André Moraes Nicola

15 **Dissecting Tuberculosis Through Transcriptomic Studies**.................... 289
    Rodrigo Ferracine Rodrigues, Rogério Silva Rosada, Thiago Malardo,
    Wendy Martin Rios and Celio Lopes Silva

**16 Understanding Chagas Disease by Genome
    and Transcriptome Exploration** 311
    Ludmila Rodrigues P. Ferreira and Edecio Cunha-Neto

**17 Expression Tests in Actual Clinical Practice: How Medically
    Useful is the Transcriptome?** 327
    Bertrand R. Jordan

Concluding Remarks and Perspectives... 341

Index.. 343

Contributors

Daniel Paiva Agustinho Laboratório de Biologia Molecular, CEL/IB (Pós-graduação em Patologia Molecular/FM/UnB), Universidade de Brasília, Brasília, DF, Brazil

Patrícia Albuquerque Universidade de Brasília, Faculdade de Ceilândia, Brasília, DF, Brazil

Amanda F. Assis Molecular Immunogenetics Group, Department of Genetics, Ribeirão Preto Medical School, University of São Paulo, Ribeirão Preto, Brazil

Silvia Yumi Bando Departamento de Pediatria, Faculdade de Medicina da Universidade de São Paulo, São Paulo, São Paulo, Brazil

Fernanda Bernardi Bertonha Departamento de Pediatria, Faculdade de Medicina da Universidade de São Paulo, São Paulo, São Paulo, Brazil

Karina F. Bombonato-Prado Department of Morphology, Physiology and Basic Pathology, School of Dentistry of Ribeirão Preto, University of São Paulo, Ribeirão Preto, São Paulo, Brazil

Marcelo Macedo Brígido Laboratório de Biologia Molecular, CEL/IB (Pós-graduação em Biologia Molecular/CEL/IB), Universidade de Brasília, Brasília, DF, Brazil

Magda Carneiro-Sampaio Departamento de Pediatria, Faculdade de Medicina da Universidade de São Paulo, São Paulo, São Paulo, Brazil

Dirce Maria Carraro A. C. Camargo Cancer Center, São Paulo, SP, Brazil

Cristhianna V. A. Collares Department of Medicine, Division of Clinical Immunology, Ribeirão Preto Medical School, University of São Paulo, Ribeirão Preto, SP, Brazil

Luciano da Fontoura Costa Instituto de Física de São Carlos, Universidade de São Paulo, São Carlos, São Paulo, Brazil

Hugo Costa Paes Universidade de Brasília, Campus Universitário Darcy Ribeiro, Brasília, DF, Brazil

Edecio Cunha-Neto Laboratory of Immunology, Heart Institute (Incor), São Paulo, Brazil

Bruno Dallagiovanna Instituto Carlos Chagas, FIOCRUZ Paraná, Curitiba, Paraná, Brazil

Marcelo De Franco Laboratory of Immunogenetics, Butantan Institute, São Paulo, São Paulo, Brazil

Renê Donizeti Ribeiro de Oliveira Department of Clinical Medicine, Ribeirão Preto Medical School, University of São Paulo, Ribeirão Preto, São Paulo, Brazil

Janaína A. Dernowsek Department of Genetics, Ribeirão Preto Medical School, University of São Paulo, Ribeirão Preto, São Paulo, Brazil

Eduardo A. Donadi Department of Medicine, Division of Clinical Immunology, Ribeirão Preto Medical School, University of São Paulo, Ribeirão Preto, SP, Brazil

Paula B. Donate Inflammation and Pain Group, Department of Pharmacology, Ribeirão Preto Medical School, University of São Paulo, Ribeirão Preto, Brazil

Adriane F. Evangelista Department of Genetics, Ribeirão Preto Medical School, University of São Paulo, Ribeirão Preto, São Paulo, Brazil

Molecular Oncology Research Center, Barretos Cancer Hospital, Barretos, São Paulo, Brazil

Larissa Fernandes Universidade de Brasília, Faculdade de Ceilândia, Brasília, DF, Brazil

Elisa Napolitano Ferreira A. C. Camargo Cancer Center, São Paulo, SP, Brazil

Ludmila Rodrigues P. Ferreira Laboratory of Immunology, Heart Institute (Incor), São Paulo, Brazil

Vanessa Fontana Department of Genetics, Ribeirão Preto Medical School, University of São Paulo, Ribeirão Preto, São Paulo, Brazil

Silvana Giuliatti Bioinformatics Group, Department of Genetics, Ribeirão Preto Medical School, University of São Paulo, Ribeirão Preto, Brazil

Fabiola Holetz Instituto Carlos Chagas, FIOCRUZ Paraná, Curitiba, Paraná, Brazil

Bertrand R. Jordan Marseille Medical School, Aix-Marseille Université/EFS/CNRS, Marseille, France

Elza A. S. Lang Department of Genetics, Ribeirão Preto Medical School, University of São Paulo, Ribeirão Preto, São Paulo, Brazil

Anamélia Lorenzetti Bocca Universidade de Brasília, Campus Universitário Darcy Ribeiro, Brasília, DF, Brazil

Paulo Louzada-Júnior Department of Clinical Medicine, Ribeirão Preto Medical School, University of São Paulo, Ribeirão Preto, São Paulo, Brazil

Thiago Malardo Center for Tuberculosis Research, Ribeirão Preto Medical School, University of São Paulo, Ribeirão Preto, São Paulo, Brazil

Calliandra Maria de Souza Silva Laboratório de Biologia Molecular, CEL/IB (Pós-graduação em Biologia Molecular/CEL/IB), Universidade de Brasília, Brasília, DF, Brazil

Marcia M. C. Marques Molecular Oncology Research Center, Barretos Cancer Hospital, Barretos, São Paulo, Brazil

Barretos School of Health Sciences—FACISB, Barretos, São Paulo, Brazil

Olga Martinez Ibañez Laboratory of Immunogenetics, Butantan Institute, São Paulo, São Paulo, Brazil

Nilce M. Martinez-Rossi Department of Genetics, Ribeirão Preto Medical School, University of São Paulo, Ribeirão Preto, São Paulo, Brazil

André Moraes Nicola Programa de Pós-Graduação em Ciências Genômicas e Biotecnologia, Universidade Católica de Brasília, Brasília, DF, Brazil

Carlos Alberto Moreira-Filho Departamento de Pediatria, Faculdade de Medicina da Universidade de São Paulo, São Paulo, São Paulo, Brazil

Catherine Nguyen Laboratory TAGC INSERM U1090, National Institute for Health and Medical Research, Scientific Park of Luminy, Marseille, France

Ernna H. Oliveira Molecular Immunogenetics Group, Department of Genetics, Ribeirão Preto Medical School, University of São Paulo, Ribeirão Preto, Brazil

Paulo T. Oliveira Department of Morphology, Physiology and Basic Pathology, School of Dentistry of Ribeirão Preto, University of São Paulo, Ribeirão Preto, São Paulo, Brazil

Geraldo A. Passos Molecular Immunogenetics Group, Department of Genetics, Ribeirão Preto Medical School, University of São Paulo, Ribeirão Preto, Brazil

Department of Morphology, Physiology and Basic Pathology, School of Dentistry of Ribeirão Preto, University of São Paulo, Ribeirão Preto, São Paulo, Brazil

Nalu T. A. Peres Department of Genetics, Ribeirão Preto Medical School, University of São Paulo, Ribeirão Preto, São Paulo, Brazil

Gabriela F. Persinoti Department of Genetics, Ribeirão Preto Medical School, University of São Paulo, Ribeirão Preto, São Paulo, Brazil

Tainá Raiol Laboratório de Biologia Molecular, CEL/IB (Pós-graduação em Biologia Molecular/CEL/IB), Universidade de Brasília, Brasília, DF, Brazil

José Ricardo Jensen Laboratory of Immunogenetics, Butantan Institute, São Paulo, São Paulo, Brazil

Wendy Martin Rios Center for Tuberculosis Research, Ribeirão Preto Medical School, University of São Paulo, Ribeirão Preto, São Paulo, Brazil

Rodrigo Ferracine Rodrigues Center for Tuberculosis Research, Ribeirão Preto Medical School, University of São Paulo, Ribeirão Preto, São Paulo, Brazil

Adalberto L. Rosa Department of Oral and Maxillofacial Surgery and Periodontology, School of Dentistry of Ribeirão Preto, University of São Paulo, Ribeirão Preto, São Paulo, Brazil

Rogério Silva Rosada Center for Tuberculosis Research, Ribeirão Preto Medical School, University of São Paulo, Ribeirão Preto, São Paulo, Brazil

Antonio Rossi Department of Genetics, Ribeirão Preto Medical School, University of São Paulo, Ribeirão Preto, São Paulo, Brazil

Elza T. Sakamoto-Hojo Department of Biology, Faculty of Philosophy, Sciences and Letters of Ribeirão Preto, University of São Paulo, Ribeirão Preto, São Paulo, Brazil

Patricia Severino Instituto Israelita de Ensino e Pesquisa Albert Einstein, São Paulo, São Paulo, Brazil

Patricia Shigunov Instituto Carlos Chagas, FIOCRUZ Paraná, Curitiba, Paraná, Brazil

Celio Lopes Silva Center for Tuberculosis Research, Ribeirão Preto Medical School, University of São Paulo, Ribeirão Preto, São Paulo, Brazil

Filipi Nascimento Silva Instituto de Física de São Carlos, Universidade de São Paulo, São Carlos, São Paulo, Brazil

Ildinete Silva-Pereira Laboratório de Biologia Molecular, CEL/IB (Pós-graduação em Biologia Molecular/CEL/IB), Universidade de Brasília, Brasília, DF, Brazil

Universidade de Brasília, Campus Universitário Darcy Ribeiro, Brasília, DF, Brazil

Kelly Cristina Rodrigues Simi Laboratório de Biologia Molecular, CEL/IB (Pós-graduação em Biologia Molecular/CEL/IB), Universidade de Brasília, Brasília, DF, Brazil

Maria Sueli Soares Felipe Universidade de Brasília, Campus Universitário Darcy Ribeiro, Brasília, DF, Brazil

Paula Takahashi Department of Genetics, Ribeirão Preto Medical School, University of São Paulo, Ribeirão Preto, São Paulo, Brazil

Aldo Henrique Tavares Universidade de Brasília, Faculdade de Ceilândia, Brasília, DF, Brazil

Maria Emilia Walter Departamento de Ciência da Computação, Instituto de Ciências Exatas, Universidade de Brasília, Brasília, DF, Brazil

Danilo J. Xavier Department of Genetics, Ribeirão Preto Medical School, University of São Paulo, Ribeirão Preto, São Paulo, Brazil

Part I
Basic Principles of the Transcriptome and Its Analysis

Chapter 1
What Is the Transcriptome and How it is Evaluated?

Amanda F. Assis, Ernna H. Oliveira, Paula B. Donate, Silvana Giuliatti, Catherine Nguyen and Geraldo A. Passos

Abstract The concept of the transcriptome revolves around the complete set of transcripts present in a given cell type, tissue or organ and encompasses both coding and non-coding RNA molecules, although we often assume that it consists only of messenger RNAs (mRNAs) because of their importance in encoding proteins. Unlike the nuclear genome, whose composition and size are essentially static, the transcriptome often changes. The transcriptome is influenced by the phase of the cell cycle, the organ, exposure to drugs or physical agents, aging, diseases and a multitude of other variables, all of which must be considered at the time of its determination. However, it is precisely this property that makes the transcriptome useful for the discovery of gene function and as a molecular signature. In this chapter, we review the beginnings of transcriptome research, the main types of RNA molecules found in a mammalian cell, the methods of analysis, and the bioinformatics pipelines used to organize and interpret the large quantities of data generated by the two current gold-standard methods of analysis: microarrays and high-throughput RNA sequencing (RNA-Seq). Attention is also given to non-coding RNAs, using microRNAs (miRNAs) as an example because they physically interact with mRNAs and play a role in the fine control of gene expression.

G. A. Passos (✉) · A. F. Assis · E. H. Oliveira
Molecular Immunogenetics Group, Department of Genetics, Ribeirão Preto Medical School, University of São Paulo, 14049-900, Ribeirão Preto, Brazil
e-mail: passos@usp.br

P. B. Donate
Inflammation and Pain Group, Department of Pharmacology, Ribeirão Preto Medical School, University of São Paulo, 14049-900, Ribeirão Preto, Brazil

S. Giuliatti
Bioinformatics Group, Department of Genetics, Ribeirão Preto Medical School, University of São Paulo, 14049-900, Ribeirão Preto, Brazil

C. Nguyen
Laboratory TAGC INSERM U1090, National Institute for Health and Medical Research, Scientific Park of Luminy, 13100, Marseille, France

© Springer International Publishing Switzerland 2014
G. A. Passos (ed.), *Transcriptomics in Health and Disease,*
DOI 10.1007/978-3-319-11985-4_1

1.1 What is the Transcriptome, How it is Evaluated and What Types of RNA Molecules Exist?

Strictly speaking, the *transcriptome* can be conceptualized as the total set of RNA species, including coding and non-coding RNAs (ncRNAs), that are transcribed in a given cell type, tissue or organ at any given time under normal physiological or pathological conditions. This term was coined by Charles Auffray in 1996 to refer to the entire set of transcripts. Soon after, this concept was applied to the study of large-scale gene expression in the yeast *S. cerevisiae* (Velculescu et al. 1997; Dujon 1998; Pietu et al. 1999).

However, due to the importance of messenger RNAs (mRNAs), which represent protein-coding RNAs, the term transcriptome is often associated with this set of RNA and as an analogy species. Researchers later coined the analogous term *miR-Nome* to refer to the total set of miRNAs.

The *proteome* is conceptually similar to the transcriptome and refers the total set of proteins translated in a given cell type, tissue or organ at any given time during normal physiological or pathological conditions. Nevertheless, despite its importance, the proteome will not be discussed in this book, and we suggest the following reviews for further reading: Anderson 2014; Forler et al. 2014; Padron and Dormont 2014; Altelaar et al. 2013; and Ahrens et al. 2010.

Analyses of the transcriptome began well before its conceptualization. Large-scale analyses of gene expression in the murine thymus gland (Nguyen et al. 1995), the human brain and liver (Zhao et al. 1995) and human T cells (Schena et al. 1996) have been performed since the mid-1990s. These independent groups used cDNA clones arrayed on nylon membranes or glass slides to hybridize labeled tissue- or cell-derived samples. These arrayed cDNA clones represented the prototypes of the modern microarrays currently used in transcriptome research (Jordan 2012).

1.1.1 How the Transcriptome is Evaluated: The Birth of Transcriptome Methods

Although the first method used to analyze transcriptional gene expression emerged in 1980 with the development of Northern blot hybridization (Wreschner and Hersberg 1977), this method was not and still is not capable of being performed on a large scale, and thus cannot be considered a transcriptome approach. In 1990s, the human genome project, through partially automated DNA sequencing, had the ambition to identify, characterize and analyze all of the genes in the human genome (Watson 1990; Cantor 1990). This revolutionary approach led to thousands of entries that were constructed via the tag-sequencing of randomly selected cDNA clones (Adams et al. 1991, 1992, 1993a, b; Okubo et al. 1992; Takeda et al. 1993), thus opening an avenue for high-throughput approaches by making these data widely available in repositories such as the dbEST database (http://www.ncbi.nlm.nih.gov/dbEST). As more and more genes are identified, efforts are now being redirected towards understanding the precise temporal and cellular control of gene expression. The advances provided by the current progress in high-throughput technologies have enabled the simultaneous analysis of the activity of many genes in cells and tissues, essentially depicting a molecular portrait of the tested sample. The transcriptome approach, based on the large-scale measurement of mRNA, became the method of choice among the emerging technologies of so-called "functional genomics", primarily because this method was rapidly identified as one that can be performed at a reasonably large scale using highly parallel hybridization methods, and it has allowed a more holistic view of what is really happening in the cell (Sudo et al. 1994; Granjeaud et al. 1996, 1999; Botwell 1999; Jordan 1998).

As mentioned above, the first transcriptome analysis was performed on large nylon arrays using high-density filters containing colony cDNA (or PCR products) followed by quantitative measurements of the amount of hybridized probe at each spot. A common platform used spotted cDNA arrays, where cDNA clones representing genes were robotically spotted on the support surface either as bacterial colonies or as PCR products. These "macroarrays", or high-density filters, were made on nylon membranes measuring approximately 10 cm^2. Although this is now considered a dated approach, it was nonetheless effective enough to test sets of hundreds or even a few thousand genes.

DNA arrays allow the quantitative and simultaneous measurement of the mRNA expression levels of thousands of genes in a tissue or cell sample. The technology is based on the hybridization of a complex and heterogeneous RNA population derived from tissues or cells. Initially, this was referred as a "complex probe", i.e., a complex mix that contains varying amounts of many different cDNA sequences, corresponding to the number of copies of the original mRNA species extracted from the sample. This complex probe was produced via the simultaneous reverse transcription and ^{33}P labeling of mRNAs, which were then hybridized to large sets of DNA fragments, representing the target genes, arrayed on a solid support. Thus, each individual experiment provided a very large amount of information (Gress et al. 1992, Nguyen et al. 1995; Jordan 1998; Velculescu et al. 1995; Zhao et al. 1995; Bernard et al. 1996, Pietu et al. 1996, Rocha et al. 1997).

1.1.2   Miniaturization, an Obvious Technological Evolution Towards Microarrays

One of the major challenges that researchers faced was to obtain the highest possible sensitivity when working with a limited amount of sample (biopsies, sorted cells, etc.). In this regard, five parameters were taken into account: 1) the amount of DNA fixed on the array support; 2) the concentration of RNA that should be labeled with the ^{33}P isotope; 3) the specific activity of the labeling; 4) the duration of the hybridization; and 5) the duration of exposure of the array to the phosphor imager shields.

The miniaturization of this method lay in the intrinsic physical characteristics of nylon membranes, which allowed a significant increase in the amount of immobilized DNA. The feasibility of miniaturizing nylon was demonstrated in the Konan Peck (Academia Sinica, Taiwan) laboratory in 1998 using a colorimetric method as the detection system (Chen et al. 1998). A combination of nylon microarrays and ^{33}P-labeled radioactive probes was subsequently shown to provide similar levels of sensitivity compared with the other systems available at the time, making it possible to perform expression profiling experiments using submicrogram amounts of unamplified total RNA extracted from small biological samples (Bertucci et al. 1999).

These observations had important implications for basic and clinical research in that they provided a cheaper alternative approach that was particularly suitable for groups operating in academic environments and led to a large numbers of expression profiling analyses when only small amounts of biological material were available.

Microarrays based on solid supports, typically coated glass, were simultaneously developed in different academic and industrial laboratories. These arrays boasted the advantage of performing dual hybridization of a test sample and a reference sample, as they could be labeled with two different fluorescent compounds, namely the fluorochrome "Cy-dyes" cyanine-3 (Cy3) and cyanine-5 (Cy5) (Chee et al. 1996).

Around the same time, another well known DNA array platform was developed by Affymetrix (Santa Clara, CA, USA). Their array used oligonucleotide chips featuring hundreds of thousands of oligonucleotides that were directly synthesized *in situ* on silicon chips (each measuring a few cm^2) using photochemical reactions and a masking technology (Lockhart et al. 1996). This microarray platform promised a rapid evolution in miniaturization because it was based on the synthesis of short nucleic acid sequences, which could be updated on the basis of the current knowledge of the genome.

It quickly became clear in the academic community, as well as in industry, that the available microarray technologies represented the beginning of a revolution with considerable potential for applications in the various fields of biology and health because gene function is one of the key elements that researchers want to extract from a DNA sequence. Microarrays have become a very useful tool for this type of research (Gershon 2002). Therefore, the development of the microarray opened the door to various DNA chip technologies based on the same basic concept. For example, the maskless photolithography used to produce oligonucleotide arrays was originally developed in 1999 using the light-directed synthesis of high-resolution oligonucleotide microarrays with a digital micromirror array to form virtual masks (Singh-Gasson et al. 1999). However, this technology was barely accessible to academic laboratories at the time because of the high initial cost, the limited availability of equipment, non-reusability, and the need for a large amount of starting RNA (Bertucci et al. 1999).

This development formed the basis for the NimbleGen company, which in 2002 demonstrated the chemical synthesis quality of maskless arrays synthesis (MAS) and its utility in constructing arrays for gene expression analysis (Nuwaysir et al.

2002). Currently, NimbleGen is focused on products for sequencing (http://www. nimblegen.com/).

Similarly, in 2005, Edwin Southern's team developed a method for the *in situ* synthesis of oligonucleotide probes on polydimethylsiloxane (PDMS) microchannels through the use of conventional phosphoramidite chemistry (Moorcroft et al. 2005). This became the basis of the Oxford Gene Technology company (http:// www.ogt.co.uk/), which today develops array products centered on cytogenetics, molecular disorders and cancer.

It is also widely known that Affymetrix (http://www.affymetrix.com/estore/) and Agilent (http://www.home.agilent.com/agilent/home.jspx?lc=eng&cc=US) developed the most popular microarray technology for expression profiling based on ink jet technology, which is still widely available in the transcriptome market.

1.1.3 Reliable Microarray Results Depend on a Series of Complex Steps

The reliability of transcriptome results has concerned scientists since the beginning of transcriptome research, resulting in a number of studies comparing the different platforms, which was a real challenge in the early 2000s. Transcriptomic results largely depend on the technology used, which itself is dependent on several complex steps, ranging from the fabrication of the microarray to the experimental conditions, in addition to the chosen detection system, which also determines the method of analysis.

The results obtained with one microarray platform cannot necessarily be reproduced on another, and differences in the presence of different target sequences representing the same gene on different arrays can make it extremely difficult to integrate, combine and analyze the data (Järvinen et al. 2004).

The fabrication of high-quality microarrays has been a challenging task, taking a decade to reach several stabilized solutions, and has become an industry of its own. There are a large number of parameters and factors that affect the fabrication of a microarray, as performance depends on the array geometry, chemistry, and spot density, as well as on characteristics such as morphology, probe and hybridized density, background and sensitivity (Dufva 2005). Among the different methods used to fabricate DNA microarrays, *in situ* synthesis is the most powerful because a very high spot density can be achieved and because the probe sequence can be chosen for each synthesis.

To achieve a 10^5-fold dynamic range, which is an important parameter for gene expression analysis, the spots must contain at least 10^5 molecules, and the optimal spot size should be large enough to acquire the maximum hybridized density to obtain good sensitivity. Bead arrays that have different combinations of fluorescent dyes, which essentially constitute a barcode tag associated with the different immobilized probes, appeared to be the next evolution because they are in suspension and are therefore suitable for automation using standard equipment, leading to extremely high-throughput approaches. Optical microarrays that are detected via flow cytometry can use a large number of different beads because each bead can be decoded using a series of hybridization reactions following the immobilization of the beads to the optical fibers (Ferguson et al. 2000; Epstein et al. 2003). This increases the multiplex capacity to several thousands of different beads (Gunderson et al. 2004). Optical fiber microarrays have been commercialized by Illumina (http://www.illumina.com/), currently the leader in high-throughput sequencing technology, which allow the measurement of expression profiles by counting the amount of each RNA molecule expressed in a cell.

Experimental conditions also vary from lab to lab, as the preparation is dependent on the array platform. Variations in the quality of RNA preparations can be evaluated using the 2100 Bioanalyzer instrument developed by Agilent, which has become a standard, even if some slight variations have been observed from time to time. This system provides sizing, quantitation and quality control for RNA and DNA, as well as for proteins and cells, on a single platform, providing high-quality digital data (http://www.genomics.agilent.com/en/Bioanalyzer-System/2100-Bio-analyzer-Instruments/?cid=AG-PT-106) (Fig. 1.1).

The preparation of RNA prior to hybridization can affect microarray performance, particularly in terms of data accuracy, by distorting the quantitative measurement of transcript abundance. To obtain enough material from an initial nano- or picogram range of starting material, the RNA is transcribed *in vitro* and amplified using different protocols, which can introduce bias. In 2001, several publications discussed the different commercial protocols that were available. A publication from Charles Decreane's team examined the methods for amplifying picogram amounts of total RNA for whole genome profiling. The authors set up a specific experiment to compare three commercial RNA amplification protocols, Ambion messageAmp™, Arcturus RiboAmp™ and Epicentre Target Amp™, to the standard target labeling procedure proposed by Affymetrix, and all of the samples were tested on Affymetrix GeneChip microarrays (Clément-Ziza et al. 2009). The results obtained in this study indicated large variations between the different protocols, suggesting that the same amplification protocol should always be used to maximize the comparability of the results. Additionally, it was found that the RNA amplification affects the expression measurements as well, which was in agreement with earlier observations seen at the nanogram scale, as well as with other studies that were concerned with this question (Nygaard and Hovig 2006; Singh et al. 2005; Wang et al. 2003; Van Haaften et al. 2006; Degrelle et al. 2008).

In 2012, questions surrounding RNA amplification were still relevant. Indeed, even if the amplification of a small amount of RNA is reported to have a high reproducibility, there is still bias, and this can become time consuming. Even taking into account a correlation coefficient of 0.9 between microarray assays using non-amplified and qRT-PCR samples, the matter should still be reconsidered. In one study, the authors used the 3D-Gene™ microarray platform and compared samples prepared using either a conventional amplification method or a non-amplification protocol and a probe set selected from the MicroArray Quality Control (MAQC) project (http://www.fda.gov/ScienceResearch/BioinformaticsTools/

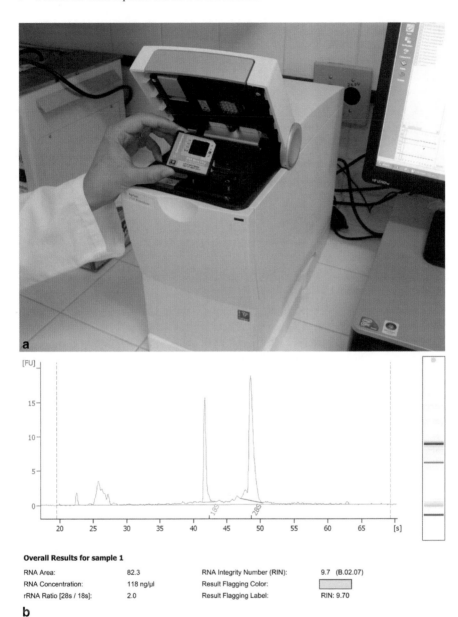

Overall Results for sample 1

RNA Area:	82.3	RNA Integrity Number (RIN):	9.7 (B.02.07)
RNA Concentration:	118 ng/µl	Result Flagging Color:	
rRNA Ratio [28s / 18s]:	2.0	Result Flagging Label:	RIN: 9.70

b

Fig. 1.1 Agilent Bioanalyzer model 2100 showing in. **a** A RNA Nano Chip and in. **b** A typical result of a microfluidic electrophoresis of a total human RNA sample extracted from leukocytes. On the *right* side of this figure appears a virtual gel with the respective bands of 28S and 18S rRNAs and 5S rRNA plus 4S tRNAs (from *top to bottom*). On the *left* side is shown the densitometry of this gel were appears the respective peaks of 28S rRNA, 18S rRNAs, 5S rRNA and 4S tRNAs. The rRNA ratio (28S/18S)=2.0 enabled a RNA integrity number (RIN=9.7), which indicated that this sample was intact (not degraded)

MicroarrayQualityControlProject/). They found that the samples from the non-amplification procedure had a higher quantitative accuracy than those from the amplification method but that the two methods exhibited comparable detection power and reproducibility (Sudo et al. 2012).

However in the above study, the researchers also used a few micrograms of RNA and a large volume of hybridization buffer. It is known that the ability to reduce the quantity of input RNA while maintaining the reaction concentration can be achieved in a device that decreases the hybridization reaction volume. Devices developed for use with beads have this characteristic; therefore, would hybridization using a bead device resolve this issue?

1.1.4   Bioinformatics and Standardization Approaches: A Possible Solution?

With regard to bioinformatics and standardization approaches, the MAQC project was initiated in 2006 to address these questions, as well as other performance and data analysis issues. The Microarray Quality Control (MAQC Consortium 2006) (http://www.fda.gov/ScienceResearch/BioinformaticsTools/MicroarrayQuality-ControlProject/) study tested a large number of laboratories, platforms and samples and found that there were notable differences in various dimensions of performance between microarray platforms. Each microarray platform has different trade-offs with respect to consistency, sensitivity, specificity and ratio compression. One interesting result was that platforms with divergent approaches for measuring expression often generated comparable results. The authors of this study concluded that the technical performance of microarrays supports their continued use for gene expression profiling in basic and applied research and may lead to the use of microarrays as a clinical diagnostic tool as well. This project has provided the microarray community with standards for data reporting, common analysis tools and useful controls that can help promote confidence in the consistency and reliability of these gene expression platforms (MAQC Consortium 2006). Similarly, in 2007, another meta-analysis of microarray results suggested several recommendations for standardization under the Standard Microarray Results Template (SMART) to facilitate the integration of microarray studies and proposed the implementation of the Minimum Information About a Microarray Experiment (MIAME) (http://www.mged.org/Workgroups/MIAME/miame.html) to facilitate the comparison of results (Cahan et al. 2007).

Given that measurement precision is critical in clinical applications, the question of the measurement precision in microarray experiments was addressed again in 2009 through an inter-laboratory protocol. In this study, the authors analyzed the results of three 2004 Expression Analysis Pilot Proficiency Test Collaborative studies using different methods. The study involved thirteen participants out of sixteen, each of whom provided triplicate microarray measurements for each of two reference RNA pools. To facilitate communication between the user and developer, this study sought to set up standardized conceptual tools, but the result of this analysis was relatively disappointing and did not allow the creation of a gold standard, though it did put forth several recommendations (Duewer et al. 2009).

All of these studies focus on the same concept that has been defended since 2001 by the Microarray Gene Expression Data Society (http://www.mged.org) – the reanalysis and reproduction of results by the scientific community. The MGED society was the first to define the MIAME, which describes the minimum information required to ensure that microarray data can be easily interpreted and that the results derived from their analysis can be independently verified. This protocol became the standard for recording and reporting microarray-based gene expression data and for inserting it in databases and public repositories (Brazma et al. 2001, Ball et al. 2002). Currently, raw and/or normalized microarray data are deposited either in the ArrayExpress databank (https://www.ebi.ac.uk/arrayexpress/) or in the Gene Expression Omnibus (GEO) (http://www.ncbi.nlm.nih.gov/geo/), providing the scientific community with data for further analysis.

1.1.5 Analysis of the Expression Data

The past two decades have seen the development of methods that allow for a nearly complete analysis of the transcriptome, in the form of microarrays and, more recently, RNA-Seq, which are the most popular technologies used in genome-scale transcriptional studies. These high-throughput gene expression analysis systems generate large and complex datasets, and the development of computational methods to obtain biological information from the generated data has been the primary challenge in bioinformatics analysis.

Even a simple microarray experiment generates a large amount of data, which places certain demands on the analysis software. Fortunately, microarrays have benefited from the availability of many commercial and open-source software packages for data manipulation that have been developed over the years. RNA-Seq, however, demands more bioinformatics expertise. There are publicly available online tools such as the Galaxy platform (Goecks et al 2010, but a basic knowledge of UNIX shell programming and Perl/Python scripting is necessary for data modification. Furthermore, similar to microarray analysis, a familiarity with the R programming environment is useful, as the software programs for many of the downstream analyses are collected in the Bioconductor (http://www.bioconductor.org/) (Gentleman et al 2004) suite of the R package. Other important considerations regarding the choice for RNA-Seq include the need for data storage resources and computing systems with large memories and/or many cores to run parallel, sophisticated algorithms efficiently and faster.

In this section, we present the main steps for analyzing multi-dimensional genomic data derived from the application of microarray or RNA-Seq assays based on a common pipeline illustrated in Fig. 1.2.

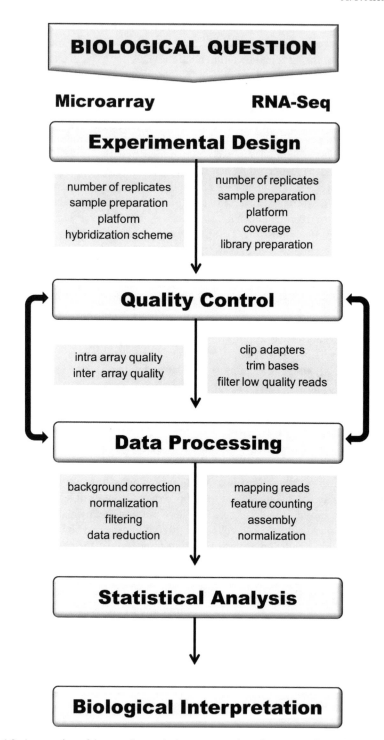

Fig. 1.2 An overview of the steps in a typical gene expression microarray or RNA-Seq experiment

1.1.5.1   Experimental Design

The aim of the experimental design is to make the experiment maximally informative given a certain amount of samples and resources and to ensure that the questions of interest can be answered. All of the decisions made at this initial step will affect the results of all the subsequent steps. The consequences of an incorrect or poor design range from a loss of statistical power and an increased number of false negatives to the inability to answer the primary scientific question (Stekel 2003).

The basic principles of experimental design rely on three fundamental aspects formalized by Fisher (1935), namely, replication, randomization and blocking.

Randomization dictates that the experimental subjects should be randomly assigned to the treatments or conditions to be studied to eliminate unknown factors that may potentially affect the results (Fang and Cui 2011).

Replication is essential for estimating and decreasing the experimental error and, thus, to detect the biological effect more precisely. A true replicate is an independent repetition of the same experimental process and an independent acquisition of the observations. There are different levels of replication in gene expression experiments: (1) a technical replicate provides measurement-level error estimates and (2) a biological replicate provides estimates of the population-level variability. If the goal is to evaluate the technology, technical replicates alone are sufficient. Otherwise, if the goal is to investigate the biological differences between tissues/conditions/treatments, biological replicates are essential (Alison et al 2006; Fang and Cui 2011). Replication is widely used in microarray experiments, though technical replicates are generally no longer performed, as analyses have shown that the results will be relatively consistent overall (Slonin and Yanai 2009). However, in RNA-Seq studies, replication is still neglected primarily due to the current high costs of these experiments. Studies conducted on the variability of this technology, both technical (Marioni et al. 2008) and biological (Bullard et al. 2010), underscore the importance of including replicates in the study design. The fundamental problem with generalizing the results gathered from unreplicated data is a complete lack of knowledge about the biological variation. Without an estimate of variability (i.e., within the treatment group), there is no basis for inference (i.e., between the treatment groups) (Auer and Doerge 2010).

As with microarray studies, RNA-Seq experiments can be affected by the variability coming from nuisance factors, often called technical effects, such as the processing date, technician, reagent batch and the hybridization/library preparation effect. In addition to these effects, in RNA-Seq experiments, there are also other technology-specific effects. For example, there is variation from one flow cell to another, resulting in a flow cell effect and variation between the individual lanes within a flow cell due to systematic variation in the sequencing cycling and/or base calling. A blocking design dictates comparisons within a block, which is a known uninteresting factor that causes variation, such as the hybridization scheme (microarray) or flow cell effect (RNA-Seq) (Fig. 1.3) (Alison et al. 2006, Slonin and Yanai 2009, Auer and Doerge 2010, Fang and Cui 2011, Luo et al 2010).

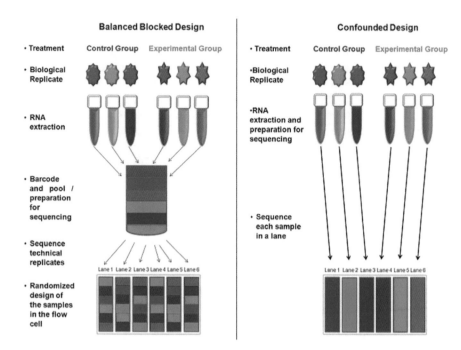

Fig. 1.3 Comparison of two methods for testing differential expression between treatments. **a** (*red*) and **b** (*blue*). In the ideal balanced block design (*left*), six samples are barcoded, pooled, and processed together. The pool is then divided into six equal portions that are input into six flow cell lanes. The confounded design (*right*) represents a typical RNA-Seq experiment and consists of the same six samples, with no barcoding, and does not permit batch and lane effects to be distinguished from the estimate of the intra-group biological variability (adapted from Auer and Doerge 2010)

In the case of microarray and RNA-Seq experiments, design issues are intrinsically dependent on hybridization and library construction, respectively. It is beyond the scope of this section to discuss and compare the different technologies available, but we recommend reading the following articles for microarray technologies: Paterson et al. (2006), Alison et al. (2006), Stekel (2003), Churchill (2002), Kerr and Churchill (2001), Jordan (2012). For RNA-Seq technologies, please see Auer and Doerge (2010) and Fang and Cui (2010), as well as chapter 2 of this book.

1.1.5.2 Quality Control

To assure the reproducibility, comparability and biological relevance of the gene expression data generated by high-throughput technologies, several research groups have provided guidelines regarding quality control (QC):

- **Minimum Information About a Microarray Experiment (MIAME):** describes the minimum information required to ensure that microarray data can be easily interpreted and that the results derived from their analysis can be independently verified (Brazma et al. 2001).

- **External RNA Control Consortium (ERCC)**: develops external RNA controls useful for evaluating the technical performance of gene expression assays performed by microarray and qRT-PCR (Baker et al. 2005).
- **MicroArray Quality Control (MAQC) Consortium**: a community-wide effort, spearheaded by the Food and Drug Administration (FDA), that seeks to experimentally address the key issues surrounding the reliability of DNA microarray data. Now in its third phase (MAQC-III), also known as Sequencing Quality Control (SEQC), the MAQC project aims to assess the technical performance of next-generation sequencing platforms by generating benchmark datasets using reference samples and evaluating the advantages and limitations of various bioinformatics strategies in RNA and DNA sequencing (Shi et al. 2006, Shi et al. 2010, (www.fda.gov/MicroArrayQC).
- **Standards, Guidelines and Best Practices for RNA-Seq**: a guideline for conducting and reporting on functional genomics experiments performed with RNA-Seq. It focuses on the best practices for creating reference-quality transcriptome measurements (The ENCODE Consortium 2011) (http://www.genome.gov/encode).

However, there are several sources of variability originating from biological and technical causes that can affect the quality of the resulting data, including biological heterogeneity in the population, sample collection, RNA quantity and quality, technical variation during sample processing, and batch effects, among others. Some of these issues can be avoided with an appropriate and carefully designed experiment that controls for the different sources of variation, but others require a quality assessment of the raw data through computational support tools. Therefore, regardless of the technology used to measure gene expression, ensuring quality control is a critical starting point for any subsequent analysis of the data (Churchill 2002, Geschwind and Gregg 2002, Cobb et al. 2005, Larkin et al. 2005, Irizarry et al. 2005, Heber and Sick 2006).

With regard to microarray technology, many tools applying diagnostic plots have been developed to visualize the spread of data and compare and contrast the probe intensity levels between the arrays of the dataset. These qualitative visualization plots include histograms, density plots, boxplots, scatter plots, MAplots, score plots of the PCA, hierarchical clustering dendrograms, and even chip pseudo plots and RNA degradation plots (Fig. 1.4). Comparing the probe intensity between samples allows us to observe if one or more of the arrays have intensity levels that are drastically different from the other arrays, which may indicate a problem with the arrays. For a better review of the use of diagnostic plots in quality control metrics, please see Gentleman et al. (2005) and Heber and Sick (2006).

In regard to RNA-Seq, several sequence artifacts are quite common, including read errors (base calling errors and small *indels*), poor quality reads and adaptor contamination. Such artifacts need to be removed before performing downstream analyses, otherwise they may lead to erroneous conclusions. Performing a quality

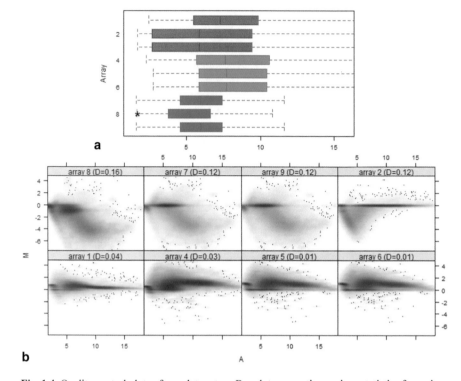

Fig. 1.4 Quality control plots of raw data sets. **a** Boxplots presenting various statistics for a given data set. The plots consist of boxes with a central line and two tails. The central line represents the median of the data, whereas the tails represent the upper (75th percentile) and lower (25th percentile) quartiles. These plots are often used to describe the range of log ratios that is associated with replicate spots. **b** *MA* plots are used to detect artifacts in the array that are intensity dependent

assessment of the reads allows us to determine the need for filtering (or cleaning) the data, removing low quality sequences, trimming bases, removing linkers, determining overrepresented sequences and identifying contamination or samples with a low sequence performance. The most important parameters used to verify the quality of the raw sequencing data are the base quality, the GC content distribution and the duplication rate (Guo et al. 2013, Patel and Jain 2012).

In addition to the QC pipelines provided commercially by the sequencing platform, there are online/standalone software packages and pipelines available as well (see: http://en.wikipedia.org/wiki/List_of_RNA-Seq_bioinformatics_tools). These packages present different features, and many are designed for a particular sequencing platform, such as NGS QC for the Illumina and Roche 454 platforms (Patel and Jain 2012) or Rolexa for Solexa sequencing data (Rougemont et al. 2009), or for a specific data storage format, such as FastQC toolkit and FastQScreen, which were both developed by the Brabaham Institute. The FastQC (http://www.bioinformatics.babraham.ac.uk/projects/fastqc) and FASTX-Tool kits (http://hannonlab.cshl.edu/fastx_toolkit/) include many of the tools used to remove indexes, barcodes and adapters and filter out the reads based on the quality metrics of the FASTQ files. For a comparison of some of the available QC tools for RNA-Seq, please refer to Patel and Jain (2012).

1.1.5.3 Data Processing

Once the quality of the data has been assessed and the applicable changes have been made, it is still necessary to perform additional processing before analyzing the differentially expressed genes. The primary objective in processing raw data is to remove unwanted sources of variation, thereby ensuring the accuracy of the final results. There are several different methods to process the data being assayed, and the specific method used depends on how the data were generated.

According to Geeleher et al. (2008), the data being assayed should be processed using several different methods, and the results should be compared to identify the most suitable method. The most appropriate method should then be used to process the raw data before the differential expression analysis.

Essentially, microarray processing involves three steps depending on the type of array: (1) background adjustment, which divides the measured hybridization intensities into a background and a signal component; (2) summarization, which combines the probe-level data into gene expression values, thereby reducing multiple probes representing a single transcript to a single measurement of expression; and (3) normalization, which aims to remove non-biological variations between arrays (Heber and Sick 2006). Other potential processing steps include transformation of the data from the raw intensities into log intensities and data filtering to remove flagged features, which are problematic features detected by the image-processing software (Stekel 2003, Allison et al. 2006).

Microarray data must also be background corrected to remove any signals arising from non-specific hybridization or spatial heterogeneity across the array. The background is a measure of the ambient signal obtained, generally, from the mean or median of the pixel intensity values surrounding each spot (Ritchie et al. 2007). The traditional correction is to subtract the local background measures from the foreground values, but the main problem with this procedure is that it can give negative corrected intensities, and there is high variability in the low-intensity log-ratios when the background is higher than the feature intensity (Stekel 2003). Instead, several different methods have been developed as alternatives. Some examples include the empirical Bayes model developed by Kooperberg et al. (2002), setting a small threshold value as suggested by Edwards (2003), the variance stabilization method (Vsn) of Huber et al. (2002), the *normexp* (normal-exponential convolution) method implemented by the RMA algorithm (Irizarry et al. 2003), and the MLE method (maximum likelihood estimation for *normexp*) (Silver et al. 2009). A detailed comparison of several of these methods can be found in the article by Ritchie et al. (2007).

The normalization of the microarray signal intensity has been widely used to adjust for experimental artifacts within the array and between all of the samples such that meaningful biological comparisons can be made (Quackenbush 2001, Lou et al. 2010). According to Stekel (2003), the methods for normalization may be broadly classified into two categories:

1. Within-array normalization (normalizes the M-values for each array separately) – these methods are applicable for two-channel arrays, in which the aim is to adjust the Cy3 and Cy5 intensities to equal levels. Methods such as the linear regression of Cy5 against Cy3 and linear or non-linear (Loess) regression of the log ratio against the average intensity can correct for the different responses of the Cy3 and Cy5 channels. However, these methods rely on the assumption that the majority of the genes on the microarray are not differentially expressed. If this assumption is not true, a different normalization method, such as using a reference sample, would be more appropriate.
2. Between-array normalization (normalizes the intensities or log-ratios to be comparable across multiple arrays) – this method is used for one- and two-channel arrays. Various methods have been proposed for this approach, such as scaling to the mean or median, centering and quantiles. Bolstad et al. (2003) presented a review of several methods and found quantile normalization to be the most reliable method.

After processing, it is strongly recommended to verify the performance of the chosen method. This can be achieved by applying the aforementioned diagnostic plots during a Quality Control session. Several studies have been published on the performance of the various processing methods (Bolstad et al. 2003, Ploner et al. 2005), but most studies have found the Robust Multichip Average method (RMA) (Irizarry et al. 2003) to be the best method. This method applies a model-based background adjustment followed by quantile normalization and a robust summary method (median polish) on the log2 intensities to obtain the probeset summary values.

The RNA-Seq data processing steps that were considered in our pipeline are as follows: (1) mapping reads; (2) transcriptome assembly; and (3) normalization of the read counts.

A common characteristic of all high-throughput sequencing technologies is the generation of relatively short reads, which should be mapped to a reference sequence, be it a reference genome or a transcriptome database. This is a critical task for most applications of the technology because the alignment algorithm must be able to efficiently find the right location for each read from among a potentially large quantity of reference data (Fonseca et al. 2012). The assembly of the transcriptome consists of the reconstruction of the full-length transcripts, except in the case of small classes of RNAs that are shorter than the sequencing length and require no assembly. The methods used to assemble reads fall into two main classes: (1) assembly based on a reference genome and (2) *de novo* assembly (Martin and Wang 2011). The strategies used to map the reads and assemble the transcriptome, along with the available tools, will be presented in more detail in chapter 2.

Normalization should always be applied to read counts due to two main sources of systematic variability: (1) RNA fragmentation during library construction causes the longer transcripts to generate more reads compared with the shorter transcripts that are present at the same abundance in the sample, and (2) the variability in the number of reads produced for each run causes fluctuations in the number of fragments mapped across the samples. Proper normalization enables accurate comparison of the expression levels between and within samples (Garber et al. 2011, Dillies et al. 2013). The RPKM (reads per kilobase of transcript per million mapped reads) is the most widely used normalization metric. It normalizes a transcript read count by both its length and the total number of mapped reads in the sample (Mortazavi et al. 2008). This approach facilitates comparisons between genes within a sample and combines the inter- and intra-sample normalization. When data originate from paired-end sequencing, the FPKM (fragments per kilobase of transcript per million mapped reads) metric is used (Garber et al. 2011, Dillies et al. 2013).

In previous years, other methods for the normalization of RNA-Seq data have been proposed as well. These methods also applied inter-sample normalization using scaling factors and include the following: (1) Total count (TC), in which the gene counts are divided by the total number of mapped reads (or library size) associated with their lane and multiplied by the mean total count across all of the samples in the dataset; (2) Upper Quartile, which has a very similar principle to TC and in which the total counts are replaced by the upper quartile of counts different from 0 in the computation of the normalization factors; (3) Median, which is similar to TC, in which the total counts are replaced by the median counts different from 0 in the computation of the normalization factors; (4) DESeq, which is the normalization method included in the DESeq Bioconductor package (version 1.6.0) (http://bioconductor.org/packages/release/bioc/html/DESeq.html) and is based on the hypothesis that most genes are not differentially expressed; (5) Trimmed Mean of M-values (TMM), which is the normalization method implemented in the edgeR Bioconductor package (version 2.4.0) (http://www.bioconductor.org/packages/release/bioc/html/edgeR.html) and is also based on the hypothesis that most genes are not differentially expressed; and (6) Quantile, which was first proposed in the context of microarray data and consists of matching the distributions of the gene counts across lanes. These proposed normalization methods, in addition to the RPKM method, were comprehensively compared and evaluated by members of The French StatOmique Consortium. Based on this comparative study, the authors proposed practical recommendations for the appropriate normalization method to be used and its impact on the differential analysis of RNA-Seq data (Dillies et al. 2013).

1.1.5.4   Statistical Analysis and Interpretation

The primary goal of gene expression studies is to identify genes that are differentially expressed between RNA samples from two types of biological conditions. Differential gene expression can provide insights into biological mechanisms or pathways and form the basis for further experiments by determining the sample and gene similarity via clustering analyses or testing a gene set for enrichment.

Differential expression analysis searches for genes whose abundance has changed significantly across the experimental conditions. In general, this means taking the quantified and normalized expression values for each library and performing statistical testing between samples of interest. In theory, the transcript abundance of the mRNA would be directly proportional to the number of reads, thereby determining the expression level (Oshlack et al. 2010).

Many methods have been developed for the analysis of differential expression using microarray data. In the early days of microarrays, only the simple fold-change method was used (Chen et al. 1997). However, the evolution of the technology called for more accurate analytical methods, and many more sophisticated statistical methods have been proposed.

In addition to the traditional t-test and ANOVA approaches used to access differential gene expression in microarray assays, variations on these tests have been developed for the purpose of overcoming the problem of a small sample size when accessing such a large dataset: dealing with many genes but only a few replicates may lead to large fold-changes driven by outliers, as well as to small error variances (Lönnstedt and Speed 2002). SAM (Significant Analysis of Microarrays) (Tusher et al. 2001) is a very popular differential expression method that uses a modified t-statistic to identify significant genes using non-parametric statistics.

Other statistical approaches for microarray data analysis have introduced linear models. The Bioconductor package Limma, developed by Smyth (2005), applies a gene-wise linear model that allows for the analysis of complex experiments (comparing many RNA samples), as well as more simple replicated experiments using only two RNA samples. Empirical Bayes and other shrinkage methods are used to borrow information across genes, making the analyses stable even for experiments with small numbers of arrays. Another powerful method to detect differentially expressed genes in microarray experiments is based on calculating the rank products (RP) from replicate experiments, while at the same time providing a straightforward and statistically stringent way to determine the significance level for each gene and allow flexible control of the false-detection rate and familywise error rate in the multiple testing situation of a microarray experiment (Breitling and Herzyk 2005).

Differential expression analysis methods that use probability distributions have also been proposed for use in modeling the count data from RNA-Seq studies, including Poisson and negative binomial (NB) distributions. The Poisson distribution forms the basis for modeling RNA-Seq counts. However, when there are biological replicates, the RNA-Seq data may exhibit more variability than expected by the Poisson distribution because it assumes that the variance is equal to the mean. If this occurs, the Poisson distribution will predict a smaller variation than that observed in the data, and the analysis will be prone to high false-positive rates that result from an underestimation of the sampling error (Anders and Huber 2010). Therefore, the NB model is the better method to address this so-called overdispersed problem because an NB distribution specifies that the variance is greater than the mean (Oshlack et al. 2010, Anders and Huber 2010, Garber et al. 2011).

Statistical analyses of RNA-Seq data will be discussed in more detail in chapter 2. There are also several reviews that discuss and compare the statistical methods used to compute differential expression. For further information, please refer to Seyednasrollah et al. (2013) and Soneson and Delorenzi (2013).

1.1.5.5 Classification and Enrichment Analysis

Classification can be performed either before or after the differential expression analysis. This process entails either placing the objects (in this case, the samples, genes or both) into pre-existing categories (known as a supervised classification) or developing a set of categories into which the objects can subsequently be placed (unsupervised classification) (Allison et al. 2006). Class discovery, or clustering analysis, is an unsupervised classification method that is widely used in the study of transcriptomic data because it allows us to identify co-regulated genes and/or samples with similar patterns of expression (biological classes). Various clustering techniques have been applied to identify patterns in gene-expression data. Most cluster analysis techniques are hierarchical: the resultant classification has an increasing number of nested classes, and the result resembles a phylogenetic classification. Non-hierarchical clustering techniques also exist, such as k-means clustering, which simply partition objects into different clusters without trying to specify the relationship between the individual elements (Quackenbush 2001). Eisen et al. (1998) is a classical reference for the use of hierarchical clustering with microarray data. In this study, the authors developed an integrated pair of open-source programs, Cluster and TreeView, for analyzing and visualizing clusters and heat maps (http://rana.lbl.gov/EisenSoftware.htm).

Biological insights into an experimental system can be gained by looking at the expression changes of sets of genes. Many tools focusing on gene set testing, network inference and knowledge databases have been designed for analyzing lists of differentially expressed genes from microarray datasets. Examples include Gene Set Enrichment Analysis (http://www.broadinstitute.org/gsea/index.jsp) (Subramanian et al. 2005) and DAVID (http://david.abcc.ncifcrf.gov/tools.jsp) (Dennis et al. 2003), which combine functional themes, such as those defined by the Gene Ontology consortium, (Ashburner et al. 2000), and metabolic and signaling pathways, such as KEGG pathways (http://www.genome.jp/kegg/pathway.html) (Kanehisa and Goto 2000) and Biocarta (http://www.biocarta.com/), with statistical enrichment analyses to determine whether specific pathways are overrepresented in a given list of differentially expressed genes. These approaches can also be applied to RNA-Seq, but the biases presented by this type of data should be taken into account (Oshlack et al. 2010). Therefore, specialized approaches (Bullard et al. 2010) and tools to perform enrichment analyses of RNA-Seq data are being developed, for example, GO-seq (http://www.bioconductor.org/packages/release/bioc/html/goseq.html) (Young et al. 2010), SeqGSA (http://www.bioconductor.org/packages/release/bioc/html/SeqGSEA.html) (Wang and Cairns 2013) and generally applicable gene set enrichment for pathway analysis (GAGE) (Luo et al. 2009).

1.2   The Diversity of the Transcriptome

Unlike the genome, which is essentially static in terms of its composition and size (barring the rare occurrence of somatic and germline mutations or the rearrangement of immunoglobulin and T cell receptor genes), the transcriptome (and similarly, the miRNome) is extremely variable and depends on the phase of the cell cycle, the organ, exposure to drugs or physical agents, aging, diseases such as cancer and autoimmune diseases and a multitude of other variables, which must be considered at the time that the transcriptome is determined. This variability arises from the fact that RNAs are differentially transcribed (or transcribed at different rates) depending on the cell type and status, though this excludes ribosomal RNAs, as they are considered housekeeping molecules.

For many years, the central dogma of molecular biology stated that RNAs molecules were intermediates between DNA and protein. This idea presupposed that the function of RNA was primarily linked to the translation of the genetic material into polypeptide chains (proteins). The genetic material was interpreted as being involved in the synthesis of these RNAs, which were termed mRNAs (Brenner et al. 1961; Jacob and Monod 1961).

During the human genome sequencing era of the 1980s and 1990s, independently led by Francis Collins and Craig Venter, the latter individual and his coworkers conceived of expressed sequence tags (ESTs), which focus on mRNAs because they encode proteins. Libraries of mRNA-derived cDNA clones were generated based on first-strand synthesis using oligonucleotide primers for that are anchored at the 3′ end of the transcript [the poly(A) tail of mRNA] (Starusberg and Riggins 2001) and then sequenced to create unique identifiers for each cDNA, with lengths ranging from 300 to 700 bp (Adams et al. 1992; Adams 2008).

ESTs were very useful for identifying new expressed genes in normal and diseased tissues (Strausberg and Riggins 2001), and transcriptome analysis at this time was largely, if not solely, based in this approach. The EST clones were distributed through the former IMAGE Consortium, whose sequences can now be retrieved via the National Center for Biotechnology Information (NCBI) dbEST Database (http:// www.ncbi.nlm.nih.gov/dbEST/). The current number of public entries for all uni- or multicellular eukaryotic organisms that have been sequenced stands at more than 74 million ESTs, including more than eight million human and nearly five million mouse ESTs.

However, as was to be expected, imaginative new strategies were emerging around the same time as well. The Serial Analysis of Gene Expression (SAGE) method (Velculescu et al. 1995), which produces short sequence tags (usually 14 nucleotides in length) positioned contiguous to defined restriction sites near the 3′ end of the cDNA strand (Strausberg and Riggins 2001), has also been widely used. At the time, the NCBI created the SAGEmap as a public repository for SAGE sequences. Currently, all of the SAGE libraries have been uploaded and accessioned through the Gene Expression Omnibus (GEO) (http://www.ncbi.nlm.nih.gov/geo/) repository.

Another novel strategy, which had yet to be tested at that time, was the generation of open reading frame (ORF) ESTs (ORESTES). This approach was jointly developed by researchers funded by the São Paulo Research Foundation (FAPESP) and by the Ludwig Institute for Cancer Research (FAPESP/LICR)-Human Cancer Genome Project (Camargo et al. 2001). Unlike ESTs, ORESTES sequences are spaced throughout the mRNA transcript, providing a scaffold to complete the full-length transcript sequences. The authors generated a substantial volume of tags (700,000 ORESTES), which at the time represented nearly 20 % of all human dbESTs (Strausberg and Riggins 2001).

The Transcript Finishing Initiative, another FAPESP/LICR project, was then undertaken for the purpose of identifying and characterizing novel human transcripts (Sogayar et al. 2004). This strategy was also novel and was based on selected EST clusters that were used for experimental validation. In this method, RT-PCR was used to fill in the gaps between paired EST clusters that were then mapped on the genome. The authors generated nearly 60,000 bp of transcribed sequences, organized into 432 exons, and ultimately defined the structure of 211 human mRNA transcripts.

However, the increasing use of modern transcriptome-wide profiling approaches, such as microarrays and whole-genome and transcriptome sequencing, allied to the precise isolation and characterization of different RNA species from eukaryotic (including mammalian) cells, led to an explosion of findings and revealed that although approximately 90 % of the mammalian genome is actively transcribed into RNA molecules, only a tiny fraction (—2 % of the total human genome) encodes mRNAs and, consequently, proteins (Maeda et al. 2006; Djebali et al. 2012).

In fact, the function of the genome can be seen from two different but complementary views. From a functional standpoint, only a fraction of the genome encodes RNA molecules (including coding and non-coding RNAs), and only a fraction of these are translated into proteins. In other words, when considering the genome in numerical terms, or rather the physical portion of DNA that is functional, we realize that only a small number of genes are transcribed specifically into mRNA molecules. However, a larger number of "variable" mRNA molecules are generated through alternative splicing, and these are translated into a greater number of proteins (including various isoforms). A large portion of the genome is then transcribed into non-coding RNAs, which play a role in the posttranscriptional control of mRNAs during their translation into proteins (Fig. 1.5).

Molecular mapping of the human genome has been largely resolved, revealing slightly more than three billion bp encompassing approximately 20–25,000 functional nuclear genes and mitochondrial DNA located in the cytoplasm. We suggest consulting the ENCODE Project (http://www.genome.gov/encode/) to follow ongoing progress in the identification of the functional elements in the human genome sequence. Nevertheless, the definition of the human transcriptome is still far from set, and it appears that most of the RNA molecules in eukaryotic cells are composed of ncRNAs that are involved in the fine control of gene expression.

Aside from knowing the exact number of mRNA molecules in a human cell, which is currently being investigated using new sequencing technologies (de Klerk

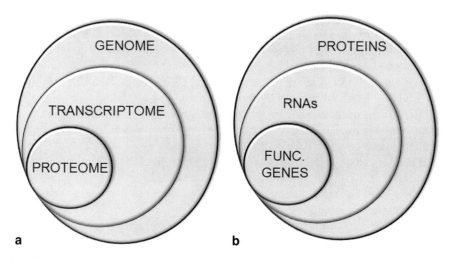

Fig. 1.5 Two ways to interpret the functioning the genome and the relative proportions of molecular entities. **a** In functional terms only a part of the genome encodes RNAs from which only a small fraction encodes proteins. **b** However, in numerical terms the set of functional genes transcribe a larger number of mRNAs from which a larger number of proteins is translated. The part A of this figure was conceived by Dr. Sven Diederichs (German Cancer Research Institute, DKFZ, Heidelberg, Germany) who allowed their use

et al. 2014; Kellis et al. 2014), one of the great challenges of the next decade will be to decipher the posttranscriptional interactions between coding and ncRNAs in the control of gene expression.

In fact, the human genome was revealed to be more than just a collection of protein-coding genes and their splice variants, rather, it displays extensive antisense, overlapping and ncRNA expression (Taft et al. 2010).

In mammals, the vast majority of the genome is transcribed into ncRNAs, which exceed the number of protein-coding genes (Liu and Taft 2013). These molecules are characterized by the absence of protein-coding capacity, but these RNAs have been described as key regulators of gene expression (Geisler and Coller 2013).

ncRNAs are grouped into two major classes based on their transcript size: small ncRNAs (19–30 nt) and long non-coding RNAs (200 nt to ~100 kilobases). These groups are distinct in their biological functions and mechanisms of gene regulation (Geisler and Coller 2013; Fatica and Bozzoni 2014; Neguembor et al. 2014).

Furthermore, ncRNAs can be grouped into a third class of housekeeping ncRNAs, which are normally constitutively expressed and include ribosomal (rRNAs), transfer (tRNAs), small nuclear (snRNAs), small nucleolar (snoRNAs) and regulatory noncoding RNAs (rnRNAs) (Ponting et al. 2009; Bratkovic and Rogelj 2014).

Small ncRNAs are primarily associated with the 5' or 3' regions of protein-coding genes, and based on their precursors and mechanism of action, they have been divided into three main classes: miRNAs, small interfering RNAs (siRNAs) and piwi-associated RNAs (piRNAs). These ncRNAs are involved in posttranscriptional gene regulation through translational repression or RNAi (Sana et al. 2012).

Interestingly, the aberrant expression of small ncRNAs has been associated with a wide variety of human diseases, including cancer, central nervous system disorders, and cardiovascular diseases (Taft et al. 2010; Sana et al. 2012) (Table 1.1).

For much of the last decade, special attention has been paid to research into long non-coding RNAs (lncRNAs), as these molecules tend to be shorter and have fewer introns than protein-coding transcripts (Ravasi et al. 2006). lncRNAs are considered to be the most numerous and functionally diverse class of RNAs (Derrien et al. 2011). Over 15,000 lncRNAs have already been identified, and this number is constantly increasing (Derrien et al. 2012; Fatica and Bozzoni 2014).

Amidst the great discoveries being made during this time of genome exploration, RNA is beginning to take center stage, and lncRNAs are a major part of this. These molecules are more abundant and functional than previously imagined, and they have been shown to be key players in gene regulation, genome stability, and chromatin modifications. Therefore, the identification and characterization of the function of lncRNAs has added a high degree of complexity to the comprehension of the structure, function and evolution of our genome.

lncRNAs can be grouped into one or more of five categories based on their position relative to protein-coding genes: (1) sense or (2) antisense, when they overlap with one or more exons of another transcript on the same or opposite strand, respectively; (3) bidirectional, when the expression of a lncRNA and a neighboring coding transcript on the opposite strand is initiated in close genomic proximity; (4) intronic, when the lncRNA is fully derived from the intron of a second transcript; or (5) intergenic, wherein a lncRNA is located within a gene (Poting et al. 2009). Most lncRNAs are transcribed by RNA Pol II and are often polyadenylated and have splice sites (Guttman et al. 2009; Mercer et al. 2013). However, they are devoid of obvious ORFs (Fatica and Bozzoni 2014).

The functional characterization of several mammalian regulatory lncRNAs has identified many biological roles, such as dosage compensation, genomic imprinting, cell cycle regulation, pluripotency, retrotransposon silencing, meiotic entry and telomerase length, and gene expression through chromatin modulation (Wery et al. 2011; Wilusz et al. 2009; Nagano and Fraser 2011).

The number of lncRNAs with described functions is steadily increasing, and many of these reports revolve around the regulatory capacity of lncRNAs. These molecules localize both to the nucleus and to the cytosol and can act at virtually every level during gene expression (Batista and Chang 2013; Van et al. 2014). Nuclear lncRNAs act as modulators of protein-coding gene expression and can be subdivided into *cis*-acting RNAs, which act in proximity to their site of transcription, or *trans*-acting lncRNAs, which work at distant loci. Both *cis*- and *trans*-acting lncRNAs can activate or repress transcription via chromatin modulation (Penny et al. 1996; Pandey et al. 2008; Nagano et al. 2008; Chu et al. 2011; Plath et al. 2003; Bertani et al. 2011).

Cytoplasmic lncRNAs can modulate translational control via sequences that are complementary to transcripts that originate from either the same chromosomal locus or independent loci. Target recognition occurs through base pairing (Batista and Chang 2013).

Table 1.1 Main RNA species found in eukaryotic cells including human

	Class	Symbol	Characteristic
Classical RNAs	**Messenger RNAs**	mRNAs	Variable in the size (average size about 2.2 kb) depending on the coded protein. Its linear structure includes a 5′ G cap, the 5′UTR, AUG start codon, coding sequence (CDS), stop codon, 3′UTR and the poly A tail. Account 1–2% of the total cellular RNA.
	Transfer RNAs	tRNAs	This class of RNAs takes the form of "clover leaf" and has variable size ranging 70-100 nt. The residues 34, 35 and 36 are complementary to the mRNA codons located at CDS. For this reason they are considered as adaptors between mRNAs and elongation peptide chains.
	Ribossomal RNAs	rRNAs	Are components of ribosomes along with ribosomal proteins. The high molecular weight rRNAs are the 28S rRNA (~5000 nt), which is present in the large subunit of the ribosome and 18S rRNA (1870 nt) present in the small ribosomal subunity. The low molecular weight rRNAs are the 5.8 rRNA (156 nt) and the 5.0S rRNA (121 nt), both present in the large ribosomal subunit.
Small non-coding RNAs	**MicroRNAs**	miRNAs	Average size about 18–25 nt; account 1–2% of the human genome; control the 50% of protein-coding genes; guide suppression of translation; Drosha and Dicer dependent small ncRNAs.
	Small interfering RNAs	siRNAs	Average size about 19–23 nt; made by Dicer processing; guide sequence specific degradation of target mRNA.
	Piwi-interacting RNAs	piRNAs	Average size about 26–30 nt; bind Piwi proteins; Dicer independent; exist in genome clusters; principally restricted to the germline and somatic cells bordering the germline.
	Small nucleolar RNAs	snoRNAs	Average size about 60–300 nt; enriched in the nucleolus; in vertebrate are excised from premRNA introns; bind snoRNP proteins.

Table 1.1 (continued)

	Class	Symbol	Characteristic
	Promoter-associated small RNAs	PASRs	Average size about 20–200 nt; modified 5′ (capped) ends; coincide with the transcriptional start sites of protein- and noncoding genes; made from transcription of short capped
	Transcription initiation RNAs	tiRNAs	Average size about 18 nt ; have the highest density just downstream of transcriptional start sites; show patterns of positional conservation; preferentially located in GC-rich promoters.
	Centromere repeat associated small interacting	crasiRNAs	Average size about 34–42 nt; processed from long dsRNAs.
	Telomere-specific small RNAs	tel-sRNAs	Average size about 24 nt; Dicer independent; 2′-O-methylated at the 3′ terminus; evolutionarily conserved from protozoa to mammals; but have not been described in human up to now.
	Pyknons		Subset of patterns of variable length; form mosaics in untranslated and protein-coding regions; more frequently in 3′ UTR.
Long non-coding RNAs	**Long intergenic Non-coding RNAs**	lincRNAs	Ranging from several hundreds to tens of thousands nts; lie within the genomic intervals between two genes; transcriptional cisregulation of neighbouring genes.
	Long intronic non-coding RNAs		Lie within the introns; evolutionary conserved; tissue and subcellular expression specified
	Telomere-associated ncRNAs	TERRAs	Average size about 100 bp to > 9 kb; conserved among eukaryotes; synthesized from C-rich strand; polyadenylated; form intermolecular G-quadruplex structure with single-stranded telomeric DNA.
	Long non-coding RNAs with dual functions		Both protein-coding and functionally regulatory RNA capacity.

Table 1.1 (continued)

Class	Symbol	Characteristic
Pseudogene RNAs		Gene copies that have lost the ability to code for a protein; potential to regulate their protein coding cousin; made through retrotransposition; tissue specific.
Transcribed-ultraconserved regions	T-UCRs	Longer than 200 bp; absolutely conserved between orthologous regions of human, rat, and mouse; located in both intra- and intergenic regions.
Circular RNAs	circRNAs	Noncoding RNAs generated during splicing through exon or intron circularization. They are transcription regulators or play their role as sponges for miRNAs.

Part of this table was adapted from Sana et al. (2012) J Transl Med 10: 103

RNA-Seq, the most powerful methodology for *de novo* sequence discovery, has been used to identify and analyze the expression of new lncRNAs in different cell types and tissues. Interestingly, sequencing experiments have shown that lncRNA expression is more cell-type specific than that of protein-coding genes (Riin and Chang 2012; Derrien et al. 2012; Guttman et al. 2012; Mercer et al. 2008; Cabili et al. 2011; Pauli et al. 2012).

The identification of lncRNAs relies on the detection of transcription from genomic regions that are not annotated as protein coding. However, other similarly robust methodologies have been used in the identification of lncRNAs, including the following: (1) Tiling arrays: this technology enables the analysis of global transcription from a specific genomic region and were initially used to both identify and analyze the expression of lncRNAs; (2) Serial analysis of gene expression (SAGE): this methodology allows both the quantification and the identification of new transcripts throughout the transcriptome; (3) Cap analysis gene expression (CAGE): this methodology is based on the isolation and sequencing of short cDNA sequence tags that originate from the 5' end of RNA transcripts; (4) Chromatin immunoprecipitation (ChIP): this method allows the isolation of DNA sequences that are associated with a chromatin component of interest, thereby allowing the indirect identification of many unknown lncRNAs; and (5) RNA-Seq: in a single sequencing run, this methodology produces billions of reads that are subsequently aligned to a reference genome (Fatica and Bozzoni 2014).

Transcriptome research began in parallel with the genome project because of Craig Venter's idea to sequence the "most important" genes, i.e., the functioning genome. This directive clearly fell upon mRNAs, as this type of RNA carries the protein code. Of course, this concept has not changed and mRNAs are still of central importance; however, what followed was the subsequent discovery of a large number of different ncRNAs whose functions are linked to the fine control of gene expression, often controlling the translation of mRNAs into proteins, i.e., posttranscriptional control as it is exerted by miRNAs. In its broadest sense, the transcriptome is undoubtedly more complex than anyone previously imagined.

1.3   The Transcriptome and miRNome are Closely Associated: The Role of MicroRNAs, a Class of Non-Coding Rnas Linked to the Fine Control of Gene Expression

Cellular gene expression is governed by a complex, multi-faceted network of regulatory interactions. In a very unique way, RNA molecules hybridize to each other. In the last decade, miRNAs have emerged as critical components of this cross-hybridization network. The miRNome was found to physically interact with the transcriptome, and this has important consequences for biological function.

The miRNA class of ncRNAs was first discovered in the worm *Caenorhabditis elegans* (Lee and Ambros 1993; Wightman and Ruykun 1993) and represents a family of small ncRNAs that posttranscriptionally regulate the stability of mRNA transcripts or their translation into proteins.

miRNAs participate in the regulation of a wide variety of biological processes, including cell differentiation and growth, development, metabolism chromosome architecture, apoptosis, and stress resistance. They are also involved in the pathogenesis of diseases as diverse as cancer and inflammation as well (Ambros 2004; Bushati and Cohen 2007; Stefani and Slack 2008). miRNAs are also promising candidates for new targeted therapeutic approaches and as biomarkers of disease. At approximately 22 nucleotides long, miRNAs are among the shortest known functional eukaryotic RNAs, and they repress most of the genes they regulate by just a small amount.

Many miRNAs are found in clusters and are transcribed from independent genes by either RNA Pol II or RNA Pol III (Chen et al. 2004; Borchert et al. 2006; Winter et al. 2009). They are normally found in three genomic locations: in the introns of protein-coding genes, in the introns of non-coding genes and in the exons of non-coding genes (Kim et al. 2006; Lin et al. 2008). Most miRNAs are derived from longer, double-stranded RNAs, which are termed primary miRNAs (pri-miRNAs).

Within these primary transcripts, miRNAs form stem-loop structures that contain the mature miRNA as part of an imperfectly paired double-stranded stem connected by a short terminal loop. pri-miRNAs are initially modified with a 5′ 7-methylguanosine cap and a 3′ poly-A tail (Cullen 2004) and contain hairpins that are further excised by the nuclear RNase III Drosha and its dsRNA-binding partner DGCR8 (DiGeorge syndrome critical region gene 8) (Gregory et al. 2004; Denli et al. 2004, Landthaler et al. 2004). The resulting pre-miRNA consists of an approximately 70-nucleotide double-stranded hairpin characterized by imperfect base-pairing in the stem-loop and a 2-nucleotide overhang at the 3′ end (Lee et al. 2003).

The stem-loop of a pre-miRNA is recognized by the nuclear transport protein exportin-5, which exports the pre-miRNA to the cytoplasm, in combination with the guanosine triphosphate (GTP) binding RAS-related nuclear protein (Ran-GTP) (Yi et al. 2003; Bohnack et al. 2004; Lund et al. 2004). In the cytoplasm, the pre-miRNAs are then cleaved by the RNAse III enzyme Dicer and the double-stranded RNA-binding protein TRBP (TAR RNA-binding protein) into duplexes of miRNA and passenger strands of approximately 22 base pairs (Hutvagner et al. 2001; Zhang et al. 2002).

After the sequential processing of the miRNA precursors, one of the two strands of the miRNA duplex is incorporated into the RNA-induced silencing complex (RISC). This complex comprises the mature miRNA strand as well as several proteins from the Argonaute and Gw182 families (Chendrimada et al. 2005; Haase et al. 2005). RISC can then find and bind to complementary mRNA sequences and perform its silencing function (Kawamata and Tomari 2010, Czech and Hannon 2011). In addition, a few miRNAs are produced by alternative pathways, independent of Drosha and/or Dicer, by exploiting diverse RNases that normally catalyze the maturation of other types of transcripts (Yang and Lai 2011).

Although miRNAs typically function in the cytoplasm, there is increasing evidence that they can play important roles in the nucleus as well (McCarthy 2008; Politz et al. 2009). They can also be found in the mitochondria, where they may be involved in the regulation of apoptotic genes (Kren et al. 2009).

The regulatory roles of miRNAs have been the subject of intense research (Shimoni et al. 2007; Wang and Raghavachari 2011; Levine et al. 2007; Levine and Hwa 2008; Mehta et al. 2008; Osella et al. 2011; Mitarai et al. 2009; Bumgarner et al. 2009; Iliopoulos et al. 2009). In mammals, the majority of miRNAs are inferred to be functional on the basis of their evolutionary conservation.

The major determinant for recognition between an miRNA and a target mRNA is a region of high sequence complementary that consists of an approximately 7-nucleotide domain at the 5′ end of the miRNA known as the "seed" sequence (Bartel 2009). The remaining nucleotides are generally only partially complementary to the target sequence. Sequences that are complementary to the seed ("seed matches") trigger a modest but detectable decrease in the expression of an mRNA. Seed matches can occur in any region of an mRNA but are more likely to decrease mRNA expression when they are located in the 3′ untranslated region (3′ UTR) (Grimson et al. 2007; Forman et al. 2008, 2010; Gu et al. 2009) (Fig. 1.6). Because the region used to create the seed is so short, more than half of the protein-coding genes in mammals are regulated by miRNAs, and thousands of other mRNAs appear to have

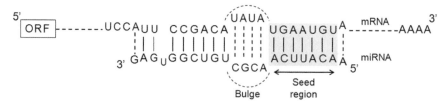

Fig. 1.6 Interaction of a miRNA with the 3′UTR of its mRNA target by base pairing. (Figure adapted from Filipowicz et al (2008) Nat Rev Genetics 9: 102–114)

undergone negative selection to avoid seed matches with miRNAs that are present in the same cell (Baek et al. 2008; Lewis et al. 2003, 2005; Farh et al. 2005, Stark 2005; Lewis 2005).

Despite the aforementioned basic features, a "seed" sequence is neither necessary nor sufficient for target silencing. It has been shown that miRNA target sites can often tolerate G:U wobble base pairs within the seed region (Miranda et al. 2006; Vella et al. 2004), and extensive base pairing at the 3' end of the miRNA may offset the absence of complementarity in the seed region (Brennecke et al. 2005; Reinhart et al. 2000). Moreover, centered sites showing 11–12 contiguous nucleotide base pairing with the central region of the miRNA without pairing to either end have also been reported (Shin et al. 2010). Adding to this repertoire, other studies have reported efficient silencing from sites that do not fit any of the above patterns and appear to be seemingly random (Lal et al. 2009; Tay et al. 2008), and even sites with extensive 5' complementarity can be inactive when tested in reporter constructs (Didiano et al. 2006).

How miRNAs repress or activate gene expression in animals is another important question, in addition to the high number of high-quality studies examining the biochemistry, biology and genomics of miRNA-directed mRNA regulation. The factors that determine which mRNAs will be targeted by miRNAs, or the mechanism by which they will be silenced, remain unclear. Extensive computational and experimental research over the last decade has substantially improved our understanding of the mechanisms underlying miRNA-mediated gene regulation (Ameres and Zamore 2013; Yue et al. 2009; Ripoli et al. 2010; Bartel 2009, Chekulaeva et al. 2009, Brodersen and Voinnet 2009).

miRNAs posttranscriptionally control gene expression by regulating mRNA translation or stability (Valencia-Sanchez et al. 2006, Standart et al. 2007; Jackson 2007, Nilsen 2007). What is known is that miRNAs can interfere with the initiation or elongation of translation; alternatively, the target mRNA may be affected by isolating it from the ribosomal machinery (Nottrott et al. 2006; Pillai et al. 2007). The binding of eIF4E to the cap region of an mRNA marks the initiation of initiation complex assembly. It has been demonstrated that miRNAs interfere with eIF4E and impair its function, and the function of the poly(A) tail can also be inhibited (Humphreys et al. 2005). There is additional evidence suggesting that miRNAs repress translation at the later stages of initiation as well. The miRNA lin-4 targets the lin-14 and lin-28 mRNAs, but under inhibitory conditions, lin-14 and lin-28 are not altered, indicating that miRNAs inhibit translation after the initiation stage. Interestingly, in both cap-dependent and independent translation, the mRNAs are inhibited by synthetic miRNA, suggesting post-initiation inhibition. Another mechanism by which miRNAs inhibit translation is by ribosome drop off, in which the ribosomes engaged in translation are directed to prematurely terminate translation. There are also proposed mechanisms by which miRNAs can direct the degradation of nascent polypeptides by recruiting proteolytic enzymes (Olsen and Ambros 1999; Petersen et al. 2006).

Microarray studies of transcript levels in cells and tissues in which miRNA pathways were inhibited or in which miRNA levels were altered support the role of miRNAs in mRNA destabilization (Behm-Ansmant et al. 2006; Giraldez et al. 2006; Rehwinkel et al. 2006; Schmitter et al. 2006; Eulalio et al. 2007). Reports have demonstrated the interaction of the P-body protein GW182 with Argonaute 1 is a key factor that marks mRNAs for degradation, as the depletion of these proteins leads to the upregulation of many mRNA targets. Moreover, knockdown experiments and analyses of the decay intermediates originating from repressed mRNAs in mammalian cells (Wu and Belasco 2006) support the role of decapping and $5' \rightarrow 3'$ exonucleolytic activities in these systems. Although many of the mRNAs that are targeted by miRNAs undergo substantial destabilization, it is not known what factors determine whether an mRNA follows the degradation or translational-repression pathway (Filipowicz et al. 2008).

In addition to their recognized roles in repressing gene expression, miRNAs have also surprisingly been linked to gene activation. The mechanism of activation is often indirect, with the repression of a repressor leading to the increased expression of specific transcripts. A relatively small number of studies have demonstrated that miRNAs can stimulate gene expression, indicating that these effects are mediated via gene promoters, extracellular receptors and the selective control of 3' or 5' UTRs. Below, we discuss three of the current examples of the role of miRNAs as stimulators of gene expression.

1) Promoter activation: Earlier studies have shown that the exogenous application of small duplex RNAs that are complementary to promoters activates gene expression in a manner similar to proteins and hormones, a phenomenon referred to as RNA activation (RNAa) (Li et al. 2006, Janowski et al. 2007). Soon afterwards, it was discovered that mir-373 targets sites in the promoters of e-cadherin and cold shock domain containing protein C2 (CSDC2), and its overexpression induced the transcription of both genes. Subsequently, mir-205 was discovered to bind to the promoter of the interleukin (IL) tumor suppressor genes IL-24 and IL-32 and, similar to mir-373, induce gene expression (Place et al. 2008; Majid et al. 2010).

2) Target activation: Several reports have shown that miRNAs can induce translation by binding to the 5' or 3' UTR of an mRNA. In the brain, a target sequence of mir-346 was found in the 5' UTR of a splice variant of receptor-interacting protein 140 (RIP140). Gain- and loss-of-function studies established that mir-346 elevated the RIP140 protein levels by facilitating the association of its mRNA with the polysome fraction. This activity did not require Ago2, indicating that other proteins in complex with the miRNA or a different RIP140 mRNA conformation induced by the miRNA mediated the effect (Tsai et al. 2009). In another study, mir-145 was shown to regulate smooth muscle cell fate and plasticity by upregulating the myocardin gene (Cordes et al. 2009). Along with this, miR-466l, a miRNA discovered in mouse embryonic stem cells, upregulated IL-10 expression in TLR-triggered macrophages by antagonizing IL-10 mRNA degradation mediated by the RBP tristetraprolin (TTP) (Ma et al. 2010).

3) Receptor ligands: Mouse TLR7 and human TLR8, which are members of the Toll-like receptor (TLR) family that are expressed on dendritic cells and B lymphocytes, physiologically recognize and bind to and are activated by ~20-nucleotide

viral single-stranded RNAs (Heil et al. 2004; Lund et al. 2004). Because miRNAs can be secreted in exosomes and are of similar size, it was predicted that they may also serve as TLR7/8 ligands. It was also found that the tumor-secreted mir-21 and mir-29a were ligands for TLR7/8 and were capable of triggering a TLR-mediated prometastatic inflammatory response (Fabbri et al. 2012).

1.3.1   Control of miRNA Expression

Despite the substantial advances in our understanding of miRNA-mediated gene regulation, the mechanisms that control the expression of the miRNAs themselves are less well understood. Homeostatic and feedback mechanisms coordinate the levels of miRNAs with their effector proteins or harmonize the levels of the biogenesis factors that function within the complexes. Often we have the impression that these processes are constitutive and inflexible.

However, diverse mechanisms that regulate the biogenesis and function of small RNAs have been uncovered (Bronevetsky and Ansel 2013; Heo and Kim 2009). Notably, many of these mechanisms provide homeostatic control over the levels of biogenesis factors and/or the resultant miRNAs. Both transcriptional and post-transcriptional mechanisms regulate miRNA biogenesis (Carthew and Sontheimer 2009; Siomi 2010; Schanen and Li 2011).

The first and one of the most important mechanisms controlling miRNA abundance is the regulation of pri-miRNA transcription. pri-miRNAs can be positively or negatively regulated by different factors such as transcription factors, enhancers, silencers and epigenetic modification of the miRNA promoter (Ruegger et al. 2012; Macedo et al. 2013). Investigations in this area have been slowed by limitations in the methods used to define the promoters and measure the transcripts. pri-miRNAs are unstable, as they are processed by the nuclear microprocessor complex very soon after transcription. Therefore, they generally do not accumulate in great abundance in cells and are underrepresented in EST and RNA-Seq libraries.

Recently, these challenges have been overcome by epigenomic and transcriptomic experiments. One study took advantage of the fact that many pri-miRNAs accumulate in cells lacking Drosha to map pri-miRNAs using RNA-Seq (Kirigin et al. 2012).

It has long been known that the levels of mature miRNAs are not determined solely by their transcription. Measurements of pri-miRNAs and their corresponding mature miRNAs were poorly correlated, suggesting that specific miRNAs are subject to developmental regulation of their processing and/or stability (Thomson et al. 2006). Additionally, the expression of these miRNAs continues to be regulated after biogenesis is complete. Mature miRNA homeostasis can be influenced by signals that modulate the stability of the miRISC complex, by nucleases that degrade miRNAs, and/or by the abundance of their mRNA targets. It is estimated that 5–10 % of mammalian miRNAs are epigenetically regulated (Breving and Esquela-Kerscher 2010, Brueckner et al. 2007, Han et al. 2007, Toyota et al. 2008).

Despite early reports indicating that miRNAs are often surprisingly stable in cells, displaying half-lives up to 12 days (van Rooij et al. 2007), cell differentiation and cell-fate decisions are frequently marked by dramatic changes in the expression of mature miRNAs.

The Argonaute proteins are limiting factors that determine the total abundance of cellular miRNAs. The deletion of these proteins, specifically Ago1 and Ago 2, was sufficient to drastically reduce miRNA expression (Bronevetsky et al. 2013; Diederichs and Haber 2007; Lund et al. 2011). Conversely, overexpressing Ago2, but not the other proteins in the miRNA biogenesis pathway, increases miRNA expression in HEK293 cells. Thus, changes in the expression and stability of Ago proteins can have dramatic effects on the expression of mature miRNAs within cells.

The action of miRNA nucleases in the regulation of miRNAs is not well understood, especially in mammals. At least two ribonucleases have been shown to negatively regulate the expression of mature miRNAs. IRE1a, an endoplasmic reticulum (ER) transmembrane RNase activated in response to ER stress, cleaves precursors corresponding to miR-17, miR-34a, miR-96, and miR-125b and mediates the rapid decay of their expression in response to sustained cellular stress (Upton et al. 2012). Additionally, Eri1, a 3'-to-5' exoribonuclease with a double-stranded RNA-binding SAP domain, was discovered to limit miRNA abundance in CD4 + T cells and natural killer (NK) cells (Thomas et al. 2012).

The sequence-specific degradation of miRNAs has also been observed with the addition of RNA targets. miRNA "antagomirs" and "miRNA sponges" are two technologies used to specifically knockdown miRNA expression, and both rely on miRNA degradation induced by high levels of miRNA-to-target complementarity (Krutzfeldt et al. 2005; Ebert et al. 2007; Plank et al. 2013). Further work is still needed to determine the extent to which miRNA expression is regulated by target mRNAs, as well as the molecular mechanisms that mediate this final step in the control of miRNA expression.

The posttranscriptional regulatory mechanisms that affect miRNA processing at different stages have recently been investigated (Siomi 2010). For example, p53 can form a complex with Drosha, which increases the processing of pri-miRNAs to pre-miRNAs (Suzuki et al. 2009). Histone deacetylase I can also enhance pri-miRNA processing by deacetylating the microprocessor complex protein DGCR8 (Wada et al. 2012). Additionally, cytokines such as interferons have been shown to inhibit Dicer expression and decrease the processing of pre-miRNAs (Wiesen and Tomasi 2009).

1.3.2   Extracellular miRNAs

RISC components and miRNAs have also been found in exosomes (Valadi et al. 2007). Exosomes isolated from the culture supernatant of many hematopoietic cells, including cytotoxic T lymphocytes, mast cells, and dendritic cells (DCs), as well as DC-derived exosomes, have been shown to stimulate CD4 + T-cell activation and induce tolerance (Zitvogel et al. 1998). Experimentally, vesicles containing both Ago2 and miRNAs, including miR-150, miR-21, and miR-26b, as well as the vesicle-derived miR-150, could be delivered to recipient HMEC-1 human endothelial cells and repress the target mRNAs in the recipient cells. These findings illustrate another mechanism by which immune cell stimulation/activation can lead to significant changes in mature miRNA levels. Interest in extracellular miRNAs in various body fluids has increased substantially as early findings indicated their utility as readily accessible biomarkers.

Circulating miRNAs have been studied in patient samples and animal models in the context of cardiovascular disease, liver injury, sepsis, cancer, and various other physiological and pathophysiological states (Cortez et al. 2011). The origin of extracellular miRNAs is still poorly understood, with blood cells appearing to be a major contributor to circulating miRNAs (Pritchard et al. 2012).

It has also become clear that extracellular miRNAs exist in several distinct forms in human plasma. In addition to miRNAs encapsulated in vesicles such as exosomes, there are stable non-vesicular miRNAs that can be copurified with Ago proteins, which are accessible for direct immunoprecipitation from plasma samples (Arroyo et al. 2011). Further research is needed to clarify the cellular sources of miRNAs, the forms in which they are released, and whether this process is regulated during biological processes.

1.3.3   An Example of the Biological Consequence of miRNAs: Their Role in the Immune System

The role of miRNAs in the immune system has been extensively investigated. Both innate and adaptive immune responses are highly regulated by miRNAs. By targeting the signal transduction proteins involved in the transmission of intracellular signals following initial pathogen recognition and by directly targeting mRNAs that encode specific inflammatory cytokines, miRNAs can have a significant impact on the innate immune response. In addition to their role in regulating the innate immune system, miRNAs have been implicated in adaptive immunity, wherein they control the development, activation and plasticity of T and B cells (Lu and Liston 2009; Xiao and Rajewsky 2009; O' Connell et al. 2010; O' Neill et al. 2011; Plank et al. 2013; Baumjohann and Ansel 2013; Donate et al. 2013).

Furthermore, the central role of miRNAs across many important aspects of innate and adaptive immunity strongly supports their potential in regulating inflammatory diseases. The identification of a broad range of miRNAs that play pathogenic roles is growing. To date, a relatively small number of miRNAs has been associated with specific inflammatory diseases, and most of the identified miRNAs are expressed across multiple tissues and cell types, and many have been shown to play roles in other disease settings, particularly in cancer. Despite the limited numbers of verified targets in inflammatory diseases, many of the targets that were verified in other experimental settings may also be relevant in inflammatory diseases (Plank et al. 2013).

1.4   Conclusion

Early on, transcriptome research was intertwined with the genome. Much of this was due to the mapping of ESTs, and sequencing dominated the scene. Through the use of EST clones and the application of technical concepts such as nucleic acid hybridization, researchers began to use arrayed filters to explore the transcriptional expression of a large number of genes in a single experiment.

The constant improvement of these DNA arrays led to the fabrication of high-density arrays and, finally, microarrays.

At the same time, sequencing also underwent significant changes involving automation and the endless quest to increase the number of reads, and this contributed substantially to a better understanding of the diversity of the transcriptome. Indeed, transcriptome research was rooted in these two major technological approaches (i.e., large-scale hybridization and sequencing).

What made microarrays robust and increased their popularity was the increase in the number of sequences deposited on the slides (currently, these slides contain the entire human or mouse functional genome), the sensitivity of the method (currently, experiments are being performed with nanogram amounts of total RNA to screen the entire functional genome), the simplicity of its use, its commercial availability and the availability of bioinformatics packages dedicated to analyzing the large amounts of data being generated.

Of key importance was the development of statistical procedures for the analysis of large amounts of data, which opened the door for biostatisticians and bioinformaticians.

All of these ongoing technological advances have contributed to the consolidation of the concept of the transcriptome. Unlike the genome, which is essentially static, the transcriptome is variable and is dependent on normal physiological, pathological or environmental conditions. Moreover, it is composed not only of mRNAs but also non-coding RNAs, including miRNAs.

This concept has provided the opportunity for all types of biomedical research to re-examine their results in light of transcriptomics.

References

Adams J (2008) Sequencing human genome: the contributions of Francis Collins and Craig Venter. Nat Educ 1(1):133

Adams MD, Kelley JM, Gocayne JD, Dubnick M, Polymeropoulos MH, Xiao H, Merril CR, Wu A, Olde B et al (1991) Complementary DNA sequencing: expressed sequence tags and human genome project. Science 252:1651–1656

Adams MD, Dubnick M, Kerlavage AR, Moreno R, Kelley JM, Utterback TR, Nagle JW, Fields C, Venter JC (1992) Sequence identification of 2,375 human brain genes. Nature 355:632–634

Adams MD, Soares MB, Kerlavage AR, Fields C, Venter JC (1993a) Rapid cDNA sequencing (expressed sequence tags) from a directionally cloned human infant brain cDNA library. Nat Genet 4:373–380

Adams MD, Kerlavage AR, Fields C, Venter JC (1993b). Initial assessment of human gene diversity and expression patterns based upon 83 million nucleotides of cDNA sequence. Nat Genet 4:256–267

Ahrens CH, Brunner E, Qeli E, Basler K, Aebersold R (2010) Generating and navigating proteome maps using mass spectrometry. Nat Rev Mol Cell Biol 11:789–801

Allison DB, Cui X, Page GP, Sabripour M (2006) Microarray data analysis: from disarray to consolidation and consensus. Nat Rev Genet 7:55–65

Altelaar AF, Munoz J, Heck AJ (2013) Next-generation proteomics: towards an integrative view of proteome dynamics. Nat Rev Genet 14:35–48

Ambros V (2004) The functions of animal microRNAs. Nature 431:350–355

Ameres SL, Zamore PD (2013) Diversifying microRNA sequence and function. Nat Rev Mol Cell Biol 14(8):475–488

Anders S, Huber W (2010) Differential expression analysis for sequence count data. Gen Biol 11: R106

Anderson L (2014) Six decades searching for meaning in the proteome. J Proteomics. doi:10.1016/j. jprot.2014

Arroyo JD, Chevillet JR, Kroh EM, Ruf IK, Pritchard CC, Gibson DF, Mitchell PS, Bennett CF, Pogosova-Agadjanyan EL et al (2011) Argonaute2 complexes carry a population of circulating microRNAs independent of vesicles in human plasma. Proc Natl Acad Sci USA 108:5003–5008

Ashburner M, Ball CA, Blake JA, Botstein D, Butler H, Cherry JM, Davis AP, Dolinski K, Dwight SS et al (2000) Gene ontology: tool for the unification of biology. Nat Genet 25:25–29

Auer PL, Doerge RW (2010) Statistical design and analysis of RNA sequencing data. Genetics 185:405–416

Baek D, Villén J, Shin C, Camargo FD, Gygi SP, Bartel DP (2008) The impact of microRNAs on protein output. Nature 455:64–71

Baker SC, Bauer SR, Beyer RP, Brenton JD, Bromley B, Burrill J, Causton H, Conley MP, Elespuru R et al (2005) The external RNA controls consortium: a progress report. Nat Methods 2 731–734

Ball CA, Sherlock G, Parkinson H, Rocca-Sera P, Brooksbank C, Causton HC, Cavalieri D, Gaasterland T, Hingamp P et al (2002) Microarray gene expression data (MGED) society. Standards for microarray data. Science 298:539

Bartel DP (2009) MicroRNAs: target recognition and regulatory functions. Cell 136, 215–233

Batista PJ and Chang HY (2013) Long noncoding RNAs: cellular address codes in development and disease. Cell 152:1298–1307

Baumjohann D, Ansel MK (2013) MicroRNA-mediated regulation of T helper cell differentiation and plasticity. Nat Rev Immunol 13:666–678

Behm-Ansmant I, Rehwinkel J, Doerks T et al (2006) mRNA degradation by miRNAs and GW182 requires both CCR4:NOT deadenylase and DCP1:DCP2 decapping complexes. Genes Dev 20:1885–1898

Bernard K, Auphan N, Granjeaud S, Victorero G, Schmitt-Verhulst AM, Jordan BR, Nguyen C (1996) Multiplex messenger assay: simultaneous, quantitative measurement of expression for many genes in the context of T cell activation. Nucleic Acids Res 24:1435–1443

Bertani S, Sauer S, Bolotin E et al (2011) The noncoding RNA Mistral activates Hoxa6 and Hoxa7 expression and stem cell differentiation by recruiting MLL1 to chromatin. Mol Cell 43:1040–1046

Bertucci F, Bernard K, Loriod B, Chang YC, Granjeaud S, Birnbaum D, Nguyen C, Peck K, Jordan BR (1999) Sensitivity issues in DNA array-based expression measurements and performance of nylon microarrays for smalls samples. Hum Mol Genet 9:1715–1722

Bohnsack MT, Czaplinski K, Gorlich D (2004) Exportin 5 is a RanGTP-dependent dsRNA-binding protein that mediates nuclear export of pre-miRNAs. RNA 10:185–191

Bolstad BM, Irizarry RA, Astrand M, Speed TP (2003) A comparison of normalization methods for high density oligonucleotide array data based on variance and bias. Bioinformatics 19(2):185–193

Borchert GM, Lanier W, Davidson BL (2006) RNA polymerase III transcribes human microRNAs. Nat Struct Mol Biol 13:1097–101

Botwell D (1999) Options available -from start to finish- for obtaining expression data by microarray. Nat Genet 21:2–32

Bratkovic T, Rogelj B (2014) The many faces of small nucleolar RNAs. Biochim Biophys Acta 1839:438–443

Brazma A, Hingamp P, Quackenbush J, Sherlock G, Spellman P, Stoeckert C, Aach J, Ansorge W, Ball CA et al (2001) Minimum information about a microarray experiment (MIAME) – toward standards for microarray data. Nat Genet 29:365–371

Breitling R, Herzyk P (2005) Rank-based methods as a non-parametric alternative of the T-statistic for the analysis of biological microarray data. J Bioinf Comp Biol 3:1171–1189

Brennecke J, Stark A, Russell RB et al (2005) Principles of microRNA-target recognition. PLoS Biol 3:e85

Brenner S, Jacob F, Meselson M (1961) An unstable intermediate carrying information from genes to ribosomes for protein synthesis. Nature 190:576–581

Breving K, Esquela-Kerscher A (2010) The complexities of microRNA regulation: miRandering around the rules. Int J BiochemCell Biol 42:1316–1329

Brodersen P, Voinnet O (2009) Revisiting the principles of microRNA target recognition and mode of action. Nat Rev Mol Cell Biol 10(2):141–1488

Bronevetsky Y, Ansel MK (2013) Regulation of miRNA biogenesis and turnover in the immune system. Immunol Rev 253:304–316

Bronevetsky Y, Villarino AV, Eisley CJ, Barbeau R, Barczak AJ, Heinz GA, Kremmer E, Heissmeyer V, McManus MT et al (2013) T cell activation induces proteasomal degradation of Argonaute and rapid remodeling of the microRNA repertoire. J Exp Med 210:417–432

Brueckner B, Stresemann C, Kuner R, Mund C, Musch T, Meister M, Sültmann H, Lyko F (2007) The human let-7a-3 locus contains an epigenetically regulated microRNA gene with oncogenic function. Cancer Res 67:1419–1423

Bullard JH, Purdom E, Hansen KD, Dudoit S (2010) Evaluation of statistical methods for normalization and differential expression in mRNA-Seq experiments. BMC Bioinform 11:94

Bumgarner SL, Dowell RD, Grisafi P, Gifford DK, Fink GR (2009) Toggle involving cis-interfering noncoding RNAs controls variegated gene expression in yeast. Proc Natinal Acad Sci USA 106:18321–18326

Bushati N, Cohen SM (2007) MicroRNA functions. Annu Rev Cell Dev Biol 23:175–205

Cahan P, Rovegno F, Mooney D, Newman JC, St. Laurent III G, McCaffrey TA (2007) Meta-analysis of microarray results: challenges, opportunities, and recommendations for standardization. Gene 401:12–18

Camargo AA, Samaia HP, Dias-Neto E, Simão DF, Migotto IA, Briones MR, Costa FF, Nagai MA, Verjovski-Almeida S et al (2001) The contribution of 700,000 ORF sequence tags to the definition of the human transcriptome. Proc Natl Acad Sci USA 98:12103–12108

Cantor CR (1990) Orchestrating the human genome project. Science 248:49–51

Carthew RW, Sontheimer EJ (2009) Origins and mechanisms of miRNAs and siRNAs. Cell 136:642–655

Chee M, Yang R, Hubbell E, Berno A, Huang XC, Stern D, Winkler J, Lockhart DJ, Morris MS, Fodor SP (1996) Accessing genetic information with high-density DNA arrays. Science 274:610–614

Chekulaeva M, Filipowicz W (2009) Mechanisms of miRNA-mediated post-transcriptional regulation in animal cells. Curr Opin Cell Biol 21:452–460

Chen Y, Dougherty ER, Bittner ML (1997) Ratio-based decisions and the quantitative analysis of cdna microarray images. J Biomed Opt 2:364–374

Chen JJ, Wu R, Yang PC, Huang JY, Sher YP, Han MH, Kao WC, Lee PJ, Chiu TF et al (1998) Profiling expression patterns and isolating differentially expressed genes by cDNA microarray system with colorimetry detection. Genomics 51:313–324

Chen CZ, Li L, Lodish HF, Bartel DP (2004) MicroRNAs modulate hematopoietic lineage differentiation. Science 303:83–86

Chendrimada TP, Gregory RI, Kumaraswamy E (2005) TRBP recruits the Dicer complex to Ago2 for microRNA processing and gene silencing. Nature 436:740–744

Chu C, Qu K, Zhong FL et al (2011) Genomic maps of long noncoding RNA occupancy reveal principles of RNA–chromatin interactions. Mol Cell 44:667–678

Churchill GA (2002) Fundamentals of experimental design for cDNA microarrays. Nat Genet 32:490–495

Clément-Ziza M, Gentien D, Lyonnet S, Thiery JP, Besmond C, Decraene C (2009) Evaluation of methods for amplification of picogram amounts of total RNA for whole genome expression profiling. BMC Genomics 26:10:246

Cobb JP, Mindrinos MN, Miller-Graziano C, Calvano SE, Baker HV, Xiao W, Laudanski K, Brownstein BH, Elson CM et al (2005) Application of genome-wide expression analysis to human health and disease PNAS 102(13):4801–4806

Cordes KR, Sheehy NT, White MP, Berry EC, Morton SU, Muth AN, Lee TH, Miano JM, Ivey KN et al (2009) miR-145 and miR-143 regulate smooth muscle cell fate and plasticity. Nature 460:705–710

Cortez MA, Bueso-Ramos C, Ferdin J (2011) MicroRNAs in body fluids–the mix of hormones and biomarkers. Nat Rev Clin Oncol 8:467–477

Cullen BR (2004) Transcription and processing of human microRNA precursors. Mol Cell 16:861–865

Czech B, Hannon GJ (2011) Small RNA sorting: matchmaking for argonautes. Nat Rev Genet 12:19–31

De Klerk E den Dunnen JT t Hoen PA (2014) RNA sequencing : from tag-based profiling to resolving complete transcript structure. Cell Mol Life Sci (epub ahead of print) 71(18):3537–3551.

Degrelle SA, Hennequet-Antier C, Chiapello H, Piot-Kaminski K, Piumi F, Robin S, Renard JP, Hue I (2008) Amplification biases: possible differences among deviating gene expressions. BMC Genomics 9:46

Denli AM, Tops BB, Plasterk RH, Ketting RF, Hannon GJ (2004) Processing of primary microRNAs by the Microprocessor complex. Nature 432:231–235

Dennis G Jr, Sherman BT, Hosack DA, Yang J, Gao W, Lane HC, Lempicki RA (2003) DAVID: database for annotation, visualization, and integrated discovery. Genome Biol 4:3

Derrien T, Guigo R, Johnson R (2011) The long non-coding RNAs: a new (p)layer in the "dark matter". Front Genet 2:107

Didiano D, Hobert, O (2006) Perfect seed pairing is not a generally reliable predictor for miRNA-target interactions. Nat Struct Mol Biol 13:849–851

Diederichs S, Haber DA (2007) Dual role for argonautes in microRNA processing and posttranscriptional regulation of microRNA expression. Cell 131:1097–1108

Dillies MA, Rau A, Aubert J, Hennequet-Antier C, Jeanmougin M, Servant N, Keime C, Marot G, Castel D et al (2013) A comprehensive evaluation of normalization methods for illumine high-throughput RNA sequencing data analysis. Brief Bioinform 14(6):671–683

Djebali S, Davis CA, Merkel A et al (2012) Landscape of transcription in human cells. Nature 489:101–108

Donate PB, Fornari TA, Macedo C, Cunha TM, Nascimento DC, Sakamoto-Hojo ET, Donadi EA, Cunha FQ, Passos GA (2013) T cell post-transcriptional miRNA-mRNA interaction networks identify targets associated with susceptibility/resistance to collagen-induced arthritis. PLoS One 8(1):e54803

Duewer DL, Jones WD, Reid LH, Salit M (2009) Learning from microarray interlaboratory studies: measures of precision for gene expression. BMC Genomics 10:153

Dufva M (2005) Fabrication of high quality microarrays. Biomol Eng 22:173–184

Dujon, B (1998) European functional analysis network (EUROFAN) and the functional analysis of the Saccharomyces cerevisiae genome. Electrophoresis 19:617–624

Ebert MS, Neilson JR, Sharp PA (2007) MicroRNA sponges: competitive inhibitors of small RNAs in mammalian cells. Nat Methods 4:721–726

Edwards D (2003) Non-linear normalization and background correction in onechannel cDNA microarrays studies. Bioinformatics 19:825–833

Eisen MB, Spellman PT, Brown PO, Botstein D (1998) Cluster analysis and display of genome-wide expression patterns. PNAS 95(25):14863–14868

Epstein JR, Leung AP, Lee KH, Walt DR (2003) High-density, microsphere based fiber optic DNA microarrays. Biosen Bioeletron 18:541–546

Eulalio A, Rehwinkel J, Stricker M, Huntzinger E, Yang SF, Doerks T, Dorner S, Bork P, Boutros M et al (2007) Target-specific requirements for enhancers of decapping in miRNA-mediated gene silencing. Genes Dev 21:2558–2570

Fabbri M, Paone A, Calore F, Galli R, Gaudio E, Santhanam R, Lovat F, Fadda P, Mao C et al (2012) microRNAs bind to Toll-like receptors to induce prometastatic inflammatory response. Proc Natl Acad Sci USA 109:E2110–E2116

Fang Z, Cui X (2010) Design and validation issues in RNA-seq experiments. Brief Bioinform 12(3):280–287

Fang Z, Cui X (2011) Design and validation issues in RNA-Seq experiments. Brief Bioinformatics 12:280–287

Farh KK, Grimson A, Jan C, Lewis BP, Johnston WK, Lim LP, Burge CB, Bartel DP (2005) The widespread impact of mammalian microRNAs on mRNA repression and evolution. Science 310:1817–1821

Fatica A, Bozzoni I (2014) Long non-coding RNAs: new players in cell differentiation and development. Nat Rev Genet 15:7–21

Ferguson JA, Steemers FJ, Walt DR (2000) High-density fiber optic DNA random microsphere array. Anal Chem 72:5618–5624

Filipowicz W, Bhattacharyya SN, Sonenberg N (2008) Mechanisms of posttranscriptional regulation by microRNAs: are the answers in sight? Nat Rev Genet 9(2):102–114

Fisher RA (1935) The design of experiments. Oxford, England. Oliver & Boyd, p 251

Fonseca NA, Rung J, Brazma A, Marioni JC (2012) Tools for mapping high-troughput sequencing data. Bioinformatics 28(24):3169–3177

Forler S, Klein O, Klose J (2014) Individualized proteomics J Proteomics 107C:56–61

Forman JJ, Coller HA (2010) The code within the code: microRNAs target coding regions. Cell Cycle 9:1533–1541

Forman JJ, Legesse-Miller A, Coller HA (2008) A search for conserved sequences in coding regions reveals that the let†'7 microRNA targets Dicer within its coding sequence. Proc Natl Acad Sci USA 105:14879–14884

Garber M, Grabherr MG, Guttman M, Trapnell C (2011) Computational methods for trasncriptome annotation and quantification using RNA-sEq. Nat Methods 8:469–477

Geeleher P, Morris D, Golden A, Hinde JP (2008) Handbook: bioconductorBuntu users manual. http://www3.it.nuigalway.ie/agolden/bioconductor/version1/handbook.pdf

Geisler S, Coller J (2013) RNA in unexpected places: long non-coding RNA functions in diverse cellular contexts. Nat Rev Mol Cell Biol 14:699–672

Gentleman RC, Carey VJ, Bates DM (2004) Bioconductor: open software development for computational biology and bioinformatics. Genome Biol 5(10):R80

Gentleman RC, Carey VJ, Huber W et al (2005) Bioinformatics and computational biology solutions using R and bioconductor. Springer, New York, p 473

Gershon D (2002) Microarray technology, an array of opportunities; technology feature. Nature 416:885–891

Geschwind DH, Gregg JP (2002) Microarrays for the neurosciences: an essential guide. The MIT Press

Giraldez AJ, Mishima Y, Rihel J, Grocock RJ, Van Dongen S, Inoue K, Enright AJ, Schier AF (2006) Zebrafish MiR-430 promotes deadenylation and clearance of maternal mRNAs. Science 312:75–79

Goecks J, Nekrutenko A, Taylor J; Galaxy Team (2010) Galaxy: a comprehensive approach for supporting accessible, reproducible, and transparent computational research in the life sciences. Genome Biol 11:R86

Granjeaud S, Nguyen C, Rocha D, Luton R, Jordan BR (1996) From hybridization image to numerical values:a practical, high throughput quantification system for high density filter hybridizations. Genet Anal Biomol Eng 12:151–162

Granjeaud S, Bertucci F, Jordan BR (1999) Expression profiling: DNA arrays in many guises. Bioessays 21:781–790

Gregory RI, Yan KP, Amuthan G, Chendrimada T, Doratotaj B, Cooch N, Shiekhattar R (2004) The Microprocessor complex mediates the genesis of microRNAs. Nature 432:235–240

Gress TM, Hoheisel JD, Lennon GG, Zehetner G, Lehrach H (1992) Hybridization fingerprinting of high-density cDNA-library arrays with cDNA pools derived from whole tissues. Mamm Genome 3:609–661

Grimson A, Farh KK, Johnston WK, Garrett-Engele P, Lim LP, Bartel DP (2007) MicroRNA targeting specificity in mammals: determinants beyond seed pairing. Mol Cell 27:91–105

Gu S, Jin L, Zhang F, Sarnow P, Kay MA (2009) Biological basis for restriction of microRNA targets to the 3í´ untranslated region in mammalian mRNAs. Nat Struct Mol Biol 16:144–150

Gunderson KL, Kruglyak S, Graige MS, Garcia F, Kermani BG, Zhao C, Che D, Dickinson T, Wickham E et al (2004) Decoding randomly ordered DNA arrays. Genome Res 14:870–877

Guo Y, Ye F, Sheng Q, Clark T, Samuels DC (2013) Three-stage quality control strategies for DNA re-sequencing data. Briefings in Bioinformatics doi:10.1093/bib/bbt069

Haase AD, Jaskiewicz L, Zhang H, Lainé S, Sack R, Gatignol A, Filipowicz W (2005) TRBP, a regulator of cellular PKR and HIV-1 virus expression, interacts with Dicer and functions in RNA silencing. EMBO 6:961–967

Han L, Witmer PD, Casey E, Valle D, Sukumar S (2007) DNA methylation regulates MicroRNA expression. Cancer Biol Ther 6:1284–1288

Heber S, Sick B (2006) Quality assessment of Affymetrix GeneChip data. OMICS 10(3):358–368

Heil F, Hemmi H, Hochrein H, Ampenberger F, Kirschning C, Akira S, Lipford G, Wagner H, Bauer S (2004) Species-specific recognition of single-stranded RNA via Toll-like receptor 7 and 8. Science 303:1526–1529

Heo I, Kim VN (2009) Regulating the regulators: posttranslational modifications of RNA silencing factors. Cell 139:28–31

Huber W, von Heydebreck A, Sültmann H, Poustka A, Vingron M (2002) Variance stabilization applied to microarray data calibration and to the quantification of differential expression. Bioinformatics 18(1):S96–S104

Humphreys DT, Westman BJ, Martin DI, Preiss T (2005) MicroRNAs control translation initiation by inhibiting eukaryotic initiation factor 4E/cap and poly(A) tail function. Proc Natl Acad Sci USA 102:16961–16966

Hutvágner G, McLachlan J, Pasquinelli AE, Bálint E, Tuschl T, Zamore PD (2001) A cellular function for the RNAinterference enzyme Dicer in the maturation of the let-7 small temporal RNA. Science 293:834–838

Iliopoulos D, Hirsch HA, Struhl K (2009) An epigenetic switch involving NF-kB, Lin28, Let-7 microRNA, and IL6 links inflammation to cell transformation. Cell 139:693–706

Irizarry RA, Hobbs B, Collin F, Beazer-Barclay YD, Antonellis KJ, Scherf U, Speed TP (2003) Exploration, normalization, and summaries of high density oligonucleotide array probe level data. Bioinformatics 4(2):249–264

Irizarry RA, Warren D, Spencer F, Kim IF, Biswal S, Frank BC, Gabrielson E, Garcia JG, Geoghegan J et al (2005) Multiple-laboratory comparison of microarray platforms. Nat Methods 2:345–350

Jackson RJ, Standart N (2007) How do microRNAs regulate gene expression? Sci STKE 2007(367):re1

Jacob F, Monod J (1961) Genetic regulatory mechanisms in the synthesis of proteins. J Mol Biol 3:318–356

Janowski BA, Younger ST, Hardy DB, Ram R, Huffman KE, Corey DR (2007) Activating gene expression in mammalian cells with promoter-targeted duplex RNAs. Nat Chem Biol 3:166–173

Järvinen AK, Hautaniemi S, Edgren H, Auvinen P, Saarela J, Kallioniemi OP, Monni O (2004) Are data from different gene expression microarray platforms comparable? Genomics 83:1164–1168

Jordan B (2012) The microarray paradigm and its various implementations. In Jordan B (ed) Microarrays in diagnostics and biomarker development. Current and future applications. Springer-Verlag, Berlin Heidelberg.

Jordan BR (1998) Large scale expression measurement by hybridization methods: from high-density membranes to "DNA chips". J Biochem 124:251–258

Kanehisa M, Goto S (2000) KEGG: Kyoto encyclopedia of genes and genomes. Nucleic Acids Res 28:27–30

Kawamata T, Tomari Y (2010) Making RISC. Trends Biochem. Sci 35:368–376

Kellis M, Wold B, Snyder MP et al (2014) Defining functional DNA elements in the human genome. Proc Natl Acad Sci USA 111:6131–6138

Kerr MK, Churchill GA (2001) Experimental design for gene expression microarrays. Biostatistics 2:183–201

Kim VN, Nam JW (2006) Genomics of microRNA. Trends Genet 22:165–173

Kirigin FF, Lindstedt K, Sellars M, Ciofani M, Low SL, Jones L, Bell F, Pauli F, Bonneau R et al (2012) Dynamic microRNA gene transcription and processing during T cell development. J Immunol 188:3257–3267

Kooperberg C, Fazzio TG, Delrow JJ, Tsukiyama T (2002) Improved background correction for spotted DNA microarrays. J Comp Biol 9:55–66

Kren BT, Wong PY, Sarver A, Zhang X, Zeng Y, Steer CJ (2009) MicroRNAs identified in highly purified liver-derived mitochondria may play a role in apoptosis. RNA Biol 6:65–72

Krützfeldt J, Rajewsky N, Braich R, Rajeev KG, Tuschl T, Manoharan M, Stoffel M (2005) Silencing of microRNAs in vivo with 'antagomirs'. Nature 438:685–689

Lal A, Navarro F, Maher CA, Maliszewski LE, Yan N, O'Day E, Chowdhury D, Dykxhoorn DM, Tsai P et al (2009) miR-24 inhibits cell proliferation by targeting E2F2, MYC, and other cell-cycle genes via binding to "seedless" 3' UTR microRNA recognition elements. Mol Cell 35:610–625

Landthaler M, Yalcin A, Tuschl T (2004) The human DiGeorge syndrome critical region gene 8 and Its D. melanogaster homolog are required for miRNA biogenesis. Curr Biol 14:2162–2167

Larkin JE, Frank BC, Gavras H, Sultana R, Quackenbush J (2005) Independence and reproducibility across microarray platforms. Nat Methods 2:337–344

Lee RC, Feinbaum RL, Ambros V (1993) The C. elegans heterochronic gene lin'4 encodes small RNAs with antisense complementarity to lin'14. Cell 75:843–854

Lee Y, Ahn C, Han J, Choi H, Kim J, Yim J, Lee J, Provost P, Rådmark O et al (2003) The nuclear RNase III Drosha initiates microRNA processing. Nature 425:415–419

Levine E, Hwa T (2008) Small RNAs establish gene expression thresholds. Curr Opin Microbiol 11:574–579

Levine E, Zhang Z, Kuhlman T, Hwa T (2007) Quantitative characteristics of gene regulation by small RNA. PLoS Biol 5:e229

Lewis BP, Shih IH, Jones-Rhoades MW, Bartel DP, Burge CB (2003) Prediction of mammalian microRNA targets. Cell 115:787–798

Lewis BP, Burge CB, Bartel DP (2005) Conserved seed pairing, often flanked by adenosines, indicates that thousands of human genes are microRNA targets. Cell 120:15–20

Li LC, Okino ST, Zhao H, Pookot D, Place RF, Urakami S, Enokida H, Dahiya R (2006) Small dsRNAs induce transcriptional activation in human cells. Proc Natl Acad Sci USA 103:17337–17342

Lin SL, Kim H, Ying SY (2008) Intron-mediated RNA interference and microRNA (miRNA). Front Biosci 13:2216–2230

Liu G, Mattick JS, Taft RJ (2013) A meta-analysis of the genomic and transcriptomics composition of complex life. Cell Cycle 12:2061–2072

Lockhart DJ, Dong H, Byrne MC, Follettie MT, Gallo MV, Chee MS, Mittmann M, Wang C, Kobayashi M (1996) Expression monitoring by hybridization to high-density oligonucleotide arrays. Nat Biotechnol 14:1675–1680

Lönnstedt I, Speed T (2002) Replicated microarray data. Stat Sinica 12:31–46

Lu LF, Liston A (2009) MicroRNA in the immune system, microRNA as an immune system. Immunology 127:291–298

Lund JM, Alexopoulou L, Sato A, Karow M, Adams NC, Gale NW, Iwasaki A, Flavell RA (2004a) Recognition of single-stranded RNA viruses by Toll-like receptor 7. Proc Natl Acad Sci USA 101:5598–5603

Lund E, Güttinger S, Calado A, Dahlberg JE, Kutay U (2004b) Nuclear export of microRNA precursors. Science 303:95–98

Lund E, Sheets MD, Imboden SB, Dahlberg JE (2011) Limiting Ago protein restricts RNAi and microRNA biogenesis during early development in Xenopus laevis. Genes Dev 25:1121–1131

Luo W, Friedman MS, Shedden K, Hankenson KD, Woolf PJ (2009) GAGE: generally applicable gene set enrichment for pathway analysis. BMC Bioinform 10:161

Luo J, Schumacher M, Scherer A, Sanoudou D, Megherbi D, Davison T, Shi T, Tong W, Shi L et al (2010) A comparison of batch effect removal methods for enhancement of prediction performance using MAQC-II microarray gene expression data. Phamacogenomics J 10:278–291

Ma F, Liu X, Li D, Wang P, Li N, Lu L, Cao X (2010) microRNA-466 l upregulates IL-10 expression in TLR-triggered macrophages by antagonizing RNA-binding protein tristetraprolin-mediated IL-10 mRNA degradation. J Immunol 184:6053–6059

Macedo C, Evangelista AF, Marques MM, Octacílio-Silva S, Donadi EA, Sakamoto-Hojo ET, Passos GA (2013) Autoimmune regulator (Aire) controls the expression of microRNAs in medullary thymic epithelial cells. Immunobiol 218:554–560

Maeda N, Kasukawa T, Oyama R et al (2006) Transcript annotation in FANTOM3: mouse gene catalog based on physical cDNAs. PloS Genet 2: e62

Majid S, Dar AA, Saini S, Yamamura S, Hirata H, Tanaka Y, Deng G, Dahiya R (2010) microRNA-205-directed transcriptional activation of tumor suppressor genes in prostate cancer. Cancer 116:5637–5649

MAQC Consortium (2006) The microarray quality control (MAQC) project shows inter- and intraplatform reproducibility of gene expression measurements. Nat Biotechnol 24:1151–1161

Marioni JC, Mason CE, Mane SM, Stephens M, Gilad Y (2008) RNA-seq: an assessment of technical reproducibility and comparison with gene expression arrays. Genome Res 18(9):1509–1517

Martin JA, Wang Z (2011) Next-generation transcriptome assembly. Nat Rev Genet 12:671–682

McCarthy JJ (2008) MicroRNA-206: the skeletal muscle-specific myomiR. Biochim Biophys Acta 1779:682–691

Mehta P, Goyal S, Wingreen NS (2008) A quantitative comparison of sRNA-based and protein-based gene regulation. Mol Syst Biol 4:211

Mercer TR, Mattick JS (2013) Structure and function of long noncoding RNAs in epigenetic regulation. Nat Struct Biol 20: 300–307

Miranda KC, Huynh T, Tay Y, Ang YS, Tam WL, Thomson AM, Lim B, Rigoutsos I (2006) A pattern-based method for the identification of MicroRNA binding sites and their corresponding heteroduplexes. Cell 126:1203–1217

Mitarai N, Benjamin JA, Krishna S, Semsey S, Csiszovszki Z, Massé E, Sneppen K (2009) Dynamic features of gene expression control by small regulatory RNAs. Proc Natl Acad Sci USA 106:10655–10659

Mitchell PS, Parkin RK, Kroh EM, Fritz BR, Wyman SK, Pogosova-Agadjanyan EL, Peterson A, Noteboom J, O'Briant KC et al (2008) Circulating microRNAs as stable blood-based markers for cancer detection. Proc Natl Acad Sci USA 105:10513–10518

Moorcroft MJ, Meuleman WR, Latham SG, Nicholls TJ, Egeland RD, Edwin M., Southern EM (2005) In situ oligonucleotide synthesis on poly(dimethylsiloxane): a flexible substrate for microarray fabrication. Nucleic Acids Res 33:e75

Mortazavi A, Williams BA, McCue K, Schaeffer L, Wold B (2008) Mapping and quantifying mammalian trasncriptome by RNA-SEq. Nat Methods 5(7):621–628

Nagano T, Mitchell JA, Sanz LA et al (2008) The Air noncoding RNA epigenetically silencestranscription by targeting G9a to chromatin. Science 322:1717–1720

Neguembor MV, Jothi M, Gabellini D (2014) Long noncoding RNAs, emerging players in muscle differentiation and disease. Skelet Muscle 4:8

Nguyen C, Rocha D, Granjeaud S, Baldit M, Bernard K, Naquet P, Jordan BR (1995) Differential gene expression inthe murine thymus assayed by quantitative hybridization of arrayed cDNA clones. Genomics 29:207–216

Nilsen TW (2007) Mechanisms of microRNA-mediated gene regulation in animal cells. Trends Genet 23:243–249

Nottrott S, Simard MJ, Richter JD (2006) Human let-7a miRNA blocks protein production on actively translating polyribosomes. Nat Struct Mol Biol 13:1108–1114

Nuwaysir EF, Huang W, Albert TJ, Singh J, Nuwaysir K, Pitas A, Richmond T, Gorski T, Berg JP et al (2002) Gene expression analysis using oligonucleotide arrays produced by maskless photolithography. Genome Res 12:1749–1755

Nygaard VL, Hovig E (2006) Options available for profiling small samples: a review of sample amplification technology when combined with microarray profiling. Nucleic Acids Res 34:996–1014

O'Connell RM, Rao DS, Chaudhuri AA, Baltimore D (2010) Physiological and pathological roles for microRNAs in the immune system. Nat Rev Immunol 10(2):111–122

O'Neill LA, Sheedy FJ, McCoy CE (2011) MicroRNAs: the fine-tuners of Toll-like receptor signalling. Nat Rev Immunol 11:163–175

Okubo K, Hori N, Matoba R, Niiyama T, Fukushima A, Kojima Y, Matsubara K (1992) Large scale cDNA sequencing for analysis of quantitative and qualitative aspects of gene expression. Nat Genet 2:173–179

Olsen PH, Ambros V (1999) The lin-4 regulatory RNA controls developmental timing in Caenorhabditis elegans by blocking LIN-14 protein synthesis after the initiation of translation. Developmental Biol 216:671–680

Osella M, Bosia C, Cora` D et al (2011) The role of incoherent microRNA-mediated feedforward loops in noise buffering. PLoS Comput Biol 7:e1001101

Oshlack A, Robinson MD, Young M (2010) From RNA-seq reads to differential expression results. Genome Biol 11:220–230

Padron G, Domont GB (2014) Two decades of proteomics in Latin America: a personal view. J Proteomics 107C:83–92

Patel RK, Jain M (2012) NGS QC toolkit: a toolkit for quality control of next generation sequencing data. PLoS ONE 7:e30619.

Pandey RR, Mondal T, Mohammad F et al (2008) Kcnq1ot1antisense noncoding RNA mediates lineage-specific transcriptional silencing through chromatin-level regulation. Mol Cell 32:232–246.

Penny GD, Kay GF, Sheardown, SA et al (1996) Requirement for Xist in X chromosome inactivation. Nature 379:131–137

Petersen CP, Bordeleau ME, Pelletier J, Sharp PA (2006) Short RNAs repress translation after initiation in mammalian cells. Mol Cell 21:533–542

Pietu G, Alibert O, Guichard V, Lamy B, Bois F, Leroy E, Mariage-Samson R, Houlgatte R, Soularue P, Auffray C (1996) Novel gene transcripts preferentially expressed in human muscles revealed by quantitative hybridization of a high density cDNA array. Genome Res 6:492–503

Pietu G, Mariage-Samson R, Fayein NA, Matingou C, Eveno E, Houlgatte R, Decraene C, Vandenbrouck Y, Tahi F et al (1999) The Genexpress IMAGE Knowledge Base of the Human Brain Transcriptome: a Prototype Integrated Resource for Functional and Computational Genomics. Genome Res 9:195–209

Pillai RS, Bhattacharyya SN, Filipowicz W (2007) Repression of protein synthesis by miRNAs: how many mechanisms? Trends Cell Biol 17:118–126

Place RF, Li LC, Pookot D, Noonan EJ, Dahiya R (2008) microRNA-373 induces expression of genes with complementary promoter sequences. Proc Natl Acad Sci USA 105:1608–1613

Plank M, Maltby S, Mattes J, Foster PS (2013) Targeting translational control as a novel way to treat inflammatory disease: The emerging role of MicroRNAs. Clin Exp Allergy 43(9):981–999

Plath K, Fang J, Mlynarczyk-Evans SK et al (2003) Role of histone H3 lysine 27 methylation in X inactivation. Science 300:131–135

Ploner A, Miller LD, Hall P, Bergh J, Pawitan Y (2005) Correlation test to assess low-level processing of high-density oligonucleotide microarray data. BMC Bioinformatics 6:80

Politz JC, Hogan EM, Pederson T (2009) MicroRNAs with a nucleolar location. RNA 15:1705–1715

Ponting CP, Oliver PL, Reik W (2009) Evolution and functions of long noncoding RNAs. Cell 136:629–641

Pritchard CC, Kroh E, Wood B, Arroyo JD, Dougherty KJ, Miyaji MM, Tait JF, Tewari M (2012) Blood cell origin of circulating microRNAs: a cautionary note for cancer biomarker studies. Cancer Prev Res 5:492–497

Quackenbush J (2001) Computational analysis of microarray data. Nat Rev Genet 2:418–427

Ravasi T, Suzuki H, Pang KC et al (2006) Experimental validation of the regulated expression of large numbers of non-coding RNAs from the mouse genome. Genome Res 16:11–19

Rehwinkel J, Natalin P, Stark A, Brennecke J, Cohen SM, Izaurralde E (2006) Genome-wide analysis of mRNAs regulated by drosha and Argonaute proteins in Drosophila melanogaster. Mol Cell Biol 26:2965–2975

Reinhart BJ, Slack FJ, Basson M, Pasquinelli AE, Bettinger JC, Rougvie AE, Horvitz HR, Ruvkun G (2000) The 21-nucleotide let-7 RNA regulates developmental timing in Caenorhabditis elegans. Nature 403:901–906

Ripoli A, Rainaldi G, Rizzo M, Mercatanti A, Pitto L (2010) The Fuzzy Logic of MicroRNA Regulation: a Key to Control Cell Complexity. Curr Genomics 11:350–353

Ritchie ME, Silver J, Oshlack A, Holmes M, Diyagama D, Holloway A, Smyth GK (2007) A comparison of background corrections methods for two-color microarrays. Bioinformatics 23(20):2700–2707

Rocha D, Carrier A, Naspetti M, Victorero G, Anderson E, Botcherby M, Nguyen C, Naquet P, Jordan BR (1997) Modulation of mRNA levels in the presence of thymocytes and genome mapping for a set of genes expressed in mouse thymic epithelial cells. Immunogenetics 46:142–151

Rougemont J, Amzallag A, Iseli C, Farinelli L, Xenarios I, Naef F (2009) Rolexa: statistical analysis of Solexa sequencing data. R package version 1.20.0 Available at Bioconductor (http://bioconductor.org/packages/release/bioc/html/Rolexa.html)

Rüegger S, Großhans H (2012) MicroRNA turnover: when, how, and why. Trends Biochem Sci 37:436–446

Sana J, Faltejskova P, Svoboda M, Slaby O (2012) Novel classes of non-coding RNAs and cancer. J Translat Med 10:103–123

Schanen BC, Li X (2011) Transcriptional regulation of mammalian miRNA genes. Genomics 97:1–6

Schena M, Shanon D, Heller R et al (1996) Parallel human genome analysis: microarray-based expression monitoring of 1000 genes. Proc. Natl. Acad. Sci. USA 93:10614–10619

Schmitter D, Filkowski J, Sewer A et al (2006) Effects of Dicer and Argonaute down-regulation on mRNA levels in human HEK293 cells. Nucleic Acids Res 34:4801–4815

Seyednasrollah F, Laiho A, Elo LL (2013) Comparison of software packages for detecting differential expression in RNA-seq studies. Briefings in Bioinformatics doi:10.1093/bib/bbt086 (in press)

Shi L, Campbell G, Jones WD et al (2010) The MicroArray Quality Control (MAQC)-II study of common practices for the development and validation of microarray-based predictive models. Nat Biotechnol 28 (8):827–838

Shi L, Reid LH, Jones WD et al (2006) The MicroArray Quality Control (MAQC) project shows inter- and intraplatform reproducibility of gene expression measurements. Nat Biotechnol 24 (9):1151–1161

Shimoni Y, Friedlander G, Hetzroni G et al (2007) Regulation of gene expression by small noncoding RNAs: a quantitative view. Mol Syst Biol 3:138

Shin C, Nam JW, Farh KK et al (2010) Expanding the microRNA targeting code: functional sites with centered pairing. Mol Cell 38:789–802

Silver JD, Ritchie ME, Smyth GK (2009) Microarray bakground correction: maximum likelihood estimation for the normal-exponential convolution. Biostatistics 10(2):352–363

Singh RL, Maganti RJ, Jabba SV, Wang M, Deng G, Heath JD, Kurn N, Wangemann P (2005) Microarray-based comparison of three amplification methods for nanogram amounts of total RNA. Am J Physiol Cell Physiol 288:C1179–C1189

Singh-Gasson S, Green RD, Yue Y, Nelson C, Blattner F, Sussman MR, Cerrina F (1999) Maskless fabrication of light-directed oligonucleotide microarrays using a digital micromirror array. Nat Biotechnol 10:974–978

Siomi H, Siomi MC (2010) Posttranscriptional regulation of microRNA biogenesis in animals. Mol Cell 38:323–332

Slonin DK, Yanai I (2009) Getting Started in Gene Expression Microarray Analysis. PLoS Comput Biol 5(10):e1000543

Smyth GK (2005) Limma: linear models for microarray data. In: Gentleman R, Carey V, Dudoit S, Irizarry R, Huber W (eds) Bioinformatics and computational biology solutions using R and bioconductor. Springer, New York, 397–420

Sogayar MC, Camargo AA, Bettoni F et al (2004) A transcript finishing initiative for closing gaps in the human transcriptome. Genome Res 14:1413–1423

Soneson C, Delorenzi M (2013) A comparison of methods for differential expression analysis of RNA-seq data. BMC Bioinform 14:91–108

Standart N, Jackson RJ (2007) MicroRNAs repress translation of m7Gppp-capped target mRNAs in vitro by inhibiting initiation and promoting deadenylation. Genes Dev 21:1975–1982

Stark A, Brennecke J, Bushati N et al (2005) Animal microRNAs confer robustness to gene expression and have a significant impact on 3í´UTR evolution. Cell 123:1133–1146

Stefani G, Slack FJ (2008) Small non-coding RNAs in animal development. Nat Rev Mol Cell Biol 9:219–230

Stekel D (2003) Microarray Bioinformatics. Cambridge University Press, Cambridge. ISBN:9780521525879

Strausberg RL, Riggins GL (2001) Navigating the human transcriptome. Proc. Natl. Acad. Sci. USA 98:11837–11838

Subramanian A, Tamayo P, Mootha VK et al (2005) Gene set enrichment analysis: a knowledge-based approach for interpreting genome-wide expression profiles. PNAS 102:15545–15550

Sudo K, Chinen K, Nakamura Y (1994) 2058 expressed sequence tags (ESTs) from a human fetal lung cDNA library. Genomics 24:276–279

Sudo H, Mizoguchi A, Kawauchi J, Akiyama H, Takizawa S (2012) Use of non-amplified RNA samples for microarray analysis of gene expression. PLoS ONE 7:e31397

Suzuki HI, Yamagata K, Sugimoto K, Iwamoto T, Kato S, Miyazono K (2009) Modulation of microRNA processing by p53. Nature 460:529–533

Taft RJ, Pang KC, Mercer TR et al (2010) Non-coding RNAs: regulators of disease. J. Pathol. 220:126–139

Takeda J, Yano H, Eng S, ZengY, Bell GI (1993) Construction of a normalized directionally cloned cDNA library from adult heart and analysis of 3040 clones by partial sequencing. Hum Mol Genet 2:1793–1798

Tay Y, Zhang J, Thomson AM et al (2008) microRNAs to Nanog, Oct4 and Sox2 coding regions modulate embryonic stem cell differentiation. Nature 455:1124–1128

The ENCODE Consortium (2011) Standards, guidelines and best practices for RNA-seq. http://encodeproject.org/ENCODE/protocols/dataStandards/ENCODE_RNAseq_Standards_V1.0.pdf

Thomson JM, Newman M, Parker JS et al (2006) Extensive post-transcriptional regulation of microRNAs and its implications for cancer. Genes Dev 20:2202–2207

Thomas MF, Abdul-Wajid S, Panduro M et al. (2012) Eri1 regulates microRNA homeostasis and mouse lymphocyte development and antiviral function. Blood 120:130–142

Toyota M, Suzuki H, Sasaki Y et al (2008) Epigenetic silencing of microRNA-34b/c and B-cell translocation gene 4 is associated with CpG island methylation in colorectal cancer. Cancer Res 68:4123–4132

Tsai NP, Lin YL, Wei LN (2009) microRNA mir-346 targets the 5-untranslated region of receptor-interacting protein 140 (RIP140) mRNA and up-regulates its protein expression. Biochem J 424:411–418

Tusher VG, Tibshirani R, Chu G (2001). Significance analysis of microarrays applied to the ionizing radiation response. Proc Natl Acad Sci USA 98(9):5116–5121

Upton JP, Wang L, Hand D et al (2012) IRE1a cleaves select microRNAs during ER stress to derepress translation of proapoptotic caspase-2. Science 338:818–822

Valadi H, Ekstrom K, Bossios A et al (2007) Exosome-mediated transfer of mRNAs and microRNAs is a novel mechanism of genetic exchange between cells. Nat Cell Biol 9:654–659

Valencia-Sanchez MA, Liu J, Hannon GJ et al (2006) Control of translation and mRNA degradation by miRNAs and siRNAs. Genes Dev 20:515–524

Van Haaften RI, Schroen B, Janssen BJ, van Erk A, Debets JJ, Smeets HJ, Smits JF, van den Wijngaard A, Pinto YM, Evelo CT (2006) Biologically relevant effects of mRNA amplification on gene expression profiles. BMC Bioinformatics 7:200

Van Heesch S, Van Iterson M, Jacobi J et al (2014) Extensive localization of long noncoding RNAs to the cytosol and mono- and polyribosomal complexes. Genome Biol 15:R6

Van Rooij E, Sutherland LB, Qi X et al (2007) Control of stress-dependent cardiac growth and gene expression by a microRNA. Science 316:575–579

Velcunescu VE, Zhang L, Volgelstein B, Kinzler KW (1995) Serial analysis of gene expression. Science 270:484–487

Velculescu VE, Zhang L, Zhou W et al (1997) Characterization of the yeast transcriptome. Cell 88:243–251

Vella MC, Choi EY, Lin SY et al (2004) The C. elegans microRNA let-7 binds to imperfect let-7 complementary sites from the lin-41 3' UTR. Genes Dev 18:132–137

Wada T, Kikuchi J, Furukawa Y (2012) Histone deacetylase 1 enhances microRNA processing via deacetylation of DGCR8. EMBO Rep 13:142–149

Wang S, Raghavachari S (2011) Quantifying negative feedback regulation by microRNAs. Phys Biol 8:055002

Wang X, Cairns MJ (2013) Gene set enrichment analysis of RNA-Seq data:integrating differential expression and splicing. BMC Bioinform 14(5):S16

Wang J, Hu L, Hamilton SR, Coombes KR, Zhang W (2003) RNA amplification strategies for cDNA microarray experiments. Biotechniques 34:394–400

Watson JD (1990) The human genome project: past, present, and future. Science 248:44–49

Wery M, Kwapisz M, Morillon A (2011) Noncoding RNAs in gene regulation. Wiley Interdiscip Rev Syst Biol Med. 3:728–738

Wiesen JL, Tomasi TB (2009) Dicer is regulated by cellular stresses and interferons. Mol Immunol 46:1222–1228

Wightman B, Ha I, Ruvkun G (1993) Posttranscriptional regulation of the heterochronic gene lin‘14 by lin‘4 mediates temporal pattern formation in C. elegans. Cell 75:855–862

Winter J, Jung S, Keller S et al (2009) Many roads to maturity: microRNA biogenesis pathways and their regulation. Nat Cell Biol 11:228–234

Wreschner DH, Herzberg M (1984) A new blotting medium for the simple Isolation and Identification of highly resolved messenger RNA. Nucleic Acids Res 12:1349–1359

Wu L, Fan J, Belasco JG (2006) microRNAs direct rapid deadenylation of mRNA. Proc. Natl Acad Sci USA 103:4034–4039

Xiao C, Rajewsky K (2009) MicroRNA control in the immune system: basic principles. Cell 136:26–36

Yang JS, Lai EC (2011) Alternative miRNA biogenesis pathways and the interpretation of core miRNA pathway mutants. Mol Cell 43:892–903

Yi R, Qin Y, Macara IG et al (2003) Exportin-5 mediates the nuclear export of pre-microRNAs and short hairpin RNAs. Genes Dev 17:3011–3016

Young MD, Wakefield MJ, Smyth GK et al (2010) Gene ontology analysis for RNA-seq: accounting for selection bias. Genome Biol 11:R14

Yue D, Liu H, Huang Y (2009) Survey of Computational Algorithms for MicroRNA Target Prediction. Curr Genomics 10:478–492

Zamore PD, Haley B (2005) Ribo-gnome: the big world of small RNAs. Science 309:1519–1524

Zhang H, Kolb FA, Brondani V et al (2002) Human Dicer preferentially cleaves dsRNAs at their termini without a requirement for ATP. EMBO J 21:5875–5885

Zhao N, Hashida H, Takahashi N, Misumi Y, Sakaki Y (1995) High-density cDNA filter analysis: a novel approach for large-scale, quantitative analysis of gene expression. Gene 156:207–213

Zitvogel L, Regnault A, Lozier A et al (1998) Eradication of established murine tumors using a novel cell-free vaccine: dendritic cell-derived exosomes. Nat Med 4:594–600

Chapter 2
Transcriptome Analysis Throughout RNA-seq

Tainá Raiol, Daniel Paiva Agustinho, Kelly Cristina Rodrigues Simi, Calliandra Maria de Souza Silva, Maria Emilia Walter, Ildinete Silva-Pereira and Marcelo Macedo Brígido

Abstract Differential gene expression profile is a powerful tool to identify changes in cell or tissue trancriptomes, which allows to understanding complex biological process such as oncogenesis, cell differentiation and host immunological response to pathogens, among others. To date, the gold standard technique to compare gene expression profile is micro-array hybridization of a RNA preparation. In recent years technological advances led to a new generation of sequencing methods, which can be explored to uncover the complete content of a cell transcriptome. Such a deep sequencing of a RNA preparation, named RNA-seq, allows to virtually detect the complete RNA content, including low abundant isoforms. The RNA-seq quantitative aspect may be further explored to detect gene differential expression based on a reference genome and gene model. In contrast to micro-arrays, RNA-seq may find a broader range of RNA isoforms as well as novel RNA molecules, and has been gradually substituting micro-arrays to differential gene expression profile. In this chapter we describe how deep sequencing may be used to describe changes in the gene expression profile, its advantages and limitations.

2.1 High-Throughput Sequencing Techniques

Since the development of Sanger's technology in the 70's, DNA sequencing has been continuously improved regarding to both throughput and low cost. Next generation sequencing (NGS), also called high-throughput or deep sequencing, constitutes a new breakthrough of increasingly research power, a revolutionary advance

M. M. Brígido (✉) · T. Raiol · K. C. R. Simi · C. Maria de Souza Silva · I. Silva-Pereira
Laboratório de Biologia Molecular, CEL/IB (Pós-graduação em Biologia Molecular/CEL/IB), Universidade de Brasília, Brasília, DF 70910-900, Brazil
e-mail: brigido@unb.br

D. P. Agustinho
Laboratório de Biologia Molecular, CEL/IB (Pós-graduação em Patologia Molecular/FM/UnB), Universidade de Brasília, Brasília, DF 70910-900, Brazil

M. E. Walter
Departamento de Ciência da Computação, Instituto de Ciências Exatas, Universidade de Brasília, Brasília, DF 70910-900, Brazil

© Springer International Publishing Switzerland 2014
G. A. Passos (ed.), *Transcriptomics in Health and Disease,*
DOI 10.1007/978-3-319-11985-4_2

in molecular biology knowledge. An increasing number of biological questions may be addressed by NGS technologies, which provides a much larger comprehensive survey compared to the Sanger method, and under a system biology perspective. In particular, transcriptomics has been strongly benefited by the use of these new technologies, also called RNA-seq, allowing a complete characterization of whole transcriptome at both gene (Kvam et al. 2012) and exon (Anders et al. 2012) levels, and with an additional ability to identify rare transcripts, new genes, novel splicing junctions and gene fusions (Katz et al. 2010; Wang et al. 2009; Van Verk et al. 2013).

In this chapter, first we address a brief overview on sequencing techniques and the most common next-generation platforms, as well as computational methods for RNA-seq data analysis. After, we discuss two case studies to assess the capabilities of RNA-seq in addressing important biological issues.

2.1.1   Sanger's Sequencing Technology

In 1977, Frederick Sanger and colleagues (Sanger et al. 1977) developed the DNA sequencing method, which in 2001 allowed the first Human genome draft (Lander et al. 2001). This method called dideoxy chain-termination or simply Sanger method is based on the use of special nucleotide molecules (called ddTNP), lacking a 3'-OH at the deoxyribose, which blocks the DNA elongation. These special nucleotides are mixed in lower concentrations to the regular nucleotides and used as reagents for DNA polymerase reaction. Therefore, with the polymer synthesis stopped by the inclusion of a ddNTP, the last nucleotide can be determined. Each of the four ddNTPs was added separately in four different reactions. At the beginning, one of the regular nucleotides, most commonly dATP or dCTP, was radioactively labeled (e.g., ^{32}P or ^{35}S) in order to achieve the radioactive signal. Usually, polyacrylamide gel electrophoresis was used for separation of the DNA molecules, which diverged in length by a single nucleotide. Then the gel was dried and exposed to X-ray film.

An important modification of the method was the substitution of the radioactive label by a fluorescent dye (Smith et al. 1986). Each distinct wave length produced by the fluorescent dyes linked to dideoxynucleotides corresponds to a different nucleotide, with the four sequencing reactions performed in the same tube. With the automation of the Sanger sequencing method, the performance reached up to 96 different reactions running in parallel capillary gel electrophoresis (Marsh et al. 1997), which is considered the first-generation technology. In the top of the technology 384 samples could be sequenced at once in a single multi-well plate. The main sequencing devices using Sanger method are ABI (Applied Biosystems) and MegaBACE (GE Healthcare Life Sciences).

The main advantage of Sanger sequencing is the length of the produced sequences, about 1000 kb, which is still unreachable by the main NGS technologies nowadays. However, deep sequencing has the advantage of high coverage, i.e., a large amount of redundant data, further treated through bioinformatics analysis, generating much more informative data in a single run.

Table 2.1 Comparison of next generation sequencing technologies

	Sanger ABI 3730xl	454 GS FLX	HiSeq 2000	SOLiDv4	Ion Torrent PGM (318 chip)
Read length (bp)	900	700	150	85	100
Cost (US$/Mb)	500	12.56	0.02	0.04	0.63
Output data/run	2.88 Mb	0.7 Gb	600 Gb	30 Gb	1 Gb
Time run	3 h	1 day	8 days	7 days	3 h

2.1.2   Next Generation Sequencing

Regulatory mechanisms and gene expression profiles have been widely investigated towards elucidation of several essential cellular processes. Hybridization-based technology, e.g., microarray, has been very useful for determining global gene expression. However, the high background levels due to cross-hybridization, a limited range of quantification and a restricted detection of known genes are bottlenecks for large scale use of this technique (Shendure 2008). RNA-seq allows a genome-scale transcriptome analysis, including novel genes and splice variants, with a large range of quantification and reduced sequencing costs (Wang et al. 2009; Soon et al. 2013). These advantages make RNA-seq a better and attractive solution for whole-genome transcriptome analysis of several organisms, even for those with no sequenced reference genomes.

Nowadays, the most commonly used NGS platform for RNA-seq research is the Illumina HiSeq. A comparison of NGS technologies is shown in Table 2.1 based on data from Liu et al. (2012).

The enormous amounts of data generated by NGS create new challenges to the downstream bioinformatics analysis, which has to handle with large sequence files while searching for comprehensive and useful biological information, discussed later in this chapter.

2.1.3   454 Sequencing

In 2005, the 454 sequencing platform was formally announced by Roche as a new massive parallelized sequencer (Margulies et al. 2005). It was the first technology, among several others, considered as next generation sequencing. Since 454 produces the largest sequences among the NGS platforms, it is mainly used for transcriptome studies concerning organisms without a reference genome.

The pyrosequencing method used by 454 sequencing is based on the detection of pyrophosphate released during the nucleotide incorporation promoted by DNA polymerase (Harrington et al. 2013; Mardis 2013). In contrast to the Sanger sequencing, pyrosequencing is designated as a sequence-by-synthesis technique because DNA synthesis is monitored in real time. Single-stranded DNA library is generated after fragmentation and addition of adaptors to both fragment ends

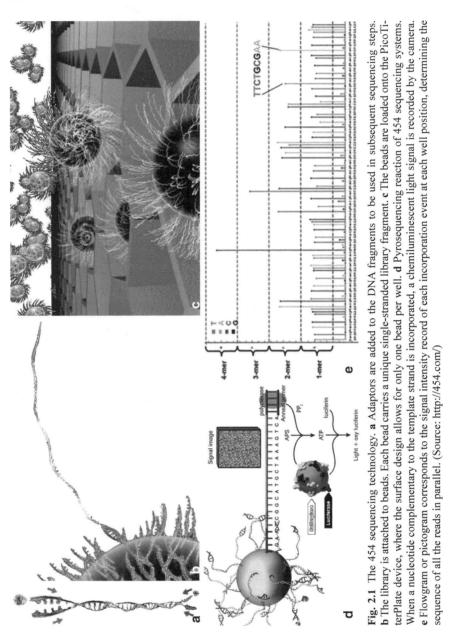

Fig. 2.1 The 454 sequencing technology. **a** Adaptors are added to the DNA fragments to be used in subsequent sequencing steps. **b** The library is attached to beads. Each bead carries a unique single-stranded library fragment. **c** The beads are loaded onto the PicoTiterPlate device, where the surface design allows for only one bead per well. **d** Pyrosequencing reaction of 454 sequencing systems. When a nucleotide complementary to the template strand is incorporated, a chemiluminescent light signal is recorded by the camera. **e** Flowgram or pictogram corresponds to the signal intensity record of each incorporation event at each well position, determining the sequence of all the reads in parallel. (Source: http://454.com/)

(Fig. 2.1a and b). One single fragment is ligated to beads covered by adaptors to proceed to the clonal amplification by emulsion PCR. Bead-ligated sequence is added, along with amplification reagents, in a water-in-oil mixture to trap individual beads in amplification micro-reactors. Next, the bead-ligated amplified sequences are added to the PicoTiterPlate device containing millions of 28 μm wells, the precise size for a single bead (Fig. 2.1c). To these wells, enzyme beads (containing sulfurylase and luciferase) are added to the device. Each nucleotide is added to the system separately during the sequencing rounds. With the incorporation of one nucleotide, a pyrophosphate is released and used by sulfurylase to convert ADP into ATP, the substrate of luciferase (Fig. 2.1d). ATP and luciferin are used by luciferase to produce luminescence, which is detected by a visible-light high-sensitivity CCD camera. Apyrase is subsequently added to remove any non-incorporated nucleotide, and, then, the next round is initiated. The signal strength is proportional to the number of added nucleotides, recorded as a pyrogram (Fig. 2.1e). Sequences are stored as standard flowgram format (SFF), a binary format that is further converted in the FASTQ format, used in the bioinformatics analysis.

2.1.4 Illumina Sequencing

Illumina sequencing uses reversible dye-terminator technique that adds a single nucleotide to the DNA template in each cycle (Bentley et al. 2008). This system was initially developed in 2007 by Solexa and was subsequently acquired by Illumina, Inc. Illumina is widely used in several whole transcriptome studies since it reaches the deepest depth among NGS technologies. However, the small sequence size (around 100 bp) hampers the assembly into contigs as normally used for Sanger and 454 sequencing. Therefore, a reference genome is usually necessary for Illumina data analysis.

As 454, Illumina sequencing is based on sequencing-by-synthesis, however, instead of clonal amplification using beads, Illumina sequencing is performed in a solid slide covered by adaptors complementary to those added to the fragmented DNA sequences (Metzker 2010). This procedure, called bridge PCR, consists in amplification of bended DNA sequences, attached by both ends to the solid surface (Fig. 2.2a). By the end of the clonal amplification, clusters of identical DNA sequences will be formed in order to amplify the fluorescence signals. In each round, one single nucleotide is added to the single-strand template sequences followed by fluorescence detection by a high-sensitivity CCD camera (Fig. 2.2b and d). As in Sanger's technology, different fluorophore molecules are attached to each nucleotide, however, these molecules hamper the polymerase to add new ones. The fluorescence emission releases the 3'OH of the recent added nucleotide allowing it to receive new monomers in the next sequencing round.

Single-end sequencing, i.e., reads generated from a single end adaptor, is being replaced by the paired-end sequencing, since the accuracy in downstream analysis is greater with a fairly price. Paired-end reads are produced from the adaptor priming sites in both template sequence ends, being the second adaptor primer used in a subsequent sequencing run (Fig. 2.2c).

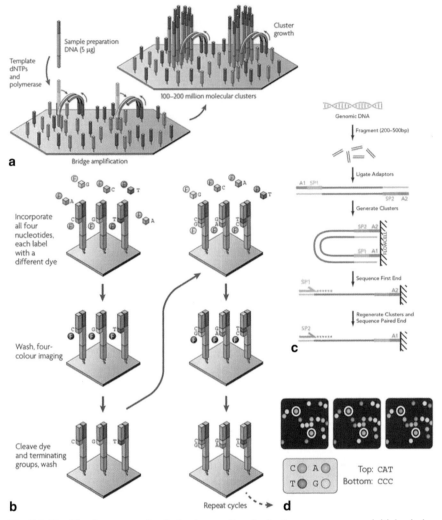

Fig. 2.2 The Illumina sequencing technology. **a** Two basic steps encompass an initial priming and extending of the single-stranded, single-molecule template, and bridge amplification of the immobilized template in a solid device with immediately adjacent primers to form clusters. **b** The four-color cyclic reversible termination (*CRT*) method uses terminator chemistry. A cleavage step removes the fluorescent dyes and regenerates the 3'-OH group. **c** Paired-end sequencing by which reads are generated from both template strand. "A" block indicates the device-ligation adaptors and "SP", sequencing primers. **d** In the images, the sequencing data is highlighted from two sequence clusters. (Source: Metzker 2010 and http://www.illumina.com/)

2.2   Bioinformatics Pipelines for Transcriptome Projects

As described previously, Illumina sequencing has been commonly used in transcriptome projects, since the volumes of sequenced reads (named *raw data*) allow to finding virtually the total of the expressed genes (transcripts). Due to the short lengths of the Illumina reads, they are usually mapped in a reference genome, the mapped regions indicating the expressed genes of the RNA-seq sample. If the organism genome is not yet sequenced, new specially developed methods to handle short reads have been used to reconstruct the transcript sequences, e.g., 454 sequencing produces sequences four times larger than those produced by Illumina. Also in this case, the original reads are usually assembled in larger sequences, in order to rebuild each (fragment of) transcript. In both cases, the metaphor for reconstructing the transcripts is like mounting a puzzle, where the pieces (the reads) have to be assembled (relative to a reference genome or not) to obtain the picture (the transcripts of the transcriptome). After this, different analyses can be performed on these reconstructed transcripts, e.g., quantitative analysis and differential expression. In transcriptome projects, the tasks of reconstructing transcripts and performing biological analyses are performed by bioinformatics pipelines, discussed next.

2.2.1   Pipelines

A bioinformatics pipeline is a computational system composed of a sequence of softwares (computer programs), sequentially executed, in which the output data from one software is the input data for the following software.

In general, transcriptome bioinformatics pipelines have the following phases, which can be combined according to the input raw data and the objectives of each project:

- *filter (or clean)* raw data for quality assessment: this is usually performed in two steps as follows. In the *clipping step*, a fragment (or the whole read) containing adapters is removed, while in the *trimming step*, reads are filtered to remove low quality sequences. This filtering phase guarantees a reliable dataset of quality short reads, to be used in the following phases of the pipeline;
- *map* short reads to reference genomes: the filtered reads are aligned to a reference genome, in order to find the genomic regions presenting matches with these reads;
- *assembly (or group)* reads: each group of reads (called *contig*), composed of reads having similar extremities (the end of a read is similar to the beginning of another read), allows to construct one larger sequence (called *consensus*), which is a predicted (fragment of) transcript;
- *analysis* of the set of (fragments of) transcripts obtained from the mapping or the assembling phase: allows to obtain relevant biological information, e.g.,
  - *quantitative analysis*: among others, coverage analysis shows the abundance of genes expressed in one RNA-seq sample, more precisely, the number of reads mapped in a certain region of the chromosome
  - *differential expression*: allows to analyze the variability of genetic expression between samples
  - *annotation*: assigns a biological function to each transcript

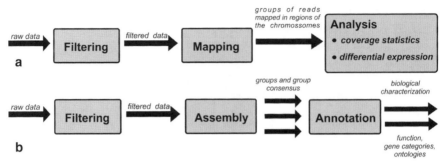

Fig. 2.3 a Pipelines for short reads, with a well-characterized reference genome, and two types of analyses—coverage statistics and differential expression. **b** Pipeline for longer reads, with no reference genome, and annotation (biological function, gene categories and ontologies)

Two generic bioinformatics pipelines for transcriptomes are discussed next, although the design of a particular pipeline depends on the objectives of the transcriptome project and other information, e.g., the sequencer (since the sequencing techniques may cause specific errors in the raw data, which have to be treated), and availability of information to be used in the analysis phase (e.g., quantitative and differential expression softwares and availability of reference genome).

Pipeline 1 The organism of interest has at least one reference genome already sequenced, with well-annotated genes and other biological characteristics, and the reads are short (about 100 bp, e.g., short reads produced by Illumina). A pipeline with three phases can be designed (Fig. 2.3a): filtering, mapping and quantitative analysis.

Pipeline 2 The organism of interest has not been sequenced before, and the reads are longer (from 400 bp to 800 bp, e.g., reads produced by 454). A pipeline can be designed with three phases (Fig. 2.3b): filtering, assembly and annotation. The assembly phase construct one consensus sequence for each group of reads presenting similar extremities. The annotation phase assigns biological functions to the consensus sequences.

A pipeline is usually implemented using a programming language (e.g., Java or Perl) that controls the execution of the softwares, which use files organized in file directories, or a database management system (e.g., MySQL (MySQL 1995) or PostgreSQL [PosGres]) that stores, retrieves and manages data. Each pipeline phase uses public (open source) or private softwares, and some of the most commonly used public ones are described next.

2.2.2 Bioinformatics Softwares

2.2.2.1 Filtering

As can be seen in Part 4.1, the high-throughput sequencers use different techniques, which may cause specific errors in the reads. These errors have to be treated to guarantee quality to the reads used in the next pipeline phases. Therefore, the *filtering*

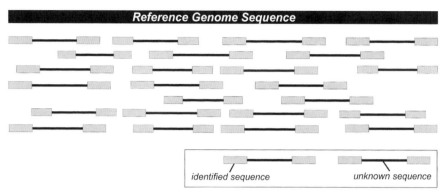

Fig. 2.4 Short reads mapped to a reference genome. (Source: http://readtiger.com/wkp/en/Genomics)

(or *cleaning*) phase performs clipping and trimming, as described before. The reads are stored using format FASTQ, which stores the nucleotides of each read together with their corresponding quality scores.

Some tools are used to assess the qualities as well as other information about the input sequences. FastQC (Andrews 2010) allows to verifying quality of raw data. FASTX-Toolkit (Gordon and Hannon 2010) provides options for performing both clipping and trimming. Other commonly used tools are Cutadapt (Martin 2011) for clipping and PRINSEQ (Schmieder and Edwards 2011) for trimming. All of them present several options such as minimum size of one read, minimum quality score and polyadenylation removal.

2.2.2.2 Softwares for Mapping

The objective of the mapping phase is to find where each filtered short read is located in a reference genome (Fig. 2.4).

There are many softwares capable to performing the mapping process. In general, these softwares are computational intensive (to process and store data), and mapping techniques use indices to accelerate the search procedure and to reduce the memory cost associated to finding the location of the short reads to the reference genome.

Bowtie (Langmead et al. 2009) is a fast short aligner that tolerates a small number of mismatches. Bowtie first concatenates all the reference genome in one single string, and performs the Burrows-Wheeler transformation to generate one index to this reference genome. Next, one character of each sequence is mapped until the entire sequence is aligned. If the sequence cannot be aligned, the program backtracks one step, substituting one character, and repeating the process. The maximum number of character substitutions is a parameter in Bowtie.

Table 2.2 Softwares and their corresponding web sites. (Adapted from Trapnell and Salzberg 2009)

Mapping software	Web site
Bowtie	http://bowtie.cbcb.umd.edu
BWA	http://bio-bwa.sourceforge.net/bwa.shtml
Maq	http://maq.sourceforge.net
Mosaik	http://bioinformatics.bc.edu/marthlab/wiki/index.php/Software
Novoalign	http://www.novocraft.com
Segemehl	http://www.bioinf.uni-leipzig.de/Software/segemehl/
SOAP2	http://soap.genomics.org.cn
TopHat	http://tophat.cbcb.umd.edu/
ZOOM	http://www.bioinfor.com

TopHat (Trapnell et al. 2009; Kim et al. 2013) first aligns the RNA-seq short reads to large genomes using Bowtie, and then analyzes the mapping results to identify splice junction between exons.

Segemehl (Hoffmann et al. 2009, 2014) maps short reads to reference genomes, detecting mismatches, insertions and deletions. Moreover, Segemehl can deal with different read lengths and is able to correctly map primer—or polyadenylation contaminated reads. Segemehl matching method is based on enhanced suffix arrays, supporting the SAM format and queries with gziped reads to save disk and memory space, and allowing both bisulfite sequencing and split read mappings.

There are many other computational methods to map short reads to a reference genome, as shown in Table 2.2.

2.2.2.3   Softwares for Assembling

The assembly phase aims to group reads with similar extremities (Fig. 2.5), i.e., the overlapping of the end of one read and the beginning of another indicates that both probably belong to the same transcript. These similar extremities enable to reconstruct larger regions of the transcripts. As said before, each of these groups is called

Fig. 2.5 Overlapping of the extremities of the reads indicates that they are parts of the same (fragment of one) transcript

contig, and the sequence resulting of the overlapping extremities of the reads in one contig, called consensus, is a predicted (fragment of) transcript.

Assembly (or *de novo* assembly) is convenient for 454, which generates sequences from 400 to 600 bp. The assembly software has to consider the sequencing errors when joining the reads produced by one sequencer. One of the main problems of the 454 sequencer is the existence of homopolymers errors.

MIRA (Mimicking Intelligent Read Assembly) (Chevreux et al. 2004) maps reads from electrophoresis sequencing (Sanger) and RNA-seq (Illumina) to contigs. In particular, miraEST is a fragment assembler for EST transcripts, capable of reconstructing mRNA transcripts while detecting and classifying single nucleotide polymorphisms (SNPs) occurring in different variations of the transcripts. The assembly system uses iterative strategies based on highly reliable regions within the sequences, with a return strategy that uses low reliable regions if needed. miraEST has: special functions to assemble numerous highly similar sequences without prior masking; an automatic editor to perform editions and analysis of alignments by looking at the underlying traces; possibilities to use incorrectly preprocessed sequences; routines to use additional sequencing information, e.g., base error probabilities and template insert sizes; and functions to detect and solve possible misassemblies. Besides, miraEST can detect and classify sequence properties, e.g., SNPs, with high specificity and sensitivity of one mutation per sequence. The assembler is commonly used for similarity analysis of transcripts between organisms, and assembly of sequences from various sources for oligonucleotide design in clinical microarray experiments. The default values for MIRA mapping should allow it to work with many ESTs and RNA-seq datasets, even from non-normalized libraries. However, for very high coverage (e.g., about 10 k coverage), some procedure to reduce data will lead to a more efficient processing with MIRA. Recent developments of MIRA allowed to perform *de novo* RNA-seq assembly of non-normalized libraries, and MIRA can be used for datasets with up to 50 million Illumina 100 bp reads.

ABySS (Assembly By Short Sequences) (Simpson et al. 2009) is a de novo, paired-end sequence assembler designed for short reads. The single-processor version is useful for assembling genomes of up to 100 Mbases, while the parallel version (implemented using MPI) is capable of assembling larger genomes.

Table 2.3 summarizes some important assemblers, together with their sites.

Table 2.3 Characteristics of assembly softwares for high-throughput sequencers	Assembly software	Web site
	Abyss (Simpson et al. 2009)	http://www.bcgsc.ca/platform/bioinfo/software/abyss
	Edena (Hernandez et al. 2008)	http://www.genomic.ch/edena
	MIRA (MIRA, Mira mapping)	http://mira-assembler.sourceforge.net/docs/DefinitiveGuideToMIRA.html#chap_mapping

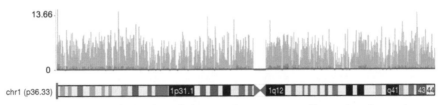

Fig. 2.6 Read coverage of transcripts relative to a reference genome. (Source: http://www.plosone. org/article/info%3Adoi%2F10.1371%2Fjournal.pone.0016266 (adapted from Twine et al. (2011) Whole transcriptome sequencing reveals gene expression and splicing differences in brain regions affected by Alzheimer's disease. Plos one 6(1):e16266))

2.2.2.4   Softwares for Analyses

In transcriptome projects, quantitative analysis, differential expression, and transcript annotation are extensively used.

Quantitative Analysis The transcript coverage is the number of reads "covering" (or the number of mapped reads in) a transcript. The larger the number, the more abundant is the expressed gene in a RNA-seq sample (Fig. 2.6).

Coverage abundance can be computed using Bioconductor (Open Source Software for Bioinformatics) (Gentleman et al. 2004), which provides tools for the analysis of high-throughput data. Bioconductor uses the R statistical programming language (Team 2005), and is open source and open development. In particular, the RNASeqMap library (Leõniewska and Okoniewski 2011) provides classes and functions to analyze the RNA-sequencing data using the coverage profiles in multiple samples at a time.

R (Team 2005) is a language and an environment for statistical computing and graphics. R provides a wide variety of statistical (e.g., linear and nonlinear modeling, classical statistical tests, classification and clustering) and graphical techniques, and is highly extensible. R is easy to use, allowing to output well-designed publication-quality plots. Among other softwares, which facilitate data manipulation, calculation and graphical display, R includes: effective data handling and storage facility; a collection of intermediate tools for data analysis; a simple and powerful programming language which includes conditionals, loops, user-defined recursive functions and input and output facilities.

Differential Expression Differential expression refers to the study of the variability of genetic expression between samples. One important objective of transcriptome projects is to identify the differentially expressed genes in two or more conditions (Rapaport et al. 2013). These genes are selected based on parameters, usually based on p-values generated by statistical modeling. The expression level is measured by the number of reads mapping to the transcript, which is expected to correlate directly with its abundance level. This measure is different from gene probe-based methods, e.g., microarrays. In RNA-seq, the expression of a transcript is limited by the sequencing depth and depends on the expression levels of other transcripts, in contrast to array-based methods, in which probe intensities are independent one of each other. This, and other technical differences, has motivated the development of many statistical algorithms, with different approaches for normalization and differential expression detection. As an example, Poisson or negative binomial distributions to model the gene count data and a variety of normalization procedures are common approaches.

Cufflinks (Trapnell et al. 2010) may be used to measure global *de novo* transcript isoform expression. It assembles transcripts, estimates their abundances, and determines differential expression (Trapnell et al. 2013) and regulation in RNA-seq samples. Moreover, Cufflinks accepts reads aligned by other mapper, and assembles the alignments to a parsimonious set of transcripts. Then, it estimates the relative abundances of these transcripts based on how many reads support each one, considering also biases in library preparation protocols.

Some articles discuss and compare statistical methods to compute differential expression. Kvam et al. (2012) make a review and compare four statistical methods—edgeR, DESeq, baySeq, and a method with a two-stage Poisson model (TSPM). Rapaport et al. (2013) describe an extensive evaluation of common methods—Cuffdiff (Trapnell et al. 2013), edgeR (Robinson et al. 2010), DESeq (Anders and Huber 2010), PoissonSeq (Li et al. 2012), baySeq (Hardcastle and Kelly 2010), and limma (Smyth 2004) adapted for RNA-seq use, using the Sequencing Quality Control (SEQC) benchmark dataset and ENCODE data.

Splice Junctions Splice junctions are nucleotide sequences at the exon–intron boundary in the pre-messenger RNA of eukaryotes, that are removed during the RNA splicing, and can generate many processed transcripts from one gene. Computationally, the problem is to recognize, given a sequence of DNA, the boundaries between exons (the parts of the DNA sequence retained after splicing) and introns (the parts of the DNA sequence that are spliced out). This problem consists of two subtasks: recognizing exon/intron boundaries (called EI sites), and recognizing intron/exon boundaries (IE sites). IE borders are called "acceptor sites" while EI borders are called "donor sites". The recognition and quantification of splice variants is one of the advances of RNA-seq over micro-array to measure differential gene expression.

TopHat (Trapnell et al. 2009) identifies splice junctions, producing the junctions. bed file, where the field score is used to indicate coverage depth. The identified splice junctions can be displayed in browsers (e.g., UCSC genome browser (Kuhn et al. 2013)) using.bed files encoding splice junctions. Junction files should be in the standard.bed format.

Pasta (Patterned Alignments for Splicing and Transcriptome Analysis) (Tang and Riva 2013) is a splice junction detection algorithm designed for RNA-seq data, based on a highly accurate alignment strategy and on a combination of heuristic and statistical methods to identify exon–intron junctions with high accuracy.

Annotation The annotation phase has the objective of assigning a biological function for each transcript, identifying genes and finding more information, e.g., biological categories and ontologies.

The annotation methods can be organized in two classes:

- softwares to find genes *ab initio*, where some structural characteristics of genes are used;
- softwares to perform pairwise comparison of one transcript against a file with known transcripts and their corresponding annotation. This can be done comparing the nucleotides, or the translated nucleotides.

The pairwise sequence comparison (or pairwise alignment), where a query sequence (transcript of the organism of interest) is compared with other sequences (and corresponding biological functions) stored in files, relies on an algorithm that computes an alignment among two transcripts. The hypothesis is based on Darwin evolution theory, which claim that living organisms evolved from ancestor organisms. Therefore, if two transcripts have similar sequences of nucleotides, they may be homologs, and probably share the same biological functions. This means that biological function may be inferred from similar sequences. Important pairwise algorithms, which produce alignments between pairs of sequences, are Smith-Waterman (Smith and Waterman 1981) and BLAST (Altschul et al. 1990).

Similarly to the assembly phase, the main difficult in annotation is due to the transcript length. The resulting genes may be fragmented, causing loss of information. Since alignment programs are error-tolerant, it is reasonable to expect that the annotation for transcripts (predicted from reads generated by high-throughput sequencers) is correct if functions of genes of other organisms have been found correctly.

In contrast, finding genes *ab initio* is not so error robust, since some kinds of errors can lead to incorrect gene prediction. In particular, sequencing errors introducing a stop codon can result in an incorrectly predicted gene.

2.3   Case Study 1

RNA-seq as an efficient tool to analyze and identify gene expression patterns related to murine bone marrow-derived macrophage's susceptibility to Candida albicans infection.

The improvements of organ transplantation techniques, as well as the rise of immune compromised diseases, like AIDS, are directly linked to the exponential growth of invasive infections in these patients. Therefore, the study of the etiological agents of these diseases, particularly fungal pathogens, together with the immune response they elicit, became paramount (Marr et al. 2002; Miceli et al. 2011; Richardson and Lass-Florl 2008). Among fungi, *Candida albicans* appears as a main cause of invasive infections, showing high rates of morbidity and mortality (Pappas 2006; Pappas et al. 2003; Chi et al. 2011; Shigemura et al. 2014).

Many studies have been done to understand the aspects of immune responses to *C. albicans* (Martinez-Alvarez et al. 2014; Miramon et al. 2013; Hunniger et al. 2014; Tierney et al. 2012). In this work, transcriptomic response of murine bone marrow-derived macrophages (BMDM) was analysed by RNA-seq to characterize the transcriptomic patterns of susceptibility. RNA-seq permits the discovery of new exons or transcripts, the identification of different alternative splicing patterns, as well as a global overview of the transcriptome, offering a more flexible experimental approach (Black et al. 2014; Zhao et al. 2014; Wang et al. 2009; Soon et al. 2013). Therefore, the main objective of this project was the identification of BMDMs gene expression patterns between resistant and susceptible mice after *C. albicans* infection, by the analysis of the resulting transcriptome profiles.

Bone marrow was extracted from the mice, and the hematopoietic stem cells were then differentiated into macrophages. 2×10^6 BMDMs were co-cultured with 4×10^6 *C. albicans* yeasts for 2 h, and the RNA was extracted using RNeasy (Qiagen). RNA quality and concentration were verified employing a Bioanalyzer (Agilent) and NanoDrop (Thermo Scientific) or Qubit® Fluorometric (Invitrogen), respectively. Three microgram of total RNA was used for the library preparation, which also included a step of rRNA depletion using Ribozero (Epicentre) before sequencing in an Illumina HiSeq platform.

The sequencing results were provided in FASTQ format. FastQC (Andrews 2010) was used to assess quality. Adaptors clipping and quality trimming were performed using Cutadapt (Martin 2011).

Two mapping softwares, NextGenMap (NGM) (Sedlazeck et al. 2013) and TopHat2 (Kim et al. 2013), were employed. Since both generate similar number of mapped reads, we chose NextGenMap due to its faster analysis. Low quality mappings were removed using Samtools (Li et al. 2009), which was also used to sort, index and convert the mapping results from.sam to.bam files. Bedtools (Quinlan and Hall 2010) was then used to count reads for both genes or exons, and generate a table of these counts, to be analyzed for differential expression. As said before, differential expression can be analyzed using different methodologies (Soneson and Delorenzi 2013; Wagner et al. 2012), and edgeR (Robinson et al. 2010) and DESeq (Anders and Huber 2010) were chosen. Both outputted very similar results. Alternative splicing can be checked by differential exons usage (Anders et al. 2012). Therefore, the resulting list of genes or transcripts differentially expressed was checked for gene onthology (GO terms) using either Biomart (Kasprzyk 2011), or the topGO (Alexa and Rahnenfuhrer 2010) Bioconductor package.

Several problems may occur in RNA-seq projects, and here we point out some of these:

- *Infection conditions:* the optimization of the protocols of co-culture conditions, as well as RNA extraction, may be hard to adjust. Setting a Multiplicity of infection (MOI—proportion of host/pathogen cells in the co-culture) that suffices to induce a transcriptomic response in the host cells is the first step. However, a very high MOI may result in host cells death and apoptosis, which may result in altered gene expression or low amounts of RNA extracted from these cells;
- *Infection time:* the definition of correct time intervals of interaction between pathogen and host cells is essential, since different genes have different kinetics of transcription during co-culture. This may vary drastically for different pathogens;

- *Biological replicates*: in transcriptomic studies, a robust statistical analysis is of very important. In this sense, the experimental design have to incorporate proper biological replicates to allow valid statistical inferences (Robles et al. 2012);
- *Library preparation and sequencing parameters*: the choice of the preparation methodologies, e.g., polyA enrichment protocols versus rRNA depletion protocols, or paired-end versus single-end sequencing, may strongly impact in the results. Improper handling of samples in this step may also result in sample degradation, or inefficient rRNA depletion, which may compromise the whole experiment if not properly adjusted. A well-defined experimental design for the sequencing step must also be taken into consideration. A final low coverage of the transcriptome can result in an inadequate analysis of differential gene expression.

2.4 Case Study 2

Differential transcriptomics from T Cell stimulated with anti-CD3 antibodies using RNAseq.

The high-throughput sequencing of a whole cell RNA can suggest changes in the genetic programming of cultured human T lymphocyte cell. The model system proposed here relies in the interaction of antibodies to plasma membrane receptors. The anti-CD3 antibodies display a profound change in cell phenotype that may lead to the control of tolerance and inflammation in one individual (Carpenter et al. 2000; Belghith et al. 2003) These antibodies induce a change in the normal signal transduction pathway of these target cells allowing a change in gene expression profile (Chatenoud 2003). The identification of up and down regulation of certain genes may help understand the fate of the antibody-stimulated lymphocyte. The use of quantitative RNA detection approach such as RNA-seq leads to accurate understating of individual RNA levels variation among samples (Zhao et al. 2014). Moreover, it may also leads to the identification of new RNA molecules, and a more wise quantification based on gene models (alternative splice forms of RNA). These last two measurements are exclusive of RNA-seq and it is not possible to achieve using other quantitative techniques such as microarrays (Sirbu et al. 2012).

The aim of this research was to analyze the transcriptome of T cells after stimulation with recombinant anti-CD3 antibodies to identify pathways involved in modulation of immune tolerance. Human peripheral mononuclear cell (PBMC) was isolated from a single individual using standard Ficoll-Hypac separation. About $1–2 \times 10^6$ cells were grown in RPMI media with 250 ng of anti-CD3 antibodies. After 72 h of stimulation T cells were isolated from PBMC culture using negative selection on magnetic beads (Invitrogen). This protocol was used to isolate unstimulated human T cells from PBMC by depleting B cells, NK cells, monocytes, platelets, dendritic cells, granulocytes and erythrocytes. After isolation, the purity of T cells obtained was more than 96%, as checked by flow cytometry (FacsVerse, Becton and Dickinson).

Total RNA was isolated from each sample utilizing the miRNeasykit (Qiagen). The RNA integrity and quantity was checked in Bioanalyzer and Qubit® Fluorometric (Invitrogen). The yield was about 3 μg total RNA for each 10^6 T-cells. Up to 1.5 μg of total RNA was sent to a commercial sequencing facility. The samples were sent in a RNA-stable tube (Biomatrica) to preserve integrity of total RNA at room temperature for a long period of time. The HiSeq Illumina platform was used due to its sequence deepness associated with a low price per base. The HiSeq 2500 can achieve 4 billion reads in a total of 1 Tb of data. The use of paired end strategy allow to a better mapping into the reference genome.

A polyA$^+$ library and paired end strategy was used for the sequencing. The sequencing results were available by the facility in FASTQ format. The sequencing output files (FASTQ) was the input for the analysis.

The reads were analyzed by first clipping (Cutadapt) (Martin 2011) and trimming (PRINSEQ) primers, besides performing quality check (FastQC) (Andrews 2010). Paired-end Illumina data was stored in two files, R1 and R2, corresponding to both ends of a single RNA fragment. The cleaning process eventually removes same bad quality sequence, and R1 and R2 files became unsynchronized. The pairing of R1 and R2 has to be performed (with a Perl script), before the mapping step. We used TopHat2 due its simply interface and speed. TopHat2 produces.bam or.sam files, which are ordered and indexed using Samtools (Li et al. 2009), to be further used for differential expression (read count) and gene model quantification. The indexed data is now used for detecting differential gene expression using two Bioconductor packages: edgeR and DESeq (Anders et al. 2013). These programs allow to uncovering fold change among samples attributing a reliability parameter (p-value) to each prediction. Cufflinks is an alternative to perform differential expression among samples using predicted gene model, suggesting differential expression of alternatively spliced isoforms.

References

Alexa A, Rahnenfuhrer J (2010) topGO: enrichment analysis for gene ontology. R package version 28

Altschul SF, Gish W, Miller W, Myers EW, Lipman DJ (1990) Basic local alignment search tool. J Mol Biol 215(3):403–410. doi:10.1016/S0022-2836(05)80360-2

Anders S, Huber W (2010) Differential expression analysis for sequence count data. Genome Biol 11(10):R106. doi:10.1186/gb-2010-11-10-r106

Anders S, Reyes A, Huber W (2012) Detecting differential usage of exons from RNA-seq data. Genome Res 22(10):2008–2017. doi:10.1101/gr.133744.111

Anders S, McCarthy DJ, Chen Y, Okoniewski M, Smyth GK, Huber W, Robinson MD (2013) Count-based differential expression analysis of RNA sequencing data using R and bioconductor. Nat Protoc 8(9):1765–1786

Andrews S (2010) FastQC: a quality control tool for high throughput sequence data. Reference Source

Belghith M, Bluestone JA, Barriot S, Megret J, Bach JF, Chatenoud L (2003) TGF-beta-dependent mechanisms mediate restoration of self-tolerance induced by antibodies to CD3 in overt autoimmune diabetes. Nat Med 9(9):1202–1208. doi:10.1038/nm924

Bentley DR, Balasubramanian S, Swerdlow HP, Smith GP et al (2008) Accurate whole human genome sequencing using reversible terminator chemistry. Nature 456(7218):53–59. doi:10.1038/nature07517

Black MB, Parks BB, Pluta L, Chu TM, Allen BC, Wolfinger RD, Thomas RS (2014) Comparison of microarrays and RNA-seq for gene expression analyses of dose-response experiments. Toxicol Sci 137(2):385–403. doi:10.1093/toxsci/kft249

Carpenter PA, Pavlovic S, Tso JY, Press OW, Gooley T, Yu XZ, Anasetti C (2000) Non-Fc receptor-binding humanized anti-CD3 antibodies induce apoptosis of activated human T cells. J Immunol 165(11):6205–6213

Chatenoud L (2003) CD3-specific antibody-induced active tolerance: from bench to bedside. Nat Rev Immunol 3(2):123–132. doi:10.1038/nri1000

Chevreux B, Pfisterer T, Drescher B, Driesel AJ, Muller WE, Wetter T, Suhai S (2004) Using the miraEST assembler for reliable and automated mRNA transcript assembly and SNP detection in sequenced ESTs. Genome Res 14(6):1147–1159. doi:10.1101/gr.1917404

Chi HW, Yang YS, Shang ST, Chen KH, Yeh KM, Chang FY, Lin JC (2011) *Candida albicans* versus non-albicans bloodstream infections: the comparison of risk factors and outcome. J Microbiol Immunol Infect 44(5):369–375. doi:10.1016/j.jmii.2010.08.010

Gentleman RC, Carey VJ, Bates DM, Bolstad B, Dettling M, Dudoit S, Ellis B, Gautier L, Ge Y, Gentry J (2004) Bioconductor: open software development for computational biology and bioinformatics. Genome Biol 5(10):R80

Gordon A, Hannon G (2010) Fastx-toolkit. FASTQ/A short-reads pre-processing tools. Available at: http://hannonlabcshledu/fastx_toolkit/

Hardcastle TJ, Kelly KA (2010) baySeq: empirical Bayesian methods for identifying differential expression in sequence count data. BMC Bioinform 11:422. doi:10.1186/1471-2105-11-422

Harrington CT, Lin EI, Olson MT, Eshleman JR (2013) Fundamentals of pyrosequencing. Arch Pathol Lab Med 137(9):1296–1303. doi:10.5858/arpa.2012-0463-RA

Hernandez D, Francois P, Farinelli L, Osteras M, Schrenzel J (2008) De novo bacterial genome sequencing: millions of very short reads assembled on a desktop computer. Genome Res 18(5):802–809. doi:10.1101/gr.072033.107

Hoffmann S, Otto C, Kurtz S, Sharma CM, Khaitovich P, Vogel J, Stadler PF, Hackermuller J (2009) Fast mapping of short sequences with mismatches, insertions and deletions using index structures. PLoS Comput Biol 5(9):e1000502. doi:10.1371/journal.pcbi.1000502

Hoffmann S, Otto C, Doose G, Tanzer A, Langenberger D, Christ S, Kunz M, Holdt L, Teupser D, Hackermueller J, Stadler PF (2014) A multi-split mapping algorithm for circular RNA, splicing, trans-splicing, and fusion detection. Genome Biol 15(2):R34. doi:10.1186/gb-2014-15-2-r34

Hunniger K, Lehnert T, Bieber K, Martin R, Figge MT, Kurzai O (2014) A virtual infection model quantifies innate effector mechanisms and *Candida albicans* immune escape in human blood. PLoS Comput Biol 10(2):e1003479. doi:10.1371/journal.pcbi.1003479

Kasprzyk A (2011) BioMart: driving a paradigm change in biological data management. Database (Oxford) 2011:bar049. doi:10.1093/database/bar049

Katz Y, Wang ET, Airoldi EM, Burge CB (2010) Analysis and design of RNA sequencing experiments for identifying isoform regulation. Nat Methods 7(12):1009–1015. doi:10.1038/nmeth.1528

Kim D, Pertea G, Trapnell C, Pimentel H, Kelley R, Salzberg SL (2013) TopHat2: accurate alignment of transcriptomes in the presence of insertions, deletions and gene fusions. Genome Biol 14(4):R36. doi:10.1186/gb-2013-14-4-r36

Kuhn RM, Haussler D, Kent WJ (2013) The UCSC genome browser and associated tools. Brief Bioinform 14(2):144–161. doi:10.1093/bib/bbs038

Kvam VM, Liu P, Si Y (2012) A comparison of statistical methods for detecting differentially expressed genes from RNA-seq data. Am J Bot 99(2):248–256. doi:10.3732/ajb.1100340

Lander ES, Linton LM, Birren B et al (2001) Initial sequencing and analysis of the human genome. Nature 409(6822):860–921. doi:10.1038/35057062

Langmead B, Trapnell C, Pop M, Salzberg SL (2009) Ultrafast and memory-efficient alignment of short DNA sequences to the human genome. Genome Biol 10(3):R25. doi:10.1186/gb-2009-10-3-r25

Lesniewska A, Okoniewski MÇJ (2011) rnaSeqMap: a bioconductor package for RNA sequencing data exploration. BMC Bioinform 12(1):200

Li H, Handsaker B, Wysoker A, Fennell T, Ruan J, Homer N, Marth G, Abecasis G, Durbin R (2009) The sequence alignment/map format and SAMtools. Bioinformatics 25(16):2078–2079. doi:10.1093/bioinformatics/btp352

Li J, Witten DM, Johnstone IM, Tibshirani R (2012) Normalization, testing, and false discovery rate estimation for RNA-sequencing data. Biostatistics 13(3):523–538. doi:10.1093/biostatistics/kxr031

Liu L, Li Y, Li S, Hu N, He Y, Pong R, Lin D, Lu L, Law M (2012) Comparison of next-generation sequencing systems. J Biomed Biotechnol 2012:251364. doi:10.1155/2012/251364

Mardis ER (2013) Next-generation sequencing platforms. Annu Rev Anal Chem 6:287–303. doi:10.1146/annurev-anchem-062012-092628

Margulies M, Egholm M, Altman WE et al (2005) Genome sequencing in microfabricated high-density picolitre reactors. Nature 437(7057):376–380. doi:10.1038/nature03959

Marr KA, Patterson T, Denning D (2002) Aspergillosis. Pathogenesis, clinical manifestations, and therapy. Infect Dis Clin North Am 16(4):875–894, vi

Marsh M, Tu O, Dolnik V, Roach D, Solomon N, Bechtol K, Smietana P, Wang L, Li X, Cartwright P, Marks A, Barker D, Harris D, Bashkin J (1997) High-throughput DNA sequencing on a capillary array electrophoresis system. J Capill Electrophor 4(2):83–89

Martin M (2011) Cutadapt removes adapter sequences from high-throughput sequencing reads. EMBnet J 17(1):10–12

Martinez-Alvarez JA, Perez-Garcia LA, Flores-Carreon A, Mora-Montes HM (2014) The immune response against Candida spp. and Sporothrix schenckii. Rev Iberoam Micol 31(1):62–66. doi:10.1016/j.riam.2013.09.015

Metzker ML (2010) Sequencing technologies—the next generation. Nat Rev Genet 11(1):31–46. doi:10.1038/nrg2626

Miceli MH, Diaz JA, Lee SA (2011) Emerging opportunistic yeast infections. Lancet Infect Dis 11(2):142–151. doi:10.1016/S1473-3099(10)70218-8

Miramon P, Kasper L, Hube B (2013) Thriving within the host: Candida spp. interactions with phagocytic cells. Med Microbiol Immunol 202(3):183–195. doi:10.1007/s00430-013-0288-z

MySQL A (1995) MySQL: the world's most popular open source database. Available at: www.mysql.com.

Pappas PG (2006) Invasive candidiasis. Infect Dis Clin North Am 20(3):485–506. doi:10.1016/j.idc.2006.07.004

Pappas PG, Rex JH, Lee J, Hamill RJ, Larsen RA, Powderly W, Kauffman CA, Hyslop N, Mangino JE, Chapman S, Horowitz HW, Edwards JE, Dismukes WE (2003) A prospective observational study of candidemia: epidemiology, therapy, and influences on mortality in hospitalized adult and pediatric patients. Clin Infect Dis 37(5):634–643. doi:10.1086/376906

Quinlan AR, Hall IM (2010) BEDTools: a flexible suite of utilities for comparing genomic features. Bioinformatics 26(6):841–842. doi:10.1093/bioinformatics/btq033

Rapaport F, Khanin R, Liang Y, Pirun M, Krek A, Zumbo P, Mason CE, Socci ND, Betel D (2013) Comprehensive evaluation of differential gene expression analysis methods for RNA-seq data. Genome Biol 14(9):R95. doi:10.1186/gb-2013-14-9-r95

Richardson M, Lass-Florl C (2008) Changing epidemiology of systemic fungal infections. Clin Microbiol Infect 14(Suppl 4):5–24. doi:10.1111/j.1469-0691.2008.01978.x

Robinson MD, McCarthy DJ, Smyth GK (2010) edgeR: a bioconductor package for differential expression analysis of digital gene expression data. Bioinformatics 26(1):139–140. doi:10.1093/bioinformatics/btp616

Robles JA, Qureshi SE, Stephen SJ, Wilson SR, Burden CJ, Taylor JM (2012) Efficient experimental design and analysis strategies for the detection of differential expression using RNA-Sequencing. BMC Genomics 13:484. doi:10.1186/1471-2164-13-484

Sanger F, Nicklen S, Coulson AR (1977) DNA sequencing with chain-terminating inhibitors. Proc Natl Acad Sci U S A 74(12):5463–5467

Schmieder R, Edwards R (2011) Quality control and preprocessing of metagenomic datasets. Bioinformatics 27(6):863–864. doi:10.1093/bioinformatics/btr026

Sedlazeck FJ, Rescheneder P, von Haeseler A (2013) NextGenMap: fast and accurate read mapping in highly polymorphic genomes. Bioinformatics 29(21):2790–2791. doi:10.1093/bioinformatics/btt468

Shendure J (2008) The beginning of the end for microarrays? Nat Methods 5(7):585–587. doi:10.1038/nmeth0708-585

Shigemura K, Osawa K, Jikimoto T, Yoshida H, Hayama B, Ohji G, Iwata K, Fujisawa M, Arakawa S (2014) Comparison of the clinical risk factors between *Candida albicans* and Candida non-albicans species for bloodstream infection. J Antibiot 67:311–314. doi:10.1038/ja.2013.141

Simpson JT, Wong K, Jackman SD, Schein JE, Jones SJ, Birol I (2009) ABySS: a parallel assembler for short read sequence data. Genome Res 19(6):1117–1123. doi:10.1101/gr.089532.108

Sirbu A, Kerr G, Crane M, Ruskin HJ (2012) RNA-seq vs dual- and single-channel microarray data: sensitivity analysis for differential expression and clustering. PLoS one 7(12):e50986. doi:10.1371/journal.pone.0050986

Smith TF, Waterman MS (1981) Identification of common molecular subsequences. J Mol Biol 147(1):195–197

Smith LM, Sanders JZ, Kaiser RJ, Hughes P, Dodd C, Connell CR, Heiner C, Kent SB, Hood LE (1986) Fluorescence detection in automated DNA sequence analysis. Nature 321(6071):674–679. doi:10.1038/321674a0

Smyth GK (2004) Linear models and empirical bayes methods for assessing differential expression in microarray experiments. Stat Appl Genet Mol Biol 3:Article 3. doi:10.2202/1544-6115.1027

Soneson C, Delorenzi M (2013) A comparison of methods for differential expression analysis of RNA-seq data. BMC Bioinform 14:91. doi:10.1186/1471-2105-14-91

Soon WW, Hariharan M, Snyder MP (2013) High-throughput sequencing for biology and medicine. Mol Syst Biol 9:640. doi:10.1038/msb.2012.61

Tang S, Riva A (2013) PASTA: splice junction identification from RNA-sequencing data. BMC Bioinform 14:116. doi:10.1186/1471-2105-14-116

Team RC (2005) R: a language and environment for statistical computing. R foundation for statistical computing, Vienna

Tierney L, Linde J, Muller S, Brunke S, Molina JC, Hube B, Schock U, Guthke R, Kuchler K (2012) An interspecies regulatory network inferred from simultaneous RNA-seq of *Candida albicans* invading innate immune cells. Front Microbiol 3:85. doi:10.3389/fmicb.2012.00085

Trapnell C, Salzberg SL (2009) How to map billions of short reads onto genomes. Nat Biotechnol 27(5):455–457. doi:10.1038/nbt0509-455

Trapnell C, Pachter L, Salzberg SL (2009) TopHat: discovering splice junctions with RNA-seq. Bioinformatics 25(9):1105–1111. doi:10.1093/bioinformatics/btp120

Trapnell C, Williams BA, Pertea G, Mortazavi A, Kwan G, van Baren MJ, Salzberg SL, Wold BJ, Pachter L (2010) Transcript assembly and quantification by RNA-seq reveals unannotated transcripts and isoform switching during cell differentiation. Nat Biotechnol 28(5):511–515. doi:10.1038/nbt.1621

Trapnell C, Hendrickson DG, Sauvageau M, Goff L, Rinn JL, Pachter L (2013) Differential analysis of gene regulation at transcript resolution with RNA-seq. Nat Biotechnol 31(1):46–53. doi:10.1038/nbt.2450

Van Verk MC, Hickman R, Pieterse CM, Van Wees SC (2013) RNA-seq: revelation of the messengers. Trends Plant Sci 18(4):175–179. doi:10.1016/j.tplants.2013.02.001

Wagner GnP, Kin K, Lynch VJ (2012) Measurement of mRNA abundance using RNA-seq data: RPKM measure is inconsistent among samples. Theory Biosci 131(4):281–285

Wang Z, Gerstein M, Snyder M (2009) RNA-seq: a revolutionary tool for transcriptomics. Nat Rev Genet 10(1):57–63. doi:10.1038/nrg2484

Zhao S, Fung-Leung WP, Bittner A, Ngo K, Liu X (2014) Comparison of RNA-seq and microarray in transcriptome profiling of activated T cells. PLoS one 9(1):e78644. doi:10.1371/journal.pone.0078644

Chapter 3
Identification of Biomarkers and Expression Signatures

Patricia Severino, Elisa Napolitano Ferreira and Dirce Maria Carraro

Abstract Recently, molecular biology has been substantially improved by the development of new technologies that allow the assessment of the genome, transcriptome and proteome on a high-throughput scale and at reasonable costs. The translation of all the information generated by these technologies into new biomarkers is an enormous challenge for the biomedical community, and vast efforts have been made in this arena. The practice of personalized medicine based on DNA/RNA information used for clinical decision-making has led to considerable advances in different areas of medicine and is now a reality at several medical centers worldwide. The aspiration is that in the near future, the medical community will have more and more available biomarkers to properly classify patients and to allow them to offer efficient and tailored treatment for a broader range of diseases, resulting in a high cure rate and minimal side effects. In this chapter, we discuss the identification of biomarkers by primarily examining gene expression. Two of the most important approaches, microarrays and RNA sequencing (RNA-Seq), and strategies for defining gene expression signatures are addressed. We also present important aspects involved in the validation of gene expression signatures as biomarkers, the bottlenecks and difficulties for their broader use in clinical practice and some good examples of signatures representing aspects of human diseases.

3.1 Biomarker Discovery and the Understanding of Human Diseases

Biological markers (biomarkers) are officially defined by the NIH (National Institutes of Health) as "a characteristic that is objectively measured and evaluated as an indicator of normal biological processes, pathogenic processes, or pharmacologic

P. Severino (✉)
Instituto Israelita de Ensino e Pesquisa Albert Einstein, Av. Albert Einstein, 627,
São Paulo, SP, 05652-000, Brazil
e-mail: psever@einstein.br

E. N. Ferreira · D. M. Carraro
A. C. Camargo Cancer Center, Rua Prof. Antonio Prudente, 211,
São Paulo, SP, 01509-010, Brazil

© Springer International Publishing Switzerland 2014
G. A. Passos (ed.), *Transcriptomics in Health and Disease,*
DOI 10.1007/978-3-319-11985-4_3

responses to a therapeutic intervention" (Biomarkers Definitions Working Group 2001).

In practical terms, biomarkers can be seen as the end result of a laboratory technique that involves the direct evaluation of a specific molecule in biological specimens such as blood, saliva, cells or tissues. The measurement of a biomarker can then be expressed as a positive or negative result, as intensities, as quantities, such as the level of a serum protein, or as high-dimensional panels of markers, also called "signatures", including those proposed by the genomic, proteomic, and metabolomic fields.

Biomarker discovery is a major topic in biomedical research and involves multidisciplinary teams with complementary expertise, including physicians, epidemiologists, and scientists whose aim is to study and understand human diseases. The fundamental idea is that it is possible to identify patterns of clinical, demographic, genomic, and other types of data that can be used alone or together to benefit individual patients. Biomarkers can be divided into two major types: biomarkers used for predicting the risk of developing a specific disease, and biomarkers used for investigating the behavior of a disease that is already established. The latter include markers for disease diagnosis, for predicting disease progression and responses to specific therapies and for monitoring treatment efficacy.

Although the applications of biomarkers in a variety of different diseases are clear, the experimental delineation, assay development and optimization, and validation of the biomarker candidates constitute a lengthy process (Rifai et al. 2006). In fact, the vast majority of biomarkers described in scientific publications has not been endorsed by a validation phase and, thus, cannot be used in clinical settings. Among the difficulties in this process, variability is a major concern. Intra-individual variability is primarily related to technical issues and to intra-disease heterogeneity. Intra-individual variability might be associated with the time of sample collection, the storage conditions, or the lack of equipment performance uniformity at different time points during a study or diagnostic follow-up. In contrast, interindividual variability includes demographic variables such as age, gender, ethnicity and the individual genetic background, which may or may not be easily associated with ethnicity. The presence of unidentified mutations associated with the response to a pharmacological therapy, for instance, will impact conclusions on the efficacy of a tested drug (Sahin et al. 2013). Other sources of variability include exposure to chemicals or toxins, as well as diet or other personal habits that may alter biomarker patterns. Thus, for reliability and reproducibility, careful consideration of the sources of variability in the measurement of a biomarker is critical to avoid potential misclassification of individuals and to minimize the high rate of failed biomarkers, especially given that less than 1 % of biomarkers introduced by literature reports make it to clinical practice, according to Kern (2012).

3.2   Gene Expression Assessment for Disclosure of Disease Biomarkers

High-throughput gene expression analysis enables the measurement of the activity of thousands of different transcripts, generating a gene expression profile of a cell or tissue. The assessment of the gene expression profile has contributed to the definition of the molecular basis of disease development, finding specific features that, in turn, can be associated with specific diseases. Similarities and disparities revealed by gene expression profiles of samples from distinct groups representing different medical conditions can be explored as potential biomarkers. DNA microarrays and RNA sequencing (RNA-Seq) will be addressed in this chapter.

Biomarker discovery studies using gene expression analysis demands robust methods for sample selection, sample processing and data analysis. The process starts with a RNA isolation procedure that recovers a representative RNA population. The term transcriptional analysis refers to the expression measurement of one specific species of RNA, the messenger RNA (mRNA). The synthesis and maturation of mRNA is part of a complex regulation network allowing fast response and adaptation under environmental changes. Thus the gene expression profile of mRNAs faithfully reflects the biology of a given sample at the time that it was collected. Problems associated with this step predominantly arise due to the susceptibility of RNA to degradation by ubiquitously found RNases. Sample handling and the subsequent procedures need to be under rigorous well-controlled conditions aiming to preserve the integrity of RNA, as the degradation of this molecule can significantly influence the quality and interpretation of gene expression data.

Microarrays have been extensively used for evaluating global gene expression profiles, revealing the complexity of the human functional transcriptome (see box 1). Typically, microarrays are able to evaluate the expression level of thousands of genes, enabling the quantitative assessment of their differential activity among distinct cell types or tissues. This characteristic has revolutionized both basic and clinical science, motivating its use for biomarker research. Despite specific details of the distinct commercially available microarray platforms used for gene expression analysis, including scale, labeling and solid substrate, the principles governing hybridization between the labeled nucleic acid that is in solution to the nucleic acid immobilized on a solid substrate are the same. Technical problems arising from hybridization are, therefore, common to every study design. Frequent complications can be related to fluorophore labeling, hybridization conditions and post-hybridization washes. Printing artifacts—spot morphology and nucleic acid concentration, for example—used to be an important issue in the past, when the majority of microarrays were made in-house. This has been mostly overcome with the advent of commercial sources of microarray slides.

Given that the most important factor for the success of a microarray experiment has been properly addressed, i.e., the quality of input material, hybridization quality assessment is performed when fluorescent signals are captured by an appropriate scanner. Scanner software usually has built-in quality control tests that can be used for filtering out spots that do not meet thresholds of saturation, signal intensity significantly above the background or standard deviation of background. The spots that do not meet the quality control criteria will be omitted from further analysis. Positive and negative control probes included in the microarray slide are also essential for quality assessment calculations and generate reports on technical variation between experiments. Notably, the ability to detect a meaningful biological distinction between the studied groups is closely related to the extent to which the technical variations have been controlled and minimized.

RNA-Seq is based on next-generation sequencing (NGS) (see box 2). RNA-Seq provides digital-counting analysis of the transcriptome, enabling quantitative and qualitative information. In contrast to microarray experiments, RNA-Seq quantification is based on absolute rather than relative values. The sequencing depth is therefore an important aspect when planning RNA-Seq experiments, and it is a critical parameter when performing comparative expression analysis. RNA-Seq data are also able to reveal the architecture of the transcriptome, such as alternatively spliced transcripts, mutations, alternative polyadenylation and transcript fusions. In addition, given that RNA-Seq does not rely on immobilized probes as do microarrays, it allows the identification of novel transcripts by the sequencing of any transcript species that is produced by the cell at a specific moment.

Typically, the first step of RNA-Seq analysis is to perform mapping of sequenced fragments, called "reads", to the reference genome. The accuracy of read mapping depends on the quality of the reads and on other specific DNA features such as GC content and the presence of repetitive elements. For quantitative measurements, the number of reads generated for each transcript must be normalized by considering the total number of reads obtained for each sample. This normalization step takes into consideration the fact that higher counts in "sample A" compared to "sample B" can be a result of the deeper sequencing of sample A instead of higher expression of the transcript in this sample. It is noteworthy that during library preparation, longer transcripts will generate a higher number of fragments for sequencing compared to shorter transcripts, impairing direct comparisons between expression measurements from different transcript lengths. Common normalization techniques for RNA-Seq are RPKM (Reads Per Kilobase of exon per Million fragments mapped), when fragments are sequenced from one end (single-end sequencing), or FPKM (Fragments Per Kilobase of exon per Million fragments mapped), when fragments are sequenced from both ends, providing two reads for each fragment (paired-end sequencing) (Mortazavi et al. 2008; Trapnell et al. 2010).

Although microarrays and RNA-Seq methodologies have a huge potential to be used for biomarkers discovery, the interpretation of such data is a major challenge. Different mathematical and statistical tools are available to detect associations between expression profiles and the clinical features of a given patient and to ultimately identify a group of genes whose expression pattern distinguishes between different groups and that is truly correlated with the clinical phenotype. Gene expression analyses by both approaches have mostly focused on identifying sets of genes that are differentially expressed or differentially co-expressed in distinct biological states, for instance, in diseased but not in disease-free individuals or vice versa (Watson 2006). Numerous studies have linked multi-gene signatures with clinical significances in human diseases as well as supported toxicological and functional studies at the systems level.

3.3 Potential and Current Clinical Applications of Gene Expression Signatures

Advances in transcriptional analysis, such as tools for the comprehensive integration of gene expression and clinical data, and technologies for assessing transcriptional information from restrict amounts of biological material have contributed to the improvement of biomarkers currently available for complex diseases.

For many cancers, for example, the standard diagnostic and prognostic evaluation methods are based on morphological characteristics with the use of a few molecular markers. Thus, it is common that patients diagnosed with the same tumor and bearing similar morphologic features have different outcomes. Research efforts in microarray profiling have addressed several aspects of cancer management, including tumor classification according to their sites of origin (Su et al. 2001; Bloom et al. 2004; Arango et al. 2013), the discovery new cancer subtypes (Alizadeh et al. 2000; Lapointe et al. 2004), and the prediction of clinical outcomes (Shipp et al. 2002; van't Veer et al. 2002; Balko et al. 2013)

Two successful implementations of gene expression microarrays as biomarkers in clinical practice are the MammaPrint and Oncotype DX assays. Both tests were developed for breast carcinomas and have been approved by the U.S. Food and Drug Administration (FDA). The MammaPrint test is available from a Dutch company, Agendia (www.agendia.com). The test is based on the expression pattern of 70 genes, evaluated by a microarray chip. Its application is to evaluate the risk of recurrence (high or low) in Estrogen Receptor (ER)-negative and -positive patients of less than 55 years of age, with invasive breast carcinomas in stage I or II, without lymph node metastases. The test was adjusted, and even in subsequent research on older breast cancer patients, the 70-gene signature turned out to be an accurate risk indicator (Drukker et al. 2014).

The Oncotype DX test is provided by the American company Genomic Health (www.genomichealth.com) and is based on the determination of the expression of 21 genes (16 genes used as the test and 5 used as a control) by quantitative RT-PCR and is indicated only for women with early-stage ER-positive breast cancer. Both MammaPrint and Oncotype DX are prognostic tests and are also used as predictors for chemotherapy results (Paik et al. 2006).

Other tests based on gene expression are in the final stages of development. For example, for colon carcinoma, there are two promising tests: ColoPrint (Maak et al. 2013), under validation by Agendia, the same company that markets MammaPrint, and the test Oncotype DX Colon Cancer, by the company Genomic Health, which predicts the risk of individual recurrence for stage II colon cancer patients who have proficient MMR (mismatch repair pathway) (Gray et al. 2011). More recently,

Oncotype DX Genomic has developed a similar test for prostate cancer, the Onco-type DX Genomic Prostate Cancer Score. The test has been shown to have broad application for men in different situations, with high or low-risk prostate tumors. Both groups seem to benefit from using the test, improving the choice between active surveillance for low-risk and aggressive treatment for high-risk patients (Klein 2013).

Although still in early stages of development, there are efforts being made in leukemia subclassification and diagnosis using gene expression signatures. Despite enormous amounts of experimental data, consensus in representative gene sets has not been met. The Microarray Innovations in Leukemia initiative (Haferlach et al. 2010), the so-called "international standardization program towards the application of gene expression profiling in routine leukemia diagnostics", exemplifies the effort and difficulties in achieving implementation within the boundaries of clinical practice. The aim of the study was to assess the clinical utility of gene expression profiling in a single test to subtype leukemias into categories of myeloid and lymphoid malignancies, complementing current diagnostic algorithms.

Gene expression signatures are also being assessed for clinical application in human neurodegenerative diseases, but the results are still very preliminary. The primary aim of these gene expression studies is to understand key molecular events associated with phenotypes, comparing patient groups, disease stages and anatomical locations. A large number of cerebral regions have been examined by microarray studies in neurodegenerative disorders such as Alzheimer's and Parkinson's to identify potential gene expression signatures associated with these diseases. However, studies addressing neurodegenerative diseases share difficulties, especially concerning sample procurement and the obtaining of high-quality tissue for reliable experiments (Cooper-Knock et al. 2012).

Blood-based gene expression profiling as a diagnostic tool has been explored for a broad range of diseases, due to the non-invasive nature of such a test. Examples include research in heart diseases, such as the identification of the presence and extent of coronary artery disease (Elashoff et al. 2011), or the early detection of cancer (Aaroe et al. 2010), Most studies evaluate the gene expression of peripheral blood cells, but they lack the corroboration and validation necessary to confirm applicability in a clinical setting. Variability due to sample collection and processing is the biggest drawback of this approach.

With respect to RNA-Seq as a biomarker, results are still preliminary. In spite of the fact that RNA-Seq exhibits great sensitivity in both qualitative and quantitative characterization of the transcriptome, the high costs associated with the technology and the complexity of data analysis are some of the obstacles that impair its broader use for discovering biomarkers. Instead, RNA-Seq is currently being used for microarray validation and as a complementary approach. However, its ability to generate an unprecedented detailing of the transcriptome promises to add substantial value to medicine and pharmacogenomics in the near future.

3.4   Final Considerations

In this chapter, we addressed characteristics and some of the major difficulties in using DNA microarrays and RNA-Seq for biomarker discovery in human diseases. Although long development periods are common due to the need of validation in different populations, several successful examples of biomarkers based on gene expression signatures are already used in clinical practice, despite the obstacles. Biomarker's applicability for populations of distinct genetic backgrounds and difficulties in the implementation of gene expression-based tests in routine laboratories are a major concern, but a more precise stratification of patients has become a reality in several clinical settings.

Box 3.1: Microarray

Microarray technology (Schena et al. 1995) is based on solid hybridization between two single-stranded nucleic acid molecules, the probes and the target mRNAs from cell or tissues. The probes are DNA molecules such as small oligonucleotides, cDNA or PCR products corresponding to the genes that are immobilized on a glass slide. The targets are mRNA molecules extracted from the tissues and/or cells of interest that are converted to complementary DNAs (cDNA) or complementary RNAs (cRNA) and labeled with reporter molecules (typically fluorescent molecules). The hybridization takes place in a hybridization chamber that maintains the appropriate temperature and humidity for 16 hours to favor the hybridizations. The non-hybridized molecules are rinsed away by successive washes, and only strongly paired complementary strands will remain hybridized on the slides. The slides are then scanned to detect the fluorescent signals of each spot. The intensity of the signal is proportional to the number of hybridized molecules, which in turn corresponds to the expression level of each gene evaluated.

Commonly, microarray experiments are based on competitive hybridizations. In a competitive hybridization, two populations of RNA, the test and reference samples, are labeled with different fluorescent molecules and co-hybridized to one single slide. The intensity of signals obtained by each fluorescent dye is computed, and the expression level of each gene is a relative measurement between the test and reference sample. This design allows comparative analysis between different samples that were co-hybridized to the same reference sample. The best reference is a sample that has the highest number of active genes and consequently generates intensity values for most spots, resulting in valid relative expression values for all genes. The most commonly used fluorophores in microarrays are cyanine 3 (Cy3)—wavelengths measurement from 550 to 610 nm—and Cyanine 5 (Cy5)—wavelengths measurement from 650 to 750 nm.

Box 3.2: RNA-Seq

RNA-Seq is a recently developed approach for transcriptome analysis based on high-throughput sequencing technologies that enables transcriptional evaluation at nucleotide resolution (Morin et al. 2008). In this approach, the entire population of mRNA that is extracted from the cell is sequenced, allowing the determination of gene expression and the identification of alternative splicing events, single-nucleotide polymorphisms, transcript fusion events and post-transcriptional RNA editing events.

Different protocols and commercially available kits have been proposed for RNA-Seq experiments. Nonetheless, they all follow a similar strategy. Briefly, total RNA is isolated from a sample of interest, and mRNAs are purified by PolyA tail enrichment or ribosomal RNA (rRNA) depletion. Obtaining high quality RNA is critical for a good RNA-Seq library.

The mRNA is then fragmented into smaller pieces by mechanical shearing (usually Covaris sonication) or chemical digestion (using the RNase III enzyme). First- and second-strand cDNA is reverse transcribed from fragmented RNA using random hexamers, oligo (dT) primers or specific linkers coupled to fragmented RNAs. The use of 5' and 3' end specific adapters is advantageous because it allows for strand-specific sequencing, enabling the investigation of sense and antisense transcripts. Finally, cDNA fragments are amplified by PCR for the enrichment of molecules that were correctly ligated to adapters. During library preparation, a size-selection step is usually performed prior to or after linker ligation or PCR amplification. The cDNA library is quantified by RT-qPCR or Bioanalyzer and is ready for sequencing.

Depending on the throughput of the sequencing platform, sample multiplexing by using barcoded adapters is an interesting option because it allows for simultaneous sequencing of multiple libraries in a single sequencing run (Fig. 3.1).

High-throughput transcriptional analysis of a group of patients representing different medical conditions (training set) are carried out using Microarray or RNA-Seq approaches, leading to the definition of a gene expression profile of each sample. Different mathematical and statistical tools are used to detect associations between expression profiles and the clinical features of a patient and to identify a gene expression signature that is able to distinguish between different patients and that is truly correlated with the clinical phenotype. In order to be validated, the gene signature must be evaluated in a large cohort of samples from an independent population. The gene expression signature that demonstrates consistent and accurate association between gene expression and the clinical feature is validated and then goes into an optimization step to determine the best evaluation method (microarray and Quantitative Real-Time PCR) and an optimal gene panel to be incorporated as a biomarker into clinical practice. The biomarker may then be used in clinical practice to help clinicians to make informed decisions on the best treatment options based on the predictive value of the biomarker concerning prognosis, recurrence or response to treatment.

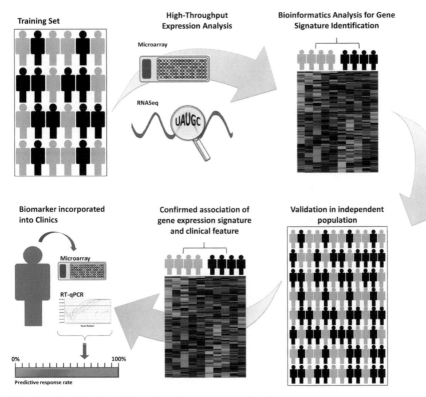

Fig. 3.1 Identification of biomarkers by gene expression signatures

References

Aaroe J, Lindahl T, Dumeaux V et al (2010) Gene expression profiling of peripheral blood cells for early detection of breast cancer. Breast Cancer Res 12(1):R7

Alizadeh AA, Eisen MB, Davis RE et al (2000) Distinct types of diffuse large B-cell lymphoma identified by gene expression profiling. Nature 403:503–511

Arango BA, Rivera CL, Glück S (2013) Gene expression profiling in breast cancer. Am J Transl Res 5:132–138

Balko JM, Giltnane J, Wang K et al (2013) Molecular profiling of the residual disease of triple-negative breast cancers after neoadjuvant chemotherapy identifies actionable therapeutic targets. Cancer Discov 4:232–245

Biomarkers Definitions Working Group (2001) Biomarkers and surrogate endpoints: preferred definitions and conceptual framework. Clin Pharmacol Ther 69:89–950

Bloom G, Yang IV, Boulware D et al (2004) Multi-platform, multi-site, microarray-based human tumor classification. Am J Pathol 164:9–16

Cooper-Knock J, Kirby J, Ferraiuolo L et al (2012) Gene expression profiling in human neurode-generative disease. Nat Rev Neurol 8:518–530

Drukker CA, van Tinteren H, Schmidt MK et al (2014) Long-term impact of the 70-gene signature on breast cancer outcome. Breast Cancer Res Treat 143:587–592

Elashoff MR, Wingrove JA, Beineke P et al (2011) Development of a blood-based gene expression algorithm for assessment of obstructive coronary artery disease in non-diabetic patients. BMC Medical Genomics 4:26

Gray RG, Quirke P, Handley K et al (2011) Validation study of a quantitative multigene reverse transcriptase-polymerase chain reaction assay for assessment of recurrence risk in patients with stage II colon cancer. J Clin Oncol 29:4611–4619

Haferlach T, Kohlmann A, Wieczorek L et al (2010) Clinical utility of microarray-based gene expression profiling in the diagnosis and subclassification of leukemia: report from the International Microarray Innovations in Leukemia Study Group. Clin Oncol 28:2529–2537

Kern SE (2012) Why your new cancer biomarker may never work: recurrent patterns and remarkable diversity in biomarker failures. Cancer Res 72:6097–6101

Klein EA (2013) A genomic approach to active surveillance: a step toward precision medicine. Asian J Androl. 15:340–341

Lapointe J, Li C, Higgins JP et al (2004) Gene expression profiling identifies clinically relevant subtypes of prostate cancer. Proc Natl Acad Sci USA 101:811–816

Maak M, Simon I, Nitsche U et al (2013) Independent validation of a prognostic genomic signature (ColoPrint) for patients with stage II colon cancer. Ann Surg 257:1053–1058

Morin R, Bainbridge M, Fejes A et al (2008) Profiling the HeLa S3 transcriptome using randomly primed cDNA and massively parallel short-read sequencing. BioTechniques 45:81–94

Mortazavi A, Williams BA, McCue K, Wold B et al (2008) Mapping and quantifying mammalian transcriptomes by RNA-SEq. Nat Methods 5:621–628

Paik S, Tang G, Shak S et al (2006) Gene expression and benefit of chemotherapy in women with node-negative, estrogen receptor-positive breast cancer. J Clin Oncol 24:3726–3734

Rifai N, Gillette MA, Carr SA (2006) Protein biomarker discovery and validation: the long and uncertain path to clinical utility. Nat Biotechnol 24:971

Sahin IH, Garrett C (2013) The heterogeneity of KRAS mutations in colorectal cancer and its biomarker implications: an ever-evolving story. Transl Gastrointestinal Cancer 2:164–166

Schena M, Shalon D, Davis RW et al (1995) Quantitative monitoring of gene expression patterns with a complementary DNA microarray. Science 270:467–470

Shipp MA, Ross KN, Tamayo P et al (2002) Diffuse large B-cell lymphoma outcome prediction by gene-expression profiling and supervised machine learning. Nat Med 8:68–74

Su AI, Welsh JB, Sapinoso LM et al (2001) Molecular classification of human carcinomas by use of gene expression signatures. Cancer Res 61:7388–7393

Trapnell C, Williams BA, Pertea G et al (2010) Transcript assembly and quantification by RNA-Seq reveals unannotated transcripts and isoform switching during cell differentiation. Nat Biotechnol 25:511–515

van't Veer LJ, Dai H, van de Vijver MJ et al (2002) Gene expression profiling predicts clinical outcome of breast cancer. Nature 415:530–536

Watson M (2006) CoXpress: differential co-expression in gene expression data. BMC Bioinformatics 7:509

Chapter 4
Methods for Gene Coexpression Network Visualization and Analysis

Carlos Alberto Moreira-Filho, Silvia Yumi Bando,
Fernanda Bernardi Bertonha, Filipi Nascimento Silva
and Luciano da Fontoura Costa

Abstract Gene network analysis is an important tool for studying the changes in steady states that characterize cell functional properties, the genome-environment interplay and the health-disease transitions. The integration of gene coexpression and protein interaction data is one current frontier of systems biology, leading, for instance, to the identification of unique and common drivers to disease conditions. In this chapter the fundamentals for gene coexpression network construction, visualization and analysis are revised, emphasizing its scale-free nature, the measures that express its most relevant topological features, and methods for network validation.

4.1 Introduction

The development of high-throughput techniques for concurrently measuring the expression levels of thousands of genes, mostly based on DNA microarrays, allowed monitoring cell's transcriptional activity across multiple conditions, opening

Electronic supplementary material The online version of this chapter (doi: 10.1007/978-3-319-11985-4_4) contains supplementary material, which is available to authorized users. Videos can also be accessed at http://www.springerimages.com/videos/978-3-319-11984-7.

C. A. Moreira-Filho (✉) · S. Y. Bando · F. B. Bertonha
Departamento de Pediatria, Faculdade de Medicina da Universidade de São Paulo,
Av. Dr. Arnaldo, 455 CEP 01246-903, São Paulo, São Paulo, Brazil
e-mail: carlos.moreira@hc.fm.usp.br

S. Y. Bando
e-mail: silvia.bando@hc.fm.usp.br

F. B. Bertonha
e-mail: bertonhafb@gmail.com

F. N. Silva · L. F. Costa
Instituto de Física de São Carlos, Universidade de São Paulo, São Carlos, São Paulo, Brazil
e-mail: filipinascimento@gmail.com

L. F. Costa
e-mail: ldfcosta@gmail.com

© Springer International Publishing Switzerland 2014
G. A. Passos (ed.), *Transcriptomics in Health and Disease,*
DOI 10.1007/978-3-319-11985-4_4

broad perspectives for functional genomics (Ideker and Krogan 2012). Hereafter, new insights were gained on the genomic mechanisms underlying relevant biological processes—such as cell cycle and development, and the genome-environment interplay—leading to a systemic approach for the identification of disease-related genes (Chuang et al. 2010; Kim et al. 2011; Sahni et al. 2013).

A single microarray generates data on the expression levels of thousands of genes and typical microarray studies encompass multiple arrays covering several distinct experimental conditions, e.g. tissue samples from patients and controls, or cultured cells submitted to different treatments (Bando et al. 2013; Herbst et al. 2014). Advanced statistical and computational tools have been developed to deal with the large amount of data derived from microarray experiments (Zhang and Horvath 2005; Lee and Tzou 2009; Faro et al. 2012). One of the most effective methods for the analysis of microarray data is based on the construction of gene coexpression networks, or GCNs: gene expression levels are pairwise compared and the pairs above a cutoff threshold are linked to create a gene-gene interaction network (Weirauch 2011). GCNs, as well as protein-protein and metabolic networks, are governed by universal laws (see Sect. 3.3.2) and the topological and dynamic properties of these networks provide important clues for understanding the functional organization of cells and tissues (Barabási and Oltvai 2004; Zhu et al. 2007). This chapter is centered in network-based methods for analyzing DNA microarray data: the fundamentals for construction, visualization, interpretation and validation of GCNs will be discussed in the next paragraphs, emphasizing the use of graph methods (Barabási and Oltvai 2004; Barabási et al. 2011; Costa et al. 2011; Villa-Vialaneix et al. 2013; Winterbach et al. 2013) for all these tasks.

A few considerations are still needed before we start to review the network-based approach to functional genomics. Firstly, one should keep in mind that gene function is not isolated: the network effect of genes is the driving force moving cell metabolism from one steady state to another, frequently in response to environmental changes (Sieberts and Schadt 2007; Sahni et al. 2013). These transitions shape what we call complex phenotypes—normal or altered by a disease state—and can be correlated with specific changes in GCNs (Benson and Breitling 2006; Carter et al. 2013; Bando et al. 2013). Secondly, in order to study these network changes is mandatory to: (i) gain access to the cells or tissues specifically involved in the physiological or pathological process under investigation; (ii) collect an adequate number of biological replicates; (iii) obtain good quality RNA samples; (iv) use a microarray platform suitable for attaining the research goals. Therefore, comments on sample quality and experimental design will be made in the following section.

4.2   Analysis of DNA microarray data

Essentially, DNA microarrays for assaying gene expression consist in grid that can contain tens of thousands of probes corresponding to known transcripts of a particular genome (human, rat, etc.). Fluorescent-labelled complementary DNA (DNA synthesized from messenger RNA) samples are hybridized to probes and the relative abundance of each sequence in a sample is quantified in microarray scanner

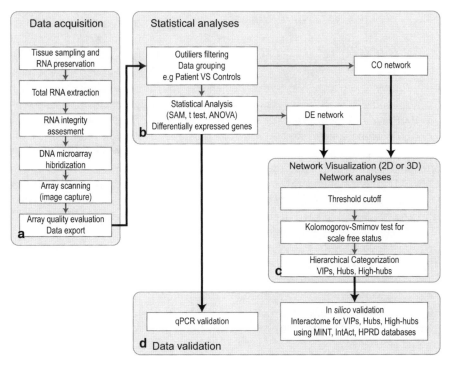

Fig. 4.1 Workflow for gene coexpression network visualization and analysis. **a** data acquisition. **b** statistical analysis. **c** network visualization and statistical analyses. **d** data validation

for fluorescence detection (image capture). The steps from RNA extraction to array scanning, data export and subsequent statistical and network analyses are outlined in the following subsections. This workflow is presented in Fig. 4.1.

4.2.1 RNA Isolation and Preservation

Messenger RNA for DNA microarray experiments must be preserved in its last physiological state and be prevented from degrading. Larger tissue samples (>50 mg) may be snap-frozen in liquid nitrogen. Smaller tissue samples (5 mm thick fragments) are usually preserved in RNA Later, a product (distributed by Ambion and Qiagen) which penetrates cell membranes and inactivates RNAses. After RNA extraction, RNA quality should be assessed in a microfluidics-based platform (e.g. BioAnalyzer) for sizing, quantification and quality control (Fig. 4.1a). The integrity of RNA molecules is estimated by using the RIN algorithm (Schroeder et al. 2006). RIN values range from 1 (total degradation) to 10 (intact). As a general rule, only RNA samples with RIN values of 7 or higher should be used in DNA microarray experiments. Irrespective of the cellular RNA extraction protocol adopted, a final column purification step (e.g. RNeasy) consistently leads to a high yielding synthesis of cDNA.

4.2.2   Gene Expression Analysis

Scanner generated data (image file) are pre-processed, filtering out probes flagged as unreliable (low intensity, saturation, restriction control probes, etc.) by the scanning software, and thereafter normalized ending up with a file of numerical values corresponding to probe's expression levels in a microarray experiment. The assessment of raw data quality and data grouping (comparison groups, e.g. patients and controls) can be done using free software packages, like R software (R Development Core Team 2012), for normalization (Lowess test for arrays normalization), outliers exclusion and exporting of valid transcript expression data.

MeV (TIGR Multiexperiment Viewer) is a popular free software for comparative analysis that can be used for clustering, visualization, classification, statistical analysis and biological theme (Gene Ontology, or GO) discovery (Saeed et al. 2003). The differentially expressed transcripts for two comparison groups are obtained using SAM test—Significance Analysis for Microarray—for parametric analysis (using non-parametric statistics) or Wilcoxon-Mann-Whitney test for non-parametric analysis. ANOVA is used for multiple comparisons across conditions. Thereafter, false discovery rate tests are applied (already included in SAM test). Finally, the differential GO annotated gene expression data can be used for gene expression analyses (Fig. 4.1b) and in the construction of coexpression networks, as described in Sect. 4.3.

4.3   Construction and Analysis of GCNs

GCNs can be obtained for a subset of genes, i.e. differentially expressed GO annotated genes (DE networks), or for all valid GO annotated genes (complete, or CO networks). These networks are constructed based on gene-gene covariance correlation, usually using Pearson's, or Spearman's rank correlation or cosine similarity measurements (Fig. 3.1c) (Prifti et al. 2008; Song et al. 2012). Genes presenting similar patterns of expression are strongly bounded together forming a weighted complete graph.

In order to construct a GCN links are removed from the initially complete graph by gradually increasing the correlation threshold (Elo et al. 2007). After link strength threshold adoption, usually above 0.80, the network is tested for scale free status (see Sect. 4.3.2) by Kolmogorov-Smirnov (K-S) statistics, i.e. power law distributions in empirical data (Clauset et al. 2009). Here we used a demonstrative example of a "patient versus control" gene coexpression analysis (Fig. 4.2). This analysis considered: 202 genes and 561 links for patients' DE network; 219 genes and 486 links for control DE networks; 6,927 genes and 12,768 links for patients' CO networks; 6,705 genes and 12,468 links for control CO networks. Link strength cut-offs were 0.998 for control CO network and 0.999 for the other three networks. Figures. 4.2 and 4.3 show K-S distribution for DE and CO networks, respectively, of patients' group (Figs. 4.2c and 4.3c) and controls' group (Figs. 4.2d and 4.3d).

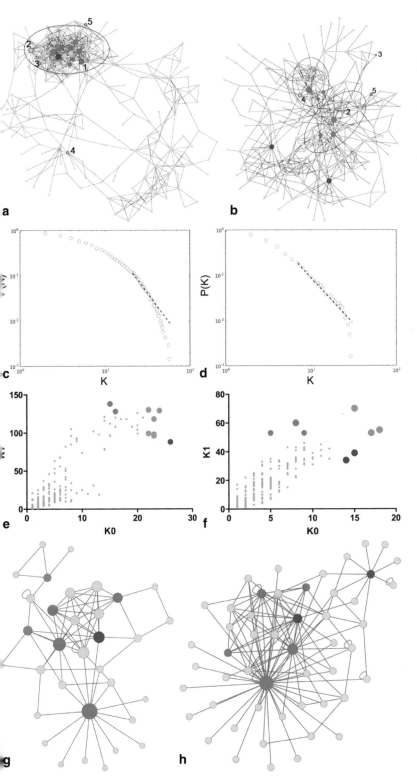

Fig. 4.2 Comparative DE network analysis for patients and control groups in a hypothetical network. DE coexpression networks for patients (**a**) and control (**b**) groups; links in blue or red indicate positive or inverse covariance correlation. It is interesting to note that the same genes (numbered nodes bordered in red or blue) have different covariance correlation between patients and control groups. Clusters are encircled in A and B networks. (**c**) and (**d**): Kolmogorov-Smirnov test for scale free status for patients and control groups, respectively. Scatter plot of node degree (k_0) vs concentric node degree (k_1) measures for patients (**e**) and control (**f**) groups. Interactome *in silico* validations for patients and control networks are depicted in (**g**) and (**h**), respectively. Hubs, VIPs and high-hubs are indicated in blue, red and green, respectively. Network analyses and visualization were accomplished through Cytoscape

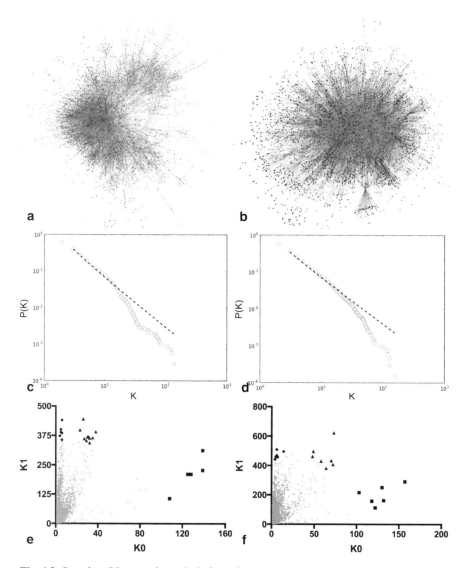

Fig. 4.3 Complete CO networks analysis for patients and control groups. CO coexpression networks for patients (**a**) and control (**b**) groups. Kolmogorov-Smirnov test for scale free status for patients (**c**) and control (**d**) groups. Scatter plot of node degree (k0) *vs* concentric node degree (k1) measures for patients (**e**) and control (**f**) groups. Hubs, VIPs and high-hubs are indicated by rectangles, diamonds and triangles, respectively. For 3D CO network visualization access the video hyperlinks (videos 4.1 and 4.2)

The number of samples available for each gene is also directly connected to the statistical significance of the generated GCN. Networks constructed from datasets with less than 5 samples per gene can lead to high adherence to the null model, where nodes are randomly connected, thus presenting degree distributions with asymptotic exponential decay behavior. This effect can occur even when considering a very large correlation threshold, such as above 0.999.

4.3.1   Network Visualization

The Cytoscape free software (Saito et al. 2012; www.cytoscape.org) is very useful for data analysis and visualization of DE networks or subnetworks (Fig. 4.2a, b). On the other hand, CO network analysis is only possible through 3D visualization (Fig. 4.3a, b and Videos 4.1 and 4.2 http://www.springerimages.com/videos/978-3-319-11984-7). Several 3D visualization softwares for gene-gene and protein-protein networks are being developed (Ishiwata et al. 2009; Pavlopoulos et al. 2008; Wang et al. 2013). One of them, under development by Luciano Costa's Research Group (http://cyvision.ifsc. usp.br/Cyvision/), Institute of Physics at São Carlos, University of São Paulo, was described in a paper by our group (Bando et al. 2013). This software—suitable for obtaining visualization of large complex networks—is based on the Fruchterman-Reingold algorithm, FR (Fruchterman and Reingold 1991), which is a force-directed technique based on molecular dynamics employing both attractive and repulsive forces between nodes (Silva et al. 2013)

4.3.2   GCNs are Scale Free Networks

GCNs, like other biological networks and similarly to social and internet networks, are not random and follow some basic principles (Newman 2010). In random networks the nodes have nearly the same number of links and, therefore, highly linked nodes are rare. In network terminology, the number of links, or edges, connected to a node is called node degree. Hence, nodes in random networks characteristically have low diversity of node degrees. Conversely, most of the "real world networks", as GCNs or protein-protein networks, are scale free, what means that the degree distribution follows a power law: the node degree distribution $P(k)$, with node degree k, follows $P(k) \sim k^{-\gamma}$, where γ is the degree exponent. Therefore, scale free networks have a limited number of highly connected nodes, or hubs, that, as we shall discuss latter, are usually associated to relevant biological functions and responsible for the network robustness, i.e., hold the whole network together (Winterbach et al. 2013).

The categorization of nodes according to their node degree encompasses two other categories besides the hubs. The VIPs (a term coined in the study of social networks) are nodes presenting low node degree but connected only with hubs (Masuda and Konno 2006; Mcauley et al. 2007). In some networks VIPs may represent the highest control hierarchy in a system and hubs may be under VIPs influence. Some nodes may present VIP status (connected with many hubs) and also present high overall number of connections, being called high-hubs (Bando et al. 2013). These hierarchical categories are all coherent with the biological role and dynamic behavior of GCNs hubs, as discussed below.

Some hubs are highly interlinked in local regions of a network thereby forming network clusters, topologically called modules or communities. Modules may be associated to specific biological processes in gene coexpression and protein-protein networks. For this reason, hubs may be sometimes classified as "party hubs", those functioning inside a module, or "date hubs", i.e., those linking different processes and organizing the network, playing a role similar to VIPs and high hubs (Zhu et al. 2007; Barabási et al. 2011; Weirauch 2011).

4.3.2.1 Concentric Characterization of Nodes

One way to classify network nodes as VIPs, hubs or high-hubs is by obtaining the node degree, k_0, and the first level concentric node degree, k_1, which takes into account all node connections leaving from its immediate neighborhood, then projecting all node values in a k_0 vs k_1 graphic. VIPs should present low k_0 but high k_1, while hubs present high k_0 and low k_1, and high-hubs present high k_0 and k_1 values. Figures. 4.2e, f and 4.3e, f show each of these node categories in scatter plots of node degree vs concentric node degree measures obtained in DE and CO networks distribution scatter plots (k_0 vs k_1) generated for distinct GCNs.

Because most real networks present scale-free distributions, there is no clear definition for setting a degree threshold for which we can classify nodes as being hubs or not (Barabási and Oltvai 2004; Barabási et al. 2011). This same is true for objectively defining VIPs and high-hubs, since the distribution of k_1 also suffers from the problem of not presenting a scale. Here we define hubs, VIPs and high-hubs by ranking them according to k_0 and k_1, and then considering a set of those presenting the highest values of each property. All calculations can be performed by using the software available at (http://cyvision.if.sc.usp.br/~bant/hierarchical/). These measures are used for nodes categorization such as Hub (high k_0 VIP (high k_1 and low k_0) and High-hub (high k_0 and k_1).

4.3.3 Betweenness Centrality

Betweenness centrality (Costa et al. 2008; Freeman 1978; Brandes 2001) is a measurement of node importance which takes into account the entire set of shortest paths between nodes and passing through a particular node in a network. Betweenness is one of the most important topological properties of a network: nodes with the highest betweenness control most of the information flow in the network (Yu et al. 2007).

4.3.4 Positive or Inverse Gene-Gene Correlation

Pearson's correlation coefficient (PCC) gives us the strength of the relationship between a pair of genes (nodes in the network) (Allen et al. 2010). PCC ranges from -1 to 1 and the closer the number to either of these boundaries the stronger the relationship: a negative number indicates an inverse correlation (e.g. expression of gene A increases as expression of gene B decreases) while a positive number indicates a positive correlation (e.g. as A increases B tends to increase). This is depicted by the blue (positive correlation) and red (negative correlation) edges in Figs. 4.2a and b.

4.3.5   Network Connectivity

This and the two next subsections will address issues on network topology. Network topology exerts a pivotal role in unravelling GCNs organization and performance under different conditions (Barabási and Oltvai 2004; Zhu et al. 2007; Costa et al. 2011). Network connectivity is an elementary network property: a pair of nodes that have just one independent path between them are more weakly connected than a pair that has many paths (Flake et al. 2002). Connectivity is commonly visualized as bottlenecks between nodes and formalized by the notion of cut set (Newman 2010). A node cut set is a set of nodes whose removal will disconnect a specific pair of nodes. Conversely, an edge cut set (or link cut set) is a set of edges whose removal will disconnect a pair of edges. A node with a higher degree of links (edges) is better connected in the network and it is supposed to play a more important role in maintaining the network structure (Barabási and Oltvai 2004; Albert 2005), what is generally associated to a relevant biological role (Langfelder et al. 2013). Connectivity is the most widely used concept for distinguishing the nodes of a network (Horvath and Dong 2008). Densely interconnected groups of nodes, or clusters (pointed out by arrows in Figs. 4.2a, b and in color in videos 4.1 and 4.2, are frequently found in most GCNs and protein-protein networks (Newman 2006) in accordance to its scale free connectivity distribution (Winterbach et al. 2013). These groups form topological modules, i.e. highly interlinked regions in a network, and have been associated, in GCNs and protein-protein networks, with highly conserved genes (Barabási and Oltvai 2004) and genes involved with complex diseases (Tuck et al. 2006; Cai et al. 2010; Barabási et al. 2011).

Robustness of complex networks is associated to the capacity of a network to preserve its topological features, such as connectivity and average path length, after the removal of a set of nodes or edges. Scale-free networks such as GCNs are found to be very resilient to random node/edge attacks. This means that random failures or perturbations in some nodes or sub-mechanisms do not seems to drive the entire system to a critical condition (Albert et al. 2008). However, attacks targeting nodes with high number of connections, i.e. hubs, present the opposite effect, thus removing a small number of such nodes in scale-free networks causes a huge impact on the network diameter and on its functionality performance.

4.3.6   Network Motifs

In biological networks it is possible to identify groups of nodes that link to each other forming a small subnetwork, or subgraph, at numbers that are significantly higher than those in randomized networks (Milo et al. 2002). These subgraphs are called motifs. Network motifs constitute smaller common patterns, or 'building blocks", of GCNs (Barabási and Oltvai 2004; Weinrauch 2011) and were found to be associated to some optimized biological functions, such as feedback and feedforward loops, related to transcriptional regulation (Shen-Orr et al. 2002; Zhang et al. 2007; Watkinson et al. 2009). Molecular components of a particular motif frequently interact with nodes in outside motifs, and aggregation of motifs into motif clusters is likely to occur in many real networks (Ravasz et al. 2002). As pointed out by Barabási and Oltvai (2004), because "the number of distinct subgraphs grows exponentially with the number of nodes that are in a subgraph, the study of larger motifs is combinatorially unfeasible". The alternative is to identify groups of highly connected nodes, called modules, directly from the network topology and manage to correlate these topological entities with their functional role (Winterbach et al. 2013).

4.3.7 Network Modules

Modules are large subgraph units, encompassing groups of densely associated nodes and connected to each other with loose links: in GCNs, for instance, modules may be hub clusters tenuously connected by VIPs. Modules serve to identify gene functions in a GCN and—as it was already observed for protein networks (Yu et al. 2007; Zhu et al. 2007)—contain "module organizer" genes, highly connected to other genes (equivalent to hubs and high-hubs) and essential to module functioning, and "connector" genes, linking different modules and relevant for intermodule communication (equivalent to VIPs) (Weirauch 2011; Bando et al. 2011; Bando et al. 2013).

There are many statistical and computational methods for identifying modules in scale free networks. One of them, the Girvan-Newman algorithm (Girvan and Newman 2002), is centered on defining the boundaries of modules by searching for those edges with high betweenness, i.e more likely to link different modules. This is an important issue: cell functions are carried out in a very modular way. Modular structure reflects a group of functionally linked nodes (genes) acting together to accomplish a specific task: it may be invariant protein-RNA complexes involved post-transcriptional control of RNAs, or temporally coregulated genes controlling processes such as cell cycle and differentiation, or bacterial response to growth and stress conditions (Costanzo et al. 2010; Wang and Zheng 2012; Rosenkrantz et al. 2013).

4.3.8 GCNs are Modular Scale Free Networks

The GCNs have a hub-dominated architecture, containing modules, or clusters, constituted by a highly connected number of nodes. The clustering coefficient C is a measure of the degree to which nodes in a graph tend to cluster together (Watts and Strogatz 1998). The average clustering coefficient $<C>$ is significantly higher in most biological networks (gene-gene, protein-protein) than in random networks of equivalent size and distribution (Barabási and Oltvai 2004). In Fig. 4.2a, b GCN gene clusters appear encircled by a solid line and in Videos 4.1 and 4.2 the clusters are identified by distinct colors.

Network modules, or clusters, are present in all cellular networks and identifiable by clustering methods based on network's topology description (Newman 2006; Li and Horwath 2009) or by combining topology and functional genomics data (Wang and Zheng 2012; Weiss et al. 2012). Therefore, to find out correspondences between cluster topology and functional properties is a main goal of GCN analysis. A large number of evidences show that modules involved in closely related biological functions tend to interact and are proximally located in the network (reviewed in Barabási et al. 2011). As we mentioned before (Sect. 4.3.6), scale free networks are robust but attacks targeting highly connected nodes may cause network disruption. There are now compelling data linking the establishment of complex diseases with the perturbation (by mutation or altered expression) of highly connected genes in GCNs (reviewed by Sahni et al. 2013; see also Bando et al. 2013). Thus, functional and disease modules overlap and the transition between health and disease can be described as a module breakdown.

Another challenging issue is to understand network controllability. Controllability analysis in complex networks, a concept introduced by Liu and Barabási (Liu et al. 2011), determines the minimum set of driver nodes necessary to (linearly) control an entire system. This also allows determining the degrees of freedom that a system can attain; therefore, it can also be understood as a measurement of the network complexity. Recently, Liu et al. (2012) introduced the control centrality measurement which considers the individual control potential of each node in a system. As GCNs may represent complex control systems, this new framework can be helpful to understand its control hierarchical structure. However, the inference of causality, i.e. who controls whom, stills presents as an open problem in network theory (Wu et al. 2012; Yuan et al. 2013).

4.4   Validation of Transcriptional Networks

The analysis of GCNs based on DNA microarray experiments has multiple applications in life sciences and medicine, ranging from the study of basic cell functions to the identification of disease markers and the molecular mechanisms underlying complex diseases. Therefore, microarray generated data need to be checked for reproducibility and biological significance. Two categories of data validation will be considered here: (i) the technical and biological validation of DNA microarray experiments (Shi et al. 2008); (ii) the validation of GNCs through interactome analysis (Wang et al. 2014). Additionally, raw microarray data and experimental design should be deposited in at least one data repository supporting MIAME (minimum information about a microarray experiment)-compliance data (Brazma et al. 2001). Two repositories commonly used for this purpose are: Gene Expression Omnibus

(GEO, www.ncbi.nlm.nih.gov/geo/) at the National Center for Biotechnology Information, and ArrayExpress—functional genomics data, at the European Bioinformatics Institute (www.ebi.ac.uk/arrayexpress/).

4.4.1 DNA Microarray Technical Validation

Good laboratory proficiency and appropriate data analysis are essential to avoid artifactual gene profiles generated from DNA microarrays experiments (Shi et al. 2008). Nevertheless, is mandatory to check for results erroneously representing either under- or over-expression of specific genes. There are several methods to quantify gene expression using RNA or gene-specific protein detection, such as quantitative real-time PCR (qPCR) and immunohistochemistry, respectively (True and Feng 2005).

A popular strategy for microarray technical validation using qPCR is to select in the gene data set those presenting the largest fold changes (statistically significant differentially expressed genes between groups). Here one can use the RNA aliquots from the same biological samples tested in the microarray experiment (Miron et al. 2006). In order to accomplish biological validations, is necessary to test additional biological samples (not those used in the experiment). This is critical, for instance, for validating certain genes as disease biomarkers. This kind of validation usually encompasses, whenever possible, immunohistochemistry validation.

4.4.2 Interactome Validation of GCNs: A Tool for Gene Function Discovery

Interactome analysis, particularly protein-protein interaction (PPI) networks (where nodes stand for proteins and edges for the physical interactions), have been used in many different areas, from the study of protein function to disease prognosis (Taylor et al. 2009), being a very useful tool for disease-gene identification (del Rio et al. 2009; Barabási et al. 2011; Carter et al. 2013; Wang et al. 2014). This kind of analysis also allows the *in silico* validation of GNC data. Protein-protein interaction (PPI) networks for GCN validation may be constructed using proteins corresponding to each of the selected hubs, VIPs and high-hubs of a particular GCN (Bando et al. 2013). Six major primary protein databases are available for PPI networks: BIND, BioGRID, DIP, HPRD, IntAct, and MINT (De La Rivas and Fontanillo 2010). In the author's laboratory these analyses are carried out by using an in house free web tool developed by LA Lima & RD Puga—Centro Internacional de Ensino e Pesquisa—Hospital A.C. Camargo and the PPI networks annotated in MINT, HPRD and IntAct databases (http://bioinfo.lbhc.hcancer.org.br/cgi-bin/interactomegraph/index.cgi). Data analysis and visualization are accomplished through Cytoscape. Figure 4.2g, h shows the interactome networks obtained for the patient and control groups DE GCNs used as a demonstrative example along this chapter. Essentially, the software helps to search for interactions among the selected GCN genes (i.e. their corresponding proteins) and their neighbors in the human interactome. Considering our patient versus control example, these neighbors could participate in some disease-related metabolic pathways, thus indicating that the selected GCN genes are involved in the molecular mechanism of that disease.

The integrative analysis of GCN and PPI data is proving to be very helpful for disclosing changes in steady states that characterize the transitions between health and disease (Sahni et al. 2013), or the common genomic drivers beyond apparently distinct pathophenotypes (Cristino et al. 2014). This approach is also advantageous for identifying disease subtypes. Our group recently showed (Bando et al. 2013)— through GNC and interactome analysis—that pathogenic and compensatory pathways differ in refractory temporal lobe epilepsy depending on the initial precipitating insult (febrile or afebrile). In this complex disease, determined by the interplay of genes and environmental factors, the role of disease genes depends substantially on their gene-gene connectivity and much less on structural gene alterations. Correlating different pathophenotypes with specific changes in GNCs and interactome may be helpful for finding novel potential therapeutic targets and design intervention strategies.

Supporting information (videos)

Video 4.1. Patients' group CO network 3D visualization Hubs, VIPs and high-hubs are indicated in blue, red and green, respectively. Clusters are identified by distinct colors.

Video 4.2. Control group CO network 3D visualization Hubs, VIPs and high-hubs are indicated in blue, red and green, respectively. Clusters are identified by distinct colors.

References

Albert R (2005) Scale-free networks in cell biology. J Cell Sci 118:4947–4957
Albert R, Jeong H, Barabási AL (2008) Error and attack tolerance of complex networks. Nature 406:378–382
Allen KD, Coffman CJ, Golightly YM et al (2010) Comparison of pain measures among patients with osteoarthritis. J Pain 11:522–527
Bando SY, Alegro MC, Amaro E Jr et al (2011) Hippocampal CA3 transcriptome signature correlates with initial precipitating injury in refractory mesial temporal lobe epilepsy. PLoS One 6(10):e26268
Bando SY, Silva FN, Costa Lda F et al (2013) Complex network analysis of CA3 transcriptome reveals pathogenic and compensatory pathways in refractory temporal lobe epilepsy. PLoS One 8(11):e79913
Barabási AL, Oltvai ZN (2004) Network biology: understanding the cell's functional organization. Nat Rev Genet 5:101–113

Barabási AL, Gulbahce N, Loscalzo J (2011) Network medicine: a network based approach to human disease. Nat Rev Genet 13:56–68

Benson M, Breitling R (2006) Network Theory to understand microarray studies of complex diseases. Curr Mol Med 6:695–701

Brandes U (2001) A Faster Algorithm for Betweenness Centrality. J Math Sociol 25:163–177

Brazma A, Hingcamp P, Quackenbush J et al (2001) Minimum information about a microarray experiment (MIAME)—toward standards for microarray data. Nat Genet 29:365–371

Cai JJ, Borenstein E, Petrov DA (2010) Broker genes in human disease. Genome Biol Evol 2:815–825

Carter H, Hofree M, Ideker T (2013) Genotype to phenotype via network analysis. Curr Opin Genet Dev 23:611–621

Clauset A, Shallizi CR, Newman MEJ (2009) Power-law distributions in empirical data. SIAM Rev 51:661–703

Chuang H-Y, Hofree M, Ideker Y (2010) A decade of systems biology. Annu Rev Cell Dev Biol 26:721–744

Costa L da F, Tognetti MAR, Silva FN (2008) Concentric characterization and classification of complex network nodes: application to an institutional collaboration network. Phys A 387:6201–6214

Costa L da F, Oliveira ON Jr, Travieso G et al (2011) Analyzing and modeling real-world phenomena with complex networks: a survey of applications. Adv Phys 60:329–412

Costanzo M, Baryshnikova A, Bellay J et al (2010) The genetic landscape of a cell. Science 327:425–431

Cristino AS, Williams SM, Hawi Z, An JY, Bellgrove MA, Schwartz CE, Costa Lda F, Claudianos C (2014) Neurodevelopmental and neuropsychiatric disorders represent an interconnected molecular system. Mol Psychiatry 19:294–301

De Las Rivas J, Fontanillo C (2010) Protein-protein interactions essentials: key concepts to building and analyzing interactome networks. PLoS Comput Biol 6:e1000807

Del Rio G, Koschutzki D, Coello G (2009) How to identify essential genes from molecular networks? BMC Syst Biol 3:102

Elo LL, Järvenpää H, Oresic M et al (2007) Systematic construction of gene coexpression networks with applications to human T helper cell differentiation process. Bioinformatics 23:2096–2103

Faro A, Giordano D, Spampinato C (2012) Combining literature text mining with microarray data: advances for system biology modeling. Brief Bioinform 13:61–82

Flake GW, Lawrence SR, Giles CL et al (2002) Self-organization and identification of Web communities. IEEE Computer 35:66–71

Freeman LC (1978) Centrality in social networks: conceptual clarification. Soc Netw 1:215–239

Fruchterman TMJ, Reingold EM (1991) Graph drawing by force-directed placement software. Pract Exp 21:1129–1164

Girvan M, Newman ME (2002) Community structure in social and biological networks. Proc Natl Acad Sci U S A 99:7821–7826

Herbst A, Jurinovic V, Krebs S et al (2014) Comprehensive analysis of β-catenin target genes in colorectal carcinoma cell lines with deregulated Wnt/β-catenin signaling. BMC Genomics 15:74

Horvath S, Dong J (2008) Geometric interpretation of gene coexpression network analysis. PLoS Comput Biol 4:e1000117

Ideker T, Krogan NJ (2012) Differential network biology. Mol Syst Biol 8:565

Ishiwata RR, Morioka MS, Ogishima S et al (2009) BioCichlid: central dogma-based 3D visualization system of time-course microarray data on a hierarchical biological network. Bioinformatics 25:543–544

Kim Y-A, Wuchty S, Przytycka TM (2011) Identifying causal genes and dysregulated pathways in complex diseases. PLoS Comput Biol 7(3):e1001095

Langfelder P, Mischel PS, Horvath S (2013) When is hub gene selection better than standard meta-analysis? PLoS ONE 8:e61505

Lee WP, Tzou WS (2009) Computational methods for discovering gene networks from expression data. Brief Bioinform 10:408–423

Li A, Horwath S (2009) Network module detection: affinity search technique with the multi-node topological overlap measure. BMC Res Notes 2:142

Liu YY, Slotine JJ, Barabási AL (2011) Controllability of complex networks. Nature 473(7346):167–173

Liu YY, Slotine JJ, Barabási AL (2012) Control centrality and hierarchical structure in complex networks. PLoS ONE 7(9):e44459

Mcauley JJ, Costa L da F, Caetano TS (2007) Rich-club phenomenon across complex network hierarchies. Appl Phy Lett 91:084103

Masuda N, Konno N (2006) VIP-club phenomenon: emergence of elites and masterminds in social networks. Soc Netw 28:297–309

Milo R, Shen-Orr S, Itzkovitz S et al (2002) Network motifs: simple building blocks of complex networks. Science 298:824–827

Miron M, Woody OZ, Marcil A et al (2006) A methodology for global validation of microarray experiments. BMC Bioinform 7:333

Newman MEJ (2006) Modularity and community structure in networks. PNAS 103:8577–8582

Newman MEJ (2010) Networks: an Introduction. Oxford University, New York

Pavlopoulos GA, O'Donoghue SI, Satagopam VP et al (2008) Arena3D: visualization of biological networks in 3D. BMC Systems Biology 2:104 (http://www.biomedcentral.com/1752–0509/2/104)

Prifti E, Zucker JD, Clement K et al (2008) Funnet: an integrative tool for exploring transcriptional interactions. Bioinformatics 24:2636–2638

R Core Team (2012) R: A language and environment for statistical computing. R Foundation for statistical computing, Vienna, Austria (http://www.R-project.org/)

Ravasz E, Somera AL, Mongru DA (2002) Hierarchical organization of modularity in metabolic networks. Science 297:1551–1555

Rosenkrantz JT, Aarts H, Abee T et al (2013) Non-essential genes form the hubs of genome scale protein function and environmental gene expression networks in Salmonella enterica serovar Typhimurium. BMC Microbiol 13:294

Saeed AS, White J et al (2003) TM4: a free, open-source system for microarray data management and analysis. Biotechniques 34:374–378

Sahni N, Yi S, Zhong Q et al (2013) Edgotype: a fundamental link between genotype and phenotype. Curr Opin Genet Dev 23:649–657

Saito R, Smoot ME, Ono K et al (2012) A travel guide to cytoscape plugins. Nat Methods 9:1069–1076

Schroeder A, Mueller O, Stocker S et al (2006) The RIN: an RNA integrity number for assigning integrity values to RNA measurements. BMC Mol Biol 7:3

Shen-Orr SS, Milo R, Mangan S et al (2002) Network motifs in the transcriptional regulation network of Escherichia coli. Nat Genet 31:64–68

Shi L, Perkins RG, Fang H et al (2008) Reproducible and reliable microarray results through quality control: good laboratory proficiency and appropriate data analysis practices are essential. Curr Opin Biotechnol 19:10–18

Sieberts SK, Schadt EE (2007) Moving toward a system genetics view of disease. Mamm Genome 18:389–401

Silva FN, Rodrigues FA, Oliveira ON Jr et al (2013) Quantifying the interdisciplinarity of scientific journals and fields. J Informetr 7:469–477

Song L, Langfelder P, Horvath S (2012) Comparison of co-expression measures: mutual information, correlation, and model based indices. BMC Bioinform 13:328

Taylor IW, Linding R, Wade-Farley D et al (2009) Dynamic modularity in protein interaction networks predicts breast cancer outcome. Nature Biotech 27:199–204

True L, Feng Z (2005) Immunohistochemical validation of expression microarray results. J Mol Diagn 7:149–151

Tuck DP, Kluger HM, Kluger Y (2006) Characterizing disease states from topological properties of transcriptional regulatory networks. BMC Bioinform 7:236

Villa-Vialaneix N, Liaubet L, Laurent T et al (2013) The structure of a gene co-expression network reveals biological functions underlying eQTLs. PLoS One 8:e60045

Wang H, Zheng H (2012) Correlation of genetic features with dynamic modularity in the yeast interactome: a view from the structural perspective. IEEE Trans Nanobiosciences 11:244–250

Wang Q, Tang B, Song L et al (2013) 3DScapeCS: application of 3 dimensional, parallel, dynamic network visualization in Cytoscape BMC Bioinformatics 14:322 (http://www.biomedcentral.com/1471-2105/14/322)

Wang XD, Huang JL, Yang L et al (2014) Identification of human disease genes from interactome network using graphlet interaction. PLoS One 9:e86142

Watkinson J, Liang KC, Wang X (2009) Inference of regulatory gene interactions from expression data using three-way mutual information. Ann NY Acad Sci 1158:302–313

Watts DJ, Strogatz SH (1998) Collective dynamics of 'small word' networks. Nature 393:440–442

Weirauch MT (2011) Gene expression network for the analysis of cDNA microarray data. In: Dehmer M, Emmert-Streib F, Graber A, Salvador A (eds) Applied statistics for network biology: methods in systems biology, vol 1. Wiley, Weinheim, pp 215–250

Weiss JM, Karma A, MacLellan WR et al (2012) "Good enough solutions" and the genetics of complex diseases. Circ Res 111:493–504

Winterbach W, Van Mieghem P, Reinders M et al (2013) Topology of molecular interaction networks. BMC Syst Biol 7:90

Wu X, Wang W, Zheng WX (2012) Inferring topologies of complex networks with hidden variables. Phys Rev E 86:046106

Yu H, Kim PM, Sprecher E et al (2007) The importance of bottlenecks in protein networks: correlation with gene essentiality and expression dynamics. PLoS Comput Biol 3:e59

Yuan Z, Zhao C, Di Z et al (2013) Exact controllability of complex networks. Nat Commun 4:2447

Zhang B, Horvath S (2005) A general framework for weighted gene co-expression network analysis. Stat Appl Genet Mol Biol 4:Article17

Zhang J, Ji Y, Zhang L (2007) Extracting three-way gene interactions from microarray data. Bioinformatics 23:2903–2909

Zhu X, Gerstein M, Snyder M (2007) Getting connected: analysis and principles of biological networks. Genes Dev 21:1010–1024

Chapter 5
Posttranscriptional Control During Stem Cells Differentiation

Bruno Dallagiovanna, Fabiola Holetz and Patricia Shigunov

Abstract Stem cells have the potential to both proliferate and self-renew. These extraordinary cells also have pluri- or multipotency rendering them the ability to differentiate into several tissue-specific lineages. These characteristics make these cells ideal candidates for use in cell therapy. However, it is essential for a successful therapy that these cells could be exclusively committed to the cell type that is needed. Commitment involves the activation of a particular genetic program and this process can be regulated at multiple levels during gene expression. An understanding of the gene regulatory networks involved in the committing of these cells to differentiation into a specific cell type is essential for the successful repair of injured tissue or even whole organogenesis.

The identity and quantity of proteins that a cell produces under a particular set of conditions, provides information about almost all cellular processes.

Gene expression can be regulated at multiple complementary levels, to obtain tight control of transcript abundance and protein synthesis. This regulation involves epigenetic and transcriptional control mechanisms and various complementary posttranscriptional steps that regulate maturation, stability and translation of an mRNA population.

Translation of mRNA into protein can be divided into three sub-process—initiation, elongation and termination. The initiation step, includes all the events that precede the formation of the first peptide bond, starting with the binding of the eukaryotic initiation complex eIF4F to the 5' cap of the mRNA. After the mRNA has been unwound by eIF4F, the pre-initiation complex 43S (which contains the 40S ribosomal subunit, eIF3, and the ternary complex eIF2, GTP and Met-tRNAi) attaches to the 5'-proximal region of the mRNA. This complex scans the 5'-UTR region to meet the first AUG codon, recruiting then the 60S ribosomal subunit to form the translational competent 80S ribosome (Hashem et al. 2013). Initiation is followed by elongation of the peptide chain, the main function of the ribosome. The final termination step includes the release of the newly synthesized protein and the dissociation of ribosomal subunits from the mRNA (Jackson et al. 2012).Structural and functional studies on translation elongation showed a considerable role of these

B. Dallagiovanna (✉) · F. Holetz · P. Shigunov
Instituto Carlos Chagas, FIOCRUZ Paraná, Rua Algacyr Munhoz Mader 3775,
Curitiba, Paraná 81350-010, Brazil
e-mail: brunod@tecpar.br

© Springer International Publishing Switzerland 2014   95
G. A. Passos (ed.), *Transcriptomics in Health and Disease,*
DOI 10.1007/978-3-319-11985-4_5

phases as targets of translation control (Dever and Green 2012; Zaretsky and Wreschner 2008). Nevertheless, in eukaryotes the initiation phase is the rate-limiting step and, thus, the main target for translational control (Fabian et al. 2010).After 80S formation, the initiation factor eIF4G interacts with poly(A)-binding protein (PABP), which is associated with the poly(A) tail, promoting mRNA circularization (Wells et al. 1998). This conformation of them RNA promotes the recycling of ribosomes for a new round of translation, resulting in the assembly of the polysome complex (Szostak and Gebauer 2013).

Regulation of protein synthesis plays a decisive role in a wide range of biological situations and is critical for maintaining homeostasis, cell proliferation, growth and development. Deregulation of translation is involved in the development of several human diseases and cancer (Gkogkas and Sonenberg 2013; Hershey et al. 2012; Livingstone et al. 2010). Translational regulation can be divided in two types, global and transcript-specific control. Modulating the activities of translation initiation factors, or the regulators that interact with them by phosphorylation, enables eukaryotic cells to regulate global rates of protein synthesis. In the mRNA specific control, translation of a defined group of mRNAs is modulated without affecting general protein biosynthesis and frequently occurs through the action of trans-acting RNA binding factors (RNA-binding proteins, miRNA and tRNA fragments). These trans-acting factors repress the association of mRNAs to translating ribosomes or the assembly of new polysomes (Gebauer et al. 2012; Hershey et al. 2012; Sobala and Hutvagner 2013).It has been suggested that the cellular abundance of proteins is predominantly controlled at the level of translation (Schwanhäusser et al. 2011). Hence, the identification of which subpopulation of mRNAs is associated to active polysomes is essential to understand the dynamics of translational control in the cell.

5.1   Stem Cells and the Control of Protein Synthesis

5.1.1   Identifying Polysome Associated mRNAs During Cellular Differentiation

Gene expression profiling has provided insight into the molecular pathways involved in stem cells self-renewal and differentiation (Ivanova et al. 2002; Song et al. 2006). Genome-wide analyses based on microarray hybridization and, more recently next generation sequencing, have been carried out to assess the global expression of gene networks. Most attempts to determine the mRNA profile of self-renewing or differentiating cells have made use of total mRNA for hybridization to microarrays or RNA-seq analysis (Jeong et al. 2007; Menssen et al. 2011). Several researches have focused on the study of the cellular transcriptome to understand gene expression regulation, assuming that the mRNA levels could reflect the final concentration of proteins in the cell (Cheadle et al. 2005; Larsson et al. 2013).

However, high-throughput analyses in eukaryotes comparing mRNA and protein levels have indicated that there is no direct correlation between transcript levels and protein synthesis, suggesting a high degree of posttranscriptional regulation in eukaryote cells (Tebaldi et al. 2012). The combination of quantitative proteomics with microarray analysis of mRNA levels in embryonic and primary hematopoietic stem cells showed very low correlation values between protein and mRNA expression during cell differentiation (Unwin et al. 2006; Williamson et al. 2008). Several lines of evidence from different organisms suggest that stem cell self-renewal and differentiation are dependent on the control of protein synthesis by posttranscriptional mechanisms (Haston et al. 2009; Kolle et al. 2011; Sampath et al. 2008). This hampers the classical transcriptome-based approach to investigate controlled expression in differentiating cells.

To overcome this limitation, in the past few years new methods were developed to compare the amount of total mRNA pool with the fraction of mRNA committed in translation. Ribosome association of an mRNA is considered as a general measure of its translational activity (Sonenberg et al. 2000). Centrifugation of the cytosolic soluble cell fraction in sucrose gradients allows the separation of polyribosome complexes from monosomes and ribosome-free transcripts or inactive mRNP particles. The different fractions of the gradient can be pooled in order to isolate the subpopulation of mRNAs that is associated to translating polysomes. Quantification of these mRNAs has been successfully used to obtain genome-wide information on translationally regulated transcripts (Arava et al. 2003; Sonenberg et al. 2000) (Fig. 5.1a). This strategy has been used by several groups to better understand the posttranscriptional regulation of the process of cell differentiation in stem cells (Bates et al. 2012; Blázquez-Domingo et al. 2005; Fromm-Dornieden et al. 2012; Kolle et al. 2009; Kolle et al. 2011; Parent and Beretta 2008; Sampath et al. 2008).

One study analyzed gene expression profiles during the differentiation of murine embryonic stem cells (ESCs) into embryoid bodies by integrating transcriptome analysis with a global assessment of ribosome loading (Sampath et al. 2008). The authors used sucrose gradient centrifugation combined with microarray analysis, also known as Translation State Array Analysis (TSAA) (Arava 2003; Arava et al. 2003; Zong et al. 1999), to obtain a genome-scale view of the effect of translation on gene expression. In TSAA changes in ribosome loading with mRNAs are measured on a genome-wide scale indicating the efficiency of translation for individual transcripts. Undifferentiated ESCs were found to be relatively polysome poor, as the result of inefficient loading of most transcripts onto ribosomes. Differentiation was accompanied by a global increase in both transcript abundance and the efficiency of mRNA translation, with almost 80 % of the transcripts showing increased ribosome loading. This study highlighted several vital genes that are exclusively regulated by translation during differentiation like the ATF5 transcription factor and Wnt1 an effector of the Wnt signaling pathway. These data indicates that protein production from a vast number of genes is limited in ESCs by both transcript and protein synthesis and that ESCs differentiation is accompanied by a notorious increase in translational efficiency (Sampath et al. 2008).

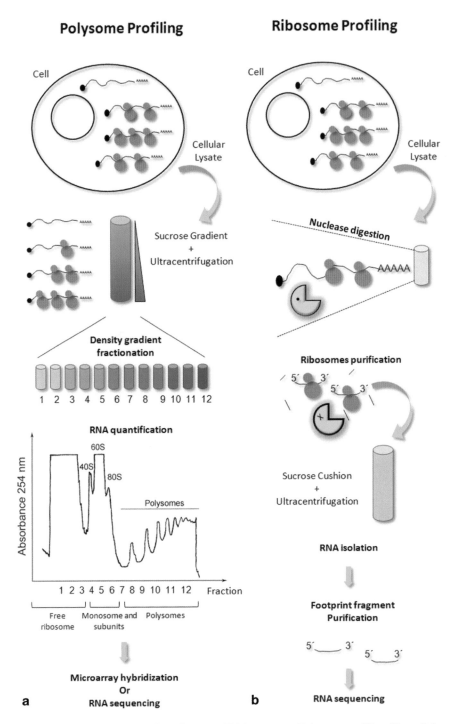

Fig. 5.1 Strategies to analyze the polysome mRNA content. **a** Polysome profiling. The cellular lysate is placed on a sucrose gradient allowing the separation of polysomal fractions. Actively

Human embryonic stem cells are isolated and characterized by surface marker expression. Identification of surface markers is essential for characterization and purification of stem cells. Kolle et al. (2009) combined immunotranscriptional profiling of human ESC lines with membrane-polysome TSAA to determine the genes encoding potential human ESC surface marker proteins (Kolle et al. 2009). This approach has been used extensively to profile transcripts encoding secreted or transmembrane proteins within a variety of cell model systems (de Jong et al. 2006; Diehn et al. 2000, 2006; Stitziel et al. 2004). The assay separates mRNAs bound to actively translating, membrane-bound polysomes from cytosolic polysome-bound and non-translated mRNAs. A total of 88 genes that encode potential cell surface markers of hESCs were identified with this approach, greatly expanding the number of protein antigens that can be used to isolate pluripotent ESCs (Kolle et al. 2009). Kolle et al. (2011) also proposed a strategy to isolate mRNAs contained in the polysome-membrane fraction of hESCs and identified these RNAs this time by large scale sequencing (Kolle et al. 2011). They found that more than 1000 genes produce transcripts that contain long 5' and/or extended 3' UTRs. Their analysis of membrane-polysome and cytosolic/untranslated fractions also identified RNAs encoding peptides destined for secretion and the extracellular space. This work highlights the efficiency of combining cellular fractionation with RNA-sequencing to characterize the transcriptome and translatome complexity in ESCs (Kolle et al. 2011).

Similar findings have been reported for differentiation of adult stem cells. Parent and Beretta (2008) used polysome profiling to investigate translational control during hepatocytic differentiation of HepaRG liver progenitor cells (Parent and Beretta 2008). They found that the vast majority of genes regulated during differentiation were contained in the polysome-bound RNA population and not in the total RNA population, suggesting a strong association between translational control and hepatocytic differentiation. They showed that hepatocytic differentiation is accompanied by a reduction in transcriptome complexity and that translational regulation is the main regulatory event. Bates et al. (2012) used a similar approach to analyze the translational regulation of genes in a model of B Cell differentiation (Bates et al. 2012). The authors used a streamlined version of traditional polysome profiling on a genomic scale during which mRNAs within sequential fractions of a linear sucrose gradient were differentially labeled and analyzed by DNA microarray (Bates et al. 2012). This procedure, called Gradient Encoding, provides an accurate and reproducible ranking of the positions of mRNAs in the gradient, allowing sensitive detection of changes in the average number of ribosome per mRNA (Hendrickson

---

translated polysome-bound mRNAs are denser than free mRNA and settle at the heavier fractions of the sucrose gradient. Translationally inactive mRNAs that are not bound to ribosomes settle at the top of the gradient. The separation of the fractions is performed by Density Gradient Fractionation (ISCO) and afterwards RNA is quantified and isolated for microarray analysis or RNA sequencing. **b** Ribosome profiling (the deep sequencing of ribosome-protected mRNA fragments). The cellular lysate is treated with an RNA nuclease and ribosomes and associated mRNA footprints are purified by ultracentrifugation through a sucrose cushion. Protected mRNA fragments from single ribosomes are purified by PAGE and sequenced

et al. 2009). The authors found that during differentiation, major changes occurred in the posttranscriptional regulation of genes with critical roles in transformation or differentiation. They also identified additional genes with potential roles in these processes based on particular changes in their translational regulation during differentiation.

An extensively studied model for adipogenesis in vitro is the mouse embryonic fibroblast cell line 3T3-L1 (MacDougald and Lane 1995). Fromm-Dornieden et al. (2012) used TSAA to analyze changes to translational control at 6 h after the induction of adipogenesis in 3T3-L1 preadipocytes. The authors detected 43 translationally up-regulated mRNAs and two translationally down-regulated mRNAs. Despite the low number of differentially expressed genes found, they conclude that a moderate reorganization of the translational activity is an important step for gene expression controls in the initial phase of adipogenesis (Fromm-Dornieden et al. 2012). Our group used polysome profiling of adult stem cells followed by RNA-seq analysis during the initial steps of adipogenesis to investigate how posttranscriptional regulation controls gene expression in human adipose stem cells (hASCs) (Spangenberg et al. 2013b). RNA-seq analysis of the total mRNA fraction and the subpopulation of mRNAs associated with translating ribosomes showed that a significant percentage of mRNAs regulated during differentiation were posttranscriptionally controlled. We demonstrated, that adipogenesis had been triggered at the molecular level after 3 days of induction with upregulation of the expression of networks of genes involved in adipocyte differentiation. Moreover, three days appears as the minimum induction time required for the initiation of adipogenesis. Our study identified 549 differentially expressed genes during initial steps of adipogenesis. Almost 60 % of these genes showed some kind of posttranscriptional regulation. In some cases, this regulation counterbalances fluctuations in total RNA levels. Thus, transcripts increasing or decreasing in abundance in the cell are recruited to polysomes in equal amounts in differentiating cells. It was also observed a subset of transcripts that showed a higher fold change in the polysomal RNA fraction, resembling a mechanism of homodirectional co-regulatory mechanism that results in the amplification or potentiation of the positive control of gene expression (Preiss et al. 2003). However, there is a subpopulation of mRNAs that is regulated solely at the translational level (Fig. 5.2a and b).Specific groups of related genes were found to display differential expression mostly in the polysomal fraction. As an example, oxidative stress response genes and a family of genes encoding proteins involved in the response to changes in the levels of reduced glutathione (Fig. 5.2c and d). Part of this regulation involved large changes in the length of untranslated regions (UTR), and the differential extension/reduction of the 3' UTR after the induction of differentiation. This was similar to what Kolle et al. described previously for ESCs. The length of 3' UTRs has been shown to differ between embryonic stem cells and somatic cells, in both humans and mice (Kolle et al. 2011). Proliferating cells produce mRNAs with shorter UTRs, which have longer half-lives. It has been suggested that transcripts may be stabilized by the loss of miRNA binding sites, which usually downregulate gene expression (Sandberg et al. 2008). By contrast, in murine stem cells, the UTRs of tissue-specific transcripts increase in length follow-

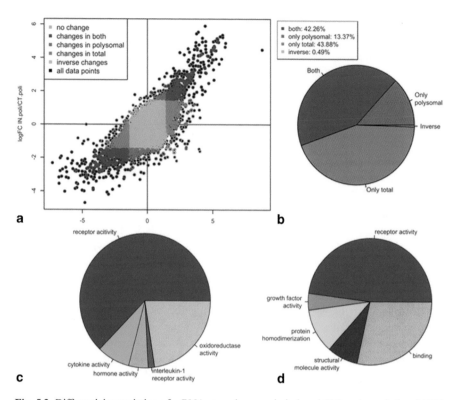

Fig. 5.2 Differential association of mRNAs to polysomes in induced (IN) and non-induced (CT) adult stem cells: **a** LogFC values from different RNA fractions were compared. The logFC values (IN vs. CT) for the polysomal fraction (y-axis) were plotted against the logFC values for the total RNA fraction (x-axis). The data points are colored according to the change in the fractions. Genes displaying changes only in the polysomal fraction include genes with a |logFC| of at least 1.5 (|logFC| > 1.5) and genes with a |logFC| < 1.5 in the total fraction (*light red*). Changes in the total RNA fraction were associated with a |logFC| > 1.5 and of |logFC| < 1.5 in the polysomal fraction (*light green*). Genes displaying changes in expression in both sets of conditions had high (or low) values of logFC (greater than 1.5/lower than − 1.5) in both RNA fractions (*violet*). Inverse changes included genes with high logFC values (logFC > 1.5) in the total fraction and low values (logFC < 1.5) in the polysomal fraction, or vice versa (*orange*). **b** The pie chart shows the percentages of genes displaying changes in expression in each of the four categories: only the changing genes are considered. GO analysis of two sets of differentially expressed genes: IN vs. CT in the polysomal RNA fraction (**c**) and IN vs. CT in total RNA (**d**). Only overrepresented Molecular Function (MF) GO terms are shown in each pie chart (**c**) and (**d**). For each over represented MF GO term its corresponding adjusted *p*-values are shown. (Spangenberg et al. 2013b)

ing cell commitment. This extension results from the use of distal polyadenylation sites and results in an increase in the half life of the mRNA, by an unknown mechanism (Ji et al. 2009). We analyzed the distribution of alternative transcripts during adipogenic differentiation and the potential role of miRNAs in post-transcriptional regulation. Our in silico analysis suggests a modest but consistent, bias in 3' UTR lengths during differentiation enabling a fine-tuned transcript regulation via small non-coding RNAs (Spangenberg et al. 2013a).

The control of translational initiation arises as a central step in the regulation of cell differentiation and commitment of stem cells. Polysome profiling is a simple and straightforward tool to analyze the flow of mRNAs between functionally distinct cell compartments, because these mRNA populations can be easily separated and isolated from a sample by centrifugation in a sucrose gradient. Its use in the study of translational regulation of stem cell commitment will help the understanding and identification of signals involved in the biology of these cells.

5.2   Ribosome Footprints on an mRNA

Several studies have used polysome profiling to study the global translation process in stem cells during a wide variety of cellular processes, from proliferation to differentiation. Nevertheless, high molecular weight ribonucleoprotein complexes not directly involved in translation could co-sediment with polysomes, making difficult the separation of those transcripts that are indeed being translated (Holetz et al. 2007; Ingolia 2014).

However, the innovative ribosome profiling technique that was described by Ingolia and colleagues in 2009, has provided a detailed view of protein synthesis mechanisms from prokaryotes to mammals. This methodology relies on the fact that ribosomes protect a stretch of bound mRNA (~ 30 nucleotides) from nuclease digestion. This protected mRNA 'footprint' can then be isolated and sequenced by deep-sequencing technologies (RNA-seq) (Fig. 5.1b). Thus, it is possible to obtain the exact location of ribosomes on mRNA, as well as a detailed overview of all translation steps, including initiation, elongation and termination by this method. Ribosome profiling measures the number and velocity of ribosomes that are translating the mRNA in vivo, instead of on the abundance of the transcript in the cell, providing measurements that closely correspond to protein abundance (Ingolia et al. 2009). This methodology makes it possible to identify sequences that are actively translated, amongst a complex array of cellular transcripts. It also enables the monitoring of translation and maturation of nascent polypeptides in vivo, and the assessment of profiles of protein synthesis (Ingolia 2014).

Ribosome profiling has emerged as a powerful technique to study several aspects of translation and is being used to unravel the mechanisms involved in the translational control of gene expression in stem cells. Ingolia and co-workers obtained genome-wide maps of protein synthesis in mouse embryonic stem cells (mESCs) and detailed information about the kinetics and mechanism of translation elongation and coupled co-translational events (Ingolia et al. 2011). The authors analyzed the cumulative distribution of footprinting counts at each codon, relative to the median density across the gene, and found thousands of pauses in the body of several transcripts. Analysis of the sequence around the pause site revealed a peptide motif associated with internal translational stalling that was not enriched in rare codons. In addition, they used a pulse-chase strategy to measure the rate of translation elongation and found that the kinetics of elongation do not depend on transcript length and protein abundance, even for transcripts that are translated at the ER surface. Furthermore, the authors suggest that translation speed does not depend on codon usage, which was consistent with the absence of pauses at rare codons. These results go against accepted biophysical models of translation, which state that elongating ribosomes translate each codon with a speed related to the features of the coding sequence and according to cellular factors, such as concentrations of elongation factors and tRNA molecules (dos Reis et al. 2004; Lu and Deutsch 2008; Tuller et al. 2011). Recently, Dana and Tuller (2012) re-analyzed ribosomal profiles of mESCs measured by Ingolia and co-workers (2011) and showed that translation elongation speed is affected by features such as the adaptation of codons to the tRNA pool, and local mRNA folding and charge (Dana and Tuller 2012).They also show that the translation elongation velocity tends to increase as translation progresses along the coding sequence.

Another intriguing result presented by Ingolia and co-workers (2011) was the presence of several unannotated near-cognate initiation sites that drive the translation of upstream open reading frames (uORFs) in mESCs, consistent with the high rate of translation observed at many 5' UTRs. Translation of uORFs was lower in differentiating cells than in mESCs, indicating that the translation of uORFs is regulated and may be part of a major program of translational control. This finding prompted the authors to search for translated regions within some lincRNAs (long intergenic non coding RNAs) which have no conserved sequence with protein-coding potential. Most putative lincRNAs showed ribosome footprints, raising the possibility that these transcripts encode small proteins. This is a striking observation, however, other findings strongly suggest that lincRNAs function as RNA molecules and not as translated proteins (Guttman et al. 2009; Guttman et al. 2010; Slavoff et al. 2013). Thus, in a recent study the same researchers developed a metric termed ribosome release score (RRS) which analyzes the pattern of ribosome occupancy across different classes of RNA and distinguishes coding from non-coding transcripts (Guttman et al. 2013). The authors categorized lincRNAs with well-established non-coding RNAs, indicating that, in general, they do not encode functional proteins.

Ingolia and colleagues also examined changes in translation when proliferative, pluripotent mESCs underwent differentiation into embryoid bodies (EBs). The abundance of ribosomal proteins was much lower in EBs than in ESCs, due 3–4-fold difference in the translational efficiency of transcripts encoding ribosomal proteins between EBs and ESCs. Translation of uORFs also declined during differentiation, and the translation rate of 5' UTRs in differentiated cells was 25 % lower than that of the CDS of individual transcripts with defined uORFs.

Ribosome profiling has also helped to characterize important proteins involved in mRNA metabolism in mESCs. One study used ribosome profiling to monitor translational efficiency after Lin28a knockdown (Cho et al. 2012). LIN28 is a conserved RNA binding protein that is highly abundant in mESCs. LIN28acts as a suppressor of let-7 micro RNA biogenesis, however many lines of evidence suggest that LIN28 carries out additional functions. The ribosome occupancy of LIN28A-bound mRNAs tended to be higher in Lin28a-depleted cells than in control siRNA-treated cells, indicating that LIN28A targets mRNAs for translation repression.

A recent study investigated the implications of canonical and non-canonical Nonsense-Mediated mRNA Decay (NMD) on the decay of endogenous mRNAs in mESCs (Hurt et al. 2013). Messenger RNAs harboring upstream open reading frames (uORFs) may be susceptible to NMD, but only a fraction of uORF-containing mRNA is actually targeted by this pathway and the influence of uORFs on mRNA stability is poorly understood (reviewed by Hurt et al. 2013). Thus, the authors carried out ribosome profiling with UPF1-depleted and control-depleted mESCs. UPF1 is a conserved protein in eukaryotes that is essential for NMD. The density of footprinting reads was used to distinguish actively translated uORFs from non-translated uORFs. The depletion of UPF1 showed that actively translated uORFs-genes are normally targeted by NMD whereas non-translated uORFs-genes escape repression (Hurt et al. 2013). The authors concluded that NMD triggered by uORF translation is an important mechanism of the regulation of gene expression in mESCs.

Adipogenic differentiation has been widely used by our group as a model to investigate the mechanisms of the posttranscriptional regulation of gene expression in hASCs (Spangenberg et al. 2013b). Polysome profiling experiments showed extensive posttranscriptional regulation 3 days after the induction of adipocyte commitment. Now, we have applied ribosomal profiling methodology to investigate differential gene expression of hASCs in the early steps of differentiation to obtain new insights into the mechanisms of translational control that may help to improve our limited understanding of stem cell differentiation. Preliminary data has confirmed extensive translational regulation during cell commitment, and shows that entire metabolic networks are regulated by modification of translational rates after induction of adipogenesis.

Overall, translational regulation is the focus of intense study and is becoming increasingly appreciated as a central step of gene expression control in both embryonic and adult stem cells. Polysome and ribosome profiling are powerful tools to analyze translational dynamics on a genome-wide scale and will enhance and improve our understanding of translational control during stem cell commitment.

Acknowledgements Supported by Ministério da Saúde Brazil, Fundação Araucária, and FIO-CRUZ. P.S. received fellowship from FIOCRUZ and B.D. from CNPq Brazil.

References

Arava Y (2003) Isolation of polysomal RNA for microarray analysis. Methods Mol Biol 224:79–87
Arava Y, Wang Y, Storey JD et al (2003) Genome-wide analysis of mRNA translation profiles in *Saccharomyces cerevisiae*. Proc Natl Acad Sci U S A 100(7):3889–3894
Bates JG, Salzman J, May D et al (2012) Extensive gene-specific translational reprogramming in a model of B cell differentiation and Abl-dependent transformation. PLoS One 7(5):e37108
Blázquez-Domingo M, Grech G, von Lindern M (2005) Translation initiation factor 4E inhibits differentiation of erythroid progenitors. Mol Cell Biol 25(19):8496–8506
Cheadle C, Fan J, Cho-Chung YS et al (2005) Stability regulation of mRNA and the control of gene expression. Ann N Y Acad Sci 1058:196–204

Cho J, Chang H, Kwon SC et al (2012) LIN28A is a suppressor of ER-associated translation in embryonic stem cells. Cell 151(4):765–777

Dana A, Tuller T (2012) Determinants of translation elongation speed and ribosomal profiling biases in mouse embryonic stem cells. PLoS Comput Biol 8(11):e1002755

de Jong M, van Breukelen B, Wittink FR et al (2006) Membrane-associated transcripts in Arabidopsis; their isolation and characterization by DNA microarray analysis and bioinformatics. Plant J 46(4):708–721

Dever TE, Green R (2012) The elongation, termination, and recycling phases of translation in eukaryotes. Cold Spring Harb Perspect Biol 4(7):a013706

Diehn M, Eisen MB, Botstein D et al (2000) Large-scale identification of secreted and membrane-associated gene products using DNA microarrays. Nat Genet 25(1):58–62

Diehn M, Bhattacharya R, Botstein D et al (2006) Genome-scale identification of membrane-associated human mRNAs. PLoS Genet 2(1):e11

dos Reis M, Savva R, Wernisch L (2004) Solving the riddle of codon usage preferences: a test for translational selection. Nucleic Acids Res 32(17):5036–5044

Fabian MR, Sonenberg N, Filipowicz W (2010) Regulation of mRNA translation and stability by microRNAs. Annu Rev Biochem 79:351–379

Fromm-Dornieden C, von der Heyde S, Lytovchenko O et al (2012) Novel polysome messages and changes in translational activity appear after induction of adipogenesis in 3T3-L1 cells. BMC Mol Biol 13:19

Gebauer F, Preiss T, Hentze MW (2012) From cis-regulatory elements to complex RNPs and back. Cold Spring Harb Perspect Biol 4(7):a012245

Gkogkas CG, Sonenberg N (2013) Translational control and autism-like behaviors. Cell Logist 3(1):e24551

Guttman M, Amit I, Garber M et al (2009) Chromatin signature reveals over a thousand highly conserved large non-coding RNAs in mammals. Nature 458(7235):223–227

Guttman M, Garber M, Levin JZ et al (2010) Ab initio reconstruction of cell type-specific transcriptomes in mouse reveals the conserved multi-exonic structure of lincRNAs. Nat Biotechnol 28(5):503–510

Guttman M, Russell P, Ingolia NT et al (2013) Ribosome profiling provides evidence that large noncoding RNAs do not encode proteins. Cell 154(1):240–251

Hashem Y, des Georges A, Dhote V et al (2013) Structure of the mammalian ribosomal 43S preinitiation complex bound to the scanning factor DHX29. Cell 153(5):1108–1119

Haston KM, Tung JY, Reijo Pera RA (2009) Dazl functions in maintenance of pluripotency and genetic and epigenetic programs of differentiation in mouse primordial germ cells in vivo and in vitro. PLoS One 4(5):e5654

Hendrickson DG, Hogan DJ, McCullough HL et al (2009) Concordant regulation of translation and mRNA abundance for hundreds of targets of a human microRNAs. PLoS Biol 7(11):e1000238

Hershey JW, Sonenberg N, Mathews MB (2012) Principles of translational control: an overview. Cold Spring Harb Perspect Biol 4(12):a011528

Holetz FB, Correa A, Avila AR et al (2007) Evidence of P-body-like structures in *Trypanosoma cruzi*. Biochem Biophys Res Commun 356(4):1062–1067

Hurt JA, Robertson AD, Burge CB (2013) Global analyses of UPF1 binding and function reveal expanded scope of nonsense-mediated mRNA decay. Genome Res 23(10):1636–1650

Ingolia NT (2014) Ribosome profiling: new views of translation, from single codons to genome scale. Nat Rev Genet 15(3):205–213

Ingolia NT, Ghaemmaghami S, Newman JR et al (2009) Genome-wide analysis in vivo of translation with nucleotide resolution using ribosome profiling. Science 324(5924):218–223

Ingolia NT, Lareau LF, Weissman JS (2011) Ribosome profiling of mouse embryonic stem cells reveals the complexity and dynamics of mammalian proteomes. Cell 147(4):789–802

Ivanova NB, Dimos JT, Schaniel C et al (2002) A stem cell molecular signature. Science 298(5593):601–604

Jackson RJ, Hellen CU, Pestova TV (2012) Termination and post-termination events in eukaryotic translation. Adv Protein Chem Struct Biol 86:45–93

Jeong JA, Ko KM, Bae S et al (2007) Genome-wide differential gene expression profiling of human bone marrow stromal cells. Stem Cells 25(4):994–1002

Ji Z, Lee JY, Pan Z et al (2009) Progressive lengthening of 3' untranslated regions of mRNAs by alternative polyadenylation during mouse embryonic development. Proc Natl Acad Sci U S A 106(17):7028–7033

Kolle G, Ho M, Zhou Q et al (2009) Identification of human embryonic stem cell surface markers by combined membrane-polysome translation state array analysis and immunotranscriptional profiling. Stem Cells 27(10):2446–2456

Kolle G, Shepherd JL, Gardiner B et al (2011) Deep-transcriptome and ribonome sequencing redefines the molecular networks of pluripotency and the extracellular space in human embryonic stem cells. Genome Res 21(12):2014–2025

Larsson O, Tian B, Sonenberg N (2013) Toward a genome-wide landscape of translational control. Cold Spring Harb Perspect Biol 5(1):a012302

Livingstone M, Atas E, Meller A et al (2010) Mechanisms governing the control of mRNA translation. Phys Biol 7(2):021001

Lu J, Deutsch C (2008) Electrostatics in the ribosomal tunnel modulate chain elongation rates. J Mol Biol 384(1):73–86

MacDougald OA, Lane MD (1995) Transcriptional regulation of gene expression during adipocyte differentiation. Annu Rev Biochem 64:345–373

Menssen A, Haupl T, Sittinger M et al (2011) Differential gene expression profiling of human bone marrow-derived mesenchymal stem cells during adipogenic development. BMC Genomics 12:461

Parent R, Beretta L (2008) Translational control plays a prominent role in the hepatocytic differentiation of HepaRG liver progenitor cells. Genome Biol 9(1):R19

Preiss T, Baron-Benhamou J, Ansorge W et al (2003) Homodirectional changes in transcriptome composition and mRNA translation induced by rapamycin and heat shock. Nat Struct Biol 10(12):1039–1047

Sampath P, Pritchard DK, Pabon L et al (2008) A hierarchical network controls protein translation during murine embryonic stem cell self-renewal and differentiation. Cell Stem Cell 2(5):448–460

Sandberg R, Neilson JR, Sarma A et al (2008) Proliferating cells express mRNAs with shortened 3' untranslated regions and fewer microRNA target sites. Science 320(5883):1643–1647

Schwanhäusser B, Busse D, Li N et al (2011) Global quantification of mammalian gene expression control. Nature 473(7347):337–342

Slavoff SA, Mitchell AJ, Schwaid AG et al (2013) Peptidomic discovery of short open reading frame-encoded peptides in human cells. Nat Chem Biol 9(1):59–64

Sobala A, Hutvagner G (2013) Small RNAs derived from the 5' end of tRNA can inhibit protein translation in human cells. RNA Biol 10(4):553–563

Sonenberg N, Hershey JWB, Matheus MB (2000) Translational control of gene expression. Cold Spring Harbor Laboratory Press, Cold Spring Harbor, NY.

Song L, Webb NE, Song Y et al (2006) Identification and functional analysis of candidate genes regulating mesenchymal stem cell self-renewal and multipotency. Stem Cells 24(7):1707–1718

Spangenberg L, Correa A, Dallagiovanna B et al (2013a) Role of alternative polyadenylation during adipogenic differentiation: an in silico approach. PLoS One 8(10):e75578

Spangenberg L, Shigunov P, Abud AP et al (2013b) Polysome profiling shows extensive posttranscriptional regulation during human adipocyte stem cell differentiation into adipocytes. Stem Cell Res 11(2):902–912

Stitziel NO, Mar BG, Liang J et al (2004) Membrane-associated and secreted genes in breast cancer. Cancer Res 64(23):8682–8687

Szostak E, Gebauer F (2013) Translational control by 3'-UTR-binding proteins. Brief Funct Genomics 12(1):58–65

Tebaldi T, Re A, Viero G et al (2012) Widespread uncoupling between transcriptome and translatome variations after a stimulus in mammalian cells. BMC Genomics 13:220

Tuller T, Veksler-Lublinsky I, Gazit N et al (2011) Composite effects of gene determinants on the translation speed and density of ribosomes. Genome Biol 12(11):R110

Unwin RD, Smith DL, Blinco D et al (2006) Quantitative proteomics reveals posttranslational control as a regulatory factor in primary hematopoietic stem cells. Blood 107(12):4687–4694

Wells SE, Hillner PE, Vale RD et al (1998) Circularization of mRNA by eukaryotic translation initiation factors. Mol Cell 2(1):135–140

Williamson AJ, Smith DL, Blinco D et al (2008) Quantitative proteomics analysis demonstrates post-transcriptional regulation of embryonic stem cell differentiation to hematopoiesis. Mol Cell Proteomics 7(3):459–472

Zaretsky JZ, Wreschner DH (2008) Protein multifunctionality: principles and mechanisms. Transl Oncogenomics 3:99–136

Zong Q, Schummer M, Hood L et al (1999) Messenger RNA translation state: the second dimension of high-throughput expression screening. Proc Natl Acad Sci U S A 96(19):10632–10636

Chapter 6
Transcriptome Analysis During Normal Human Mesenchymal Stem Cell Differentiation

Karina F. Bombonato-Prado, Adalberto L. Rosa, Paulo T. Oliveira, Janaína A. Dernowsek, Vanessa Fontana, Adriane F. Evangelista and Geraldo A. Passos

Abstract This chapter provides a review of recent advances in understanding the importance of normal mesenchymal stem cell (MSC) differentiation and the key regulators that orchestrate their fate into several cell types. Human bone marrow and umbilical cord veins are sources that achieve enough quantities of MSCs, which differentiate in vitro into osteoblasts following expansion and proper biochemical stimuli. Moreover, these cells feature fast proliferation rate and have great expansion capability. Consequently, MSCs have potential uses for clinical trials as for example in healing bone defects. However, understanding their basic processes, including the modulation of gene expression during its early differentiation, is still focus of intense investigation. Additionally, we show results suggesting that regardless the anatomical site from which stem cells are obtained, a shared set of genes is activated to trigger osteoblast differentiation.

Keywords Mesenchymal stem cells · Osteoblast differentiation · Adult stem cells · Hybridization signatures · Hierarchical clustering · Organ gene expression · Transcription profiling · Heat map

This study was approved by the local ethical committee from the School of Dentistry of Ribeirão Preto, University of São Paulo, Brazil (Protocol # 2005.1.670.58.4).

K. F. Bombonato-Prado (✉) · P. T. Oliveira · G. A. Passos
Department of Morphology, Physiology and Basic Pathology, School of Dentistry
of Ribeirão Preto, University of São Paulo, Ribeirão Preto, São Paulo 14040-900, Brazil
e-mail: karina@forp.usp.br

A. L. Rosa
Department of Oral and Maxillofacial Surgery and Periodontology, School of Dentistry of
Ribeirão Preto, University of São Paulo, Ribeirão Preto, São Paulo 14040-900, Brazil

J. A. Dernowsek · V. Fontana · A. F. Evangelista
Department of Genetics, Ribeirão Preto Medical School, University of São Paulo,
Ribeirão Preto, São Paulo 14049-900, Brazil

© Springer International Publishing Switzerland 2014
G. A. Passos (ed.), *Transcriptomics in Health and Disease,*
DOI 10.1007/978-3-319-11985-4_6

6.1 Human Mesenchymal Stem Cells Represent a Model-System for Cell Differentiation Studies

The progressive restriction of the differentiation potential from pluripotent embryonic stem cells (ESC) to different populations of multipotent adult stem cells depends on the orchestrated action of key transcription factors and changes in the profile of epigenetic modifications that ultimately lead to expression of different sets of genes. ESC are unique in their capacities to self-renew and differentiate into any somatic and germ line tissue, while, by contrast, the differentiation potential of adult stem cells is limited (Aranda et al. 2009).

Studies have shown that mesenchymal stem cells (MSC) reflect the stem cell differentiation potential and may form the basis of studies designed to provide insights into genes that confer the greatest developmental potency (Ulloa-Montoya et al. 2007). The knowledge of the fundamental processes associated with the differentiation of MSCs is still poor, and elucidation of the genetic cascade guiding these cells to become more specialized is important for both basic knowledge and clinical application (de Jeong et al. 2004).

MSCs have now been isolated from many sites throughout the body. In the bone compartment, they can be found in bone marrow, periosteum, and endosteum, thin connective tissue linings of the surface of bones, and the mineralized bone itself, and are known to be the primary sources of cells during bone repair (Knight et al. 2013).

Bone marrow is a reservoir of pluripotent stem/progenitor cells for mesenchymal tissues (Cancedda et al. 2003) and the differentiation of MSCs toward different lineage seems display different metabolism signatures (Chen et al 2014). For instance, there is a transition from glycolysis to oxidative phosphorylation in MSCs differentiation toward osteogenic lineage (Chen et al. 2008) and adipogenic lineage (Hofmann et al. 2012; Tormos et al 2011). In contrast, when MSCs differentiate toward chondrogenic lineage using pellet culture, glycolysis is enhanced (Pattappa et al. 2011). Furthermore, it has been reported that mitochondrial metabolism and reactive oxygen species (ROS) generation might be one of the causal factors rather than the merely results of adipogenic differentiation (Tormos et al. 2011). Thus, treatments altering mitochondrial metabolism and ROS generation might affect or determine MSCs fate.

Studies like the ones from Kim et al (2006) indicate that MSCs originating from specific tissues are capable of differentiation into several types of tissues. In addition to the bone marrow (BM), MSCs have been found in several other organs such as the circulating blood of preterm fetuses, hematopoietic cells (Campagnoli et al. 2001; Erices et al. 2000) and Wharton´s jelly explants (Ishige et al. 2009; Wagner et al. 2005).

Although the presence of MSCs in the umbilical cord vein (UC) of newborns was controversial some years ago (Mareschi et al. 2001; Wexler et al. 2003), this site is now being used as a standard source of these cells. Sarugaser et al. (2005) have shown that perivascular tissue from human UC vein cultivated in non-osteogenic medium contains a subpopulation of cells with an osteogenic phenotype that forms calcified nodules. The addition of osteogenic chemical supplementation to the culture medium resulted in a significant increase of these cells. Wang et al. (2004) demonstrated that mesenchymal cells from the mucous connective tissue of Wharton`s jelly express matrix receptors (CD44, CD105) and integrin markers (CD29, CD51), suggesting that these cells are similar to stem cells (SC) in that they can be differentiated into chondrogenic, adipogenic or osteogenic cell lines.

6.2 Therapeutic Potential of Human Mesenchymal Stem Cells

The therapeutic potential of stem cells is already a reality but there is still a need of understanding several aspects of their molecular biology during differentiation and induced pluripotency (Cohen and Melton 2011; Stadtfeld and Hockedlinger 2010). Considering that the control of messenger RNA (mRNA) transcription corresponds to the first step of gene regulation (Rajewsky 2011), which ultimately controls the process of differentiation, transcriptome analysis is critical for better understanding MSCs. The gene expression of pluripotency-related genes have been examined in MSCs derived from bone marrow, adipocytes, amniotic membrane and epithelial endometrium-derived stem cells as well as stroma endometrium-derived stem cells, and these studies suggest that pluripotency-related gene expression varies in different tissues (Tanabe 2014).

Investigating the genes that might act as triggers of MSC early differentiation, regardless of their tissue of origin is an interesting approach, and we hypothesized that MSCs isolated from different anatomical sites (bone marrow and umbilical cord vein) stimulated to differentiate toward a specific cell type would express a set of common genes implicated in the differentiation fate (Figs. 6.1 and 6.2).

6.3 Transcriptome Analysis During Mesenchymal Stem Cell Differentiation

To explore a larger set of genes (transcriptome profiling) in these two MSC isolates, we used microarray screening. As expected, the results showed that during early differentiation, bone marrow and umbilical cord vein cells expressed exclusive sets of genes. However, these two isolates shared expression of 25 genes, including those involved in cell-substrate junction assembly/cell-cell adhesion mediated by integrin (Integrin, alpha 5, fibronectin receptor, ITGA5), hormone-mediated signaling pathway/ossification (Thyroid hormone receptor alpha, THRA), cell differentiation (Nephronectin, NPNT) and regulation of cell growth (HtrA1 serine peptidase 1, HTRA1). Based on their involvement with the molecular/biological processes mentioned above, these could be considered key genes in driving early osteoblastic differentiation of MSCs, independent of their anatomic origin.

Fig. 6.1 Cell morphology
and ALP expression in human
umbilical cord stem cells
after 7 days of culture, **a** cells
in contact with osteogenic
medium, showing polygonal
shape and expression of ALP.
b cells in the absence of
osteogenic medium. *Green
labeling* shows actin and
blue stain labels cell nuclei
(DAPI). Magnification 400x,
fluorescence microscopy

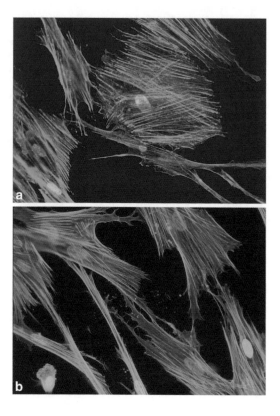

Earlier studies have compared the gene expression profile of BM stem cells, UCV cells and other types of stem cells by means of serial analysis of gene expression (SAGE) (Panepucci et al. 2004), real time PCR (Guillot et al. 2008) and microarrays (Bombonato-Prado et al. 2009; Carinci et al. 2004; Jeong et al. 2005; Schilling 2008; Shi et al. 2001; Secco et al. 2009) following an extended culture of the MSCs in osteogenic medium. Of note, the previous results of Kulterer et al. (2007) have revealed the participation of the genes ID4, CRYAB and SORT1 that were considered to be candidates as regulators of osteogenic differentiation.

In this investigation, we hypothesized that during the initial stages, as early as 24–168 h into in vitro cultivation, key genes are activated during the critical period in which the fate of MSCs is defined toward osteogenic differentiation, independent of their anatomical origin. A set of 115 specific genes were found in bone marrow MSCs, from which we highlight selected genes including Biglycan (BGN), whose coded protein is a proteoglycan of the extracellular matrix that is involved in the adhesion of collagen fibers (SOURCE Database). This protein is an extracellular matrix structural constituent, which may be involved in collagen fiber assembly (by similarity). Inkson et al. (2009) suggested that WNT1 inducible signaling pathway protein 1 (WISP-1) and BGN may functionally interact and control each other's activities, thus regulating the differentiation and proliferation of osteogenic cells.

Fig. 6.2 Cell morphology and ALP expression in mesenchymal stem cells stem cells after 7 days of culture, **a** cells in contact with osteogenic medium, showing polygonal shape and expression of ALP. **b** elongated cells in the absence of osteogenic medium with few cells positive for ALP. *Green labeling* shows actin and *blue stain labels* cell nuclei (DAPI). Magnification 400x, fluorescence microscopy

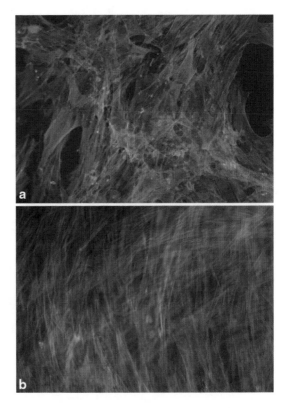

Another modulated gene was fibronectin (FN), which codes the fibronectin protein that binds cell surfaces and various compounds including collagen, fibrin, heparin and actin (SOURCE Database). Fibronectins are involved in cell adhesion, cell motility, opsonization, wound healing, and maintenance of cell shape. Ogura et al. (2004) also found that MSCs have the ability to differentiate into osteoblasts and that FN can stimulate the attachment and spreading of these cells.

The collagen, type VI, alpha 3 (COL6A3) gene codes the alpha-3 chain, one of the three alpha chains of type VI collagen, a beaded filament collagen found in most connective tissues. The alpha-3 chain of type VI collagen is much larger than the alpha-1 and -2 chains. These domains have been shown to bind extracellular matrix proteins, an interaction that explains the importance of this collagen in organizing matrix components (SOURCE Database).

A set of 178 umbilical cord specific genes was modulated, including Tafazzin (TAZ), which codes the tafazzin protein. Tafazzins compose a group of proteins that promote the differentiation and maturation of osteoblasts, while preventing adipocyte maturation (SOURCE Database). In fact, a large number of morphogens, signaling molecules, and transcriptional regulators have been implicated in regulating bone development, including transcriptional factors like TAZ, Runx2, Osterix, ATF4 and NFATc1 and the Wnt/beta-catenin, TGF-beta/BMP, FGF, Notch and Hedgehog signaling pathways (Burns et al. 2010; Deng et al. 2008).

Fig. 6.3 Venn diagram showing the specific and the 25
sharing genes during normal
differentiation of mesen-
chymal cells obtained from
human bone marrow or from
umbilical cord vein. UC:
umbilical cord vein cells;
BM: bone marrow mesenchy-
mal cells

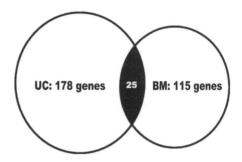

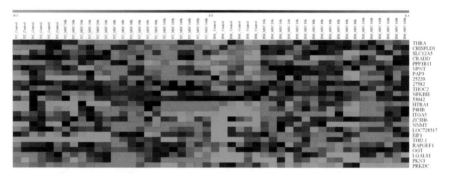

Fig. 6.4 Expression profiling of the 25 genes with shared modulation during osteoblast differentiation of bone marrow (BM) and umbilical cord vein (UC) mesenchymal stem cells [0 to 7 days
(168 h) cultured in osteogenic medium]. FDR ≤ 0.05 and fold change ≥ 2.0

Another modulated gene is the Microfibrillar-associated protein 3 (MFAP3),
which codes a microfibrillar protein important for the structure of extracellular matrix (Abrams et al. 1995), the expression of which has been correlated with bone
formation (Burns et al. 2010).

The Sprouty 2 gene (SPRY2) codes a protein associated with cell signaling and
cell fate commitment and also plays a role as a modulator of FGF signaling. Welsh
et al. (2007) demonstrated that mice carrying a deletion that removes the FGF signaling antagonist Spry2 showed cleft palate, suggesting a role for this gene in the
differentiation of MSCs into osteoblasts. Moreover, it was observed that this protein modulates tyrosine kinase signaling, regulating cell migration and proliferation
(Edwin et al. 2008).

These transcriptional profiles obtained with monolayer cultures are comparable
to those obtained with MSCs cultured in three-dimensional scaffolds (Burns et al.
2010), which mimic the in vivo bone formation. This demonstrates that the monolayer culture model-system reproduces the transcriptional modulation of three-dimensional cultures, at least for the genes above mentioned and therefore is adequate
to study gene profiling of human MSCs differentiation.

Finally, we found 25 differentially expressed genes (Figs. 6.3 and 6.4) that were
shared between the two MSC sources. Due to the biological processes in which

these genes participate, they can be considered triggers of osteoblastic differentiation of MSCs independent of their anatomical origin. Among these, we will discuss selected genes (Table 6.1). The integrin, alpha 5 (fibronectin receptor, alpha polypeptide) gene (ITGA5), which is associated with cell-matrix adhesion, was also one of these 25 genes. Integrins are cell surface receptors that interact with the extracellular matrix (ECM) and mediate various intracellular signals, defining cellular shape and mobility and regulating the cell cycle (SOURCE Database). Integrins may play significant roles in determining osteoblast function because they are signal transduction molecules. Type I collagen, fibronectin, and their integrins are critical for osteoblast function and bone development (Cowles et al. 2000; Shekaran and Garcia 2011).

The HtrA serine peptidase 1 (HTRA1) gene promotes the regulation of cell proliferation (SOURCE Database). It has been proposed that the HtrA1 protein regulates biological processes by modulating growth-factor systems other than IGF, such as the system mediated by the transforming growth factor beta 1 (TGFB1) family. Transforming growth factor beta (TGF-beta) is effective in regulating osteoblast proliferation, differentiation, bone matrix maturation and cell-specific gene expression, as well as inhibiting the expression of markers characteristic of the osteoblast phenotype such as osteocalcin (Oka et al. 2004). Hadfield et al (2008) suggested that HTRA1 may regulate matrix calcification via the inhibition of BMP-2 signaling, modulating osteoblast gene expression, and/or via the degradation of specific matrix proteins.

Nephronectin (NPNT) gene codes an extracellular matrix protein highly expressed in long bone. Kahai et al. (2009) discovered that ectopic expression of nephronectin promotes osteoblastic differentiation, thus corroborating with our results.

The Thyroid hormone receptor, alpha-1 (THRA) gene codes one of the several receptors of thyroid hormone, acting as a mediator of its biological activities. This gene is involved in the formation of bone or of a bony substance and the conversion of fibrous tissue or of cartilage into bone or a bony substance (SOURCE Database).

Protein phosphatase 1 regulatory inhibitor subunit 11 (PPP1R11) was also found to be a shared gene. Considering that phosphatase activity is important for osteoblast differentiation, as in the case of ALPL that we determined in this study, and that the PPP1R11 protein is associated with inhibition of phosphatase activity, this may be evidence for a mechanism involving phosphatase enhancement/inhibition during osteoblast differentiation. Further, genes involved in kinase activity/protein phosphorylation such as Rap guanine nucleotide exchange factor (GEF) 1 (RAPEF1) and Protein kinase 3 (PKN3) also appeared, reinforcing the importance of phosphate metabolism in osteoblast differentiation.

Genes that control apoptosis such as Lectin, galactoside-binding, soluble, 1 (LGALS1) and CASP2 and RIPK1 domain containing adaptor with death domain (CRADD) were also shared between the two sources of MSCs, providing evidence for controlled cell death during differentiation.

Genes involved in general processes such as control of transcription including Nuclear factor of kappa light polypeptide gene enhancer in B-cells inhibitor, epsi-

Table 6.1 Genes with shared modulation during osteoblast differentiation of bone marrow and umbilical cord vein mesenchymal stem cells (0 to 168 h in osteogenic medium). FDR $\leq$ 0.05, fold change $\geq$ 2.0

GenBank acc	Gene name	Symbol	Cytoband	Function
AB209346	Thyroid hormone receptor, alpha-1	THRA	17q11.2	Transcription from RNA polymerase II promoter
NM_031461	Cysteine-rich secretory protein LCCL domain containing 1	CRISPLD1	8q21.11	Function unknown
NM_020708	Solute carrier family 12, (potassium-chloride transporter) member 5	SLC12A5	20q13.12	Potassium ion transport
BC037905	CASP2 and RIPK1 domain containing adaptor with death domain	CRADD	12q21.33-q23.1	Regulation of apoptosis
NM_021959	Protein phosphatase 1, regulatory (inhibitor) subunit 11	PPP1R11	6p21.3	Protein phosphatase inhibitor activity
NM_001184691	Nephronectin	NPNT	4q24	Cell differentiation
NM_101339	Purple acid phosphatase 3	PAP3	1	Acid phosphatase activity
AF131786	Clone 25220 mRNA sequence	–	–	Function unknown
BX537526	CDNA FLJ11602 fis, clone HEMBA1003908	–	–	Function unknown
NM_001081550	THO complex 2	THOC2	Xq25-q26.3	MRNA-nucleus export
NM_004556	Nuclear factor of kappa light polypeptide gene enhancer in B-cells inhibitor, epsilon	NFKBIE	6p21.1	Cytoplasmic sequestering of transcription factor
AI220134	Transcribed locus	–	–	Function unknown
NM_002775	HtrA serine peptidase 1	HTRA1	10q26.3	Regulation of cell growth
NM_000918	Procollagen-proline, 2-oxoglutarate 4-dioxygenase beta subunit	P4HB	17q25	Electron transport

Table 6.1 (continued)

GenBank acc	Gene name	Symbol	Cytoband	Function
NM_002205	Integrin, alpha 5	ITGA5	12q11-q13	Cell-matrix adhesion
NM_198581	Zinc finger CCCH-type containing 6	ZC3H6	2q13	Function unknown
AK097984	Nicotinamide N-methyltransferase	NNMT	11q23.1	Transferase activity
BX396146	Hypothetical protein LOC728517	LOC728517	1p36.33	Function unknown
BQ278455	Eukaryotic translation initiation factor 1	EIF1	17q21.2	Regulation of translation
NM_105885	Toxin receptor binding	THI2.1	1	
NM_198679	Rap guanine nucleotide exchange factor (GEF) 1	RAPGEF1	9q34.3	Transmembrane receptor protein tyrosine kinase signaling pathway
AL050366	O-linked n-acetyl-glucosamine transferase	OGT	Xq13	Signal transduction
BF570935	Lectin, galactoside-binding, soluble, 1 (galectin 1)	LGALS1	22q13.1	Apoptosis
NM_013355	Protein kinase N3	PKN3	9q34.11	Signal transduction
NM_006904	Protein kinase, DNA-activated, catalytic polypeptide	PRKDC	8q11	Protein modification

Ion, (NFKBIE) and control of ion transport as Solute carrier family 12 (potassium/chloride transporter), member 5, (SLC12A5) were also shared.

Finally, we identified the participation of the Protein kinase, DNA-activated, catalytic polypeptide (PRKDC) gene; in addition to its role in kinase activity and osteoblast differentiation as discussed above, this gene also plays roles in DNA repair and the control of apoptosis, which are both processes that ultimately regulate cancer.

These results suggest that regardless of the anatomical site from which stem cells were obtained, a shared set of genes is activated to trigger osteoblast differentiation.

Acknowledgments This study was funded by Conselho Nacional de Desenvolvimento Científico e Tecnológico (CNPq, Brasília, Brasil), Project # 552227/2005-6.

References

Abrams WR, Ma RI, Kucich U et al (1995) Molecular cloning of the microfibrillar protein MFAP3 and assignment of the gene to human chromosome 5q32-q33.2. Genomics 26:47–54

Aranda P, Agirre X, Ballestar E et al (2009) Epigenetic signatures associated with different levels of differentiation potential in human stem cells. PLoS One 13:e7809

Bombonato-Prado KF, Bellesini LS, Junta MM et al (2009) Microarray-based gene expression analysis of human osteoblasts in response to different biomaterials. J Biomed Mater Res A 88:401–408

Burns JS, Rasmussen PL, Larsen KH et al (2010) Parameters in three-dimensional osteospheroids of telomerized human mesenchymal (stromal) stem cells grown on osteoconductive scaffolds that predict in vivo bone-forming potential. Tissue Eng Part A 16:2331–2342

Campagnoli C, Roberts IA, Kumar S et al (2001) Identification of mesenchymal stem/progenitor cells in human first-trimester fetal blood, liver, and bone marrow. Blood 98:2396–2402

Cancedda, R., G. Bianchi, A. Derubeis, R. Quarto (2003) Cell therapy for bone disease: a review of current status. Stem Cells 21:610–619

Carinci F, Piatelli A, Stabellini G et al (2004) Calcium sulfate: analysis of MG63 osteoblast-like cell response by means of a microarray technology. J Biomed Mater Res B Appl Biomater 71:260–267

Chen CT, Shih YRV, Kuo TK et al (2008) Coordinated changes of mitochondrial biogenesis and antioxidant enzymes during osteogenic differentiation of human mesenchymal stem cells. Stem Cells 26:960–968

Chen H, Liu X, Chen H et al (2014) Role of SIRT1 and AMPK in mesenchymal stem cells differentiation. Ageing Res Rev 13C:55–64

Cohen D, Melton D (2011) Turning straw into gold: directing cell fate for regenerative medicine. Nat Rev Genet 12:243–252

Cowles EA, Brailey LL, Gronowicz GA (2000) Integrin-mediated signaling regulates AP-1 transcription factors and proliferation in osteoblasts. J Biomed Mater Res 52:725–737

de Jong DS, Vaes BL, Dechering KJ et al (2004) Identification of novel regulators associated with early-phase osteoblast differentiation. J Bone Miner Res 19:947–958

Deng ZL, Sharff KA, Song WX et al (2008) Regulation of osteogenic differentiation during skeletal development. Front Biosci 13:2001–2021

Edwin F, Patel T (2008) A novel role of Sprouty 2 in regulating cellular apoptosis. J Biol Chem 283:3181–3190

Erices A, Conget P, Minguell JJ (2000) Mesenchymal progenitor cells in human umbilical cord blood. Br J Haematol 109:235–242

Guillot PV, De Bari C, Dell'accio F et al (2008) Comparative osteogenic transcription profiling of various fetal and adult mesenchymal stem cell sources. Differentiation 76:946–957

Hadfield KD, Rock CF, Inkson CA et al 2008) HtrA1 inhibits mineral deposition by osteoblasts: requirement for the protease and PDZ domains. J Biol Chem 283:5928–5938

Hofmann AD, Beyer M, Krause-Buchholz U et al (2012) OXPHOS supercomplexes as a hallmark of the mitochondrial phenotype of adipogenic differentiated human MSCs. PLoS ONE 7(4):e35160

Inkson CA, Ono M, Bi Y et al (2009) The potential functional interaction of biglycan and WISP-1 in controlling differentiation and proliferation of osteogenic cells. Cells Tissues Organs 189:153–157

Ishige I, Nagamura-Inoue T, Honda MJ et al (2009) Comparison of mesenchymal stem cells derived from arterial, venous, and Wharton´s jelly explants of human umbilical cord. Int J Hematol 90:261–269

Jeong JA, Hong SH, Gang EJ et al (2005) Differential gene expression profiling of human umbilical cord blood-derived mesenchymal stem cells by DNA microarray. Stem Cells 23:584–593

Kahai S, Lee SC, Lee DY et al (2009) MicroRNA miR-378 regulates nephronectin expression modulating osteoblast differentiation by targeting GalNT-7. Plos One 4:e7535

Kim CG, Lee JJ, Jung DY (2006) Profiling of differentially expressed genes in human stem cells by cDNA microarray. Mol Cells 21:343–355

Knight MN, Hankenson KD (2013) Mesenchymal Stem Cells in Bone Regeneration. Adv Wound Care (New Rochelle)6:306–316

Kulterer B, Friedl G, Jandrositz A et al (2007) Gene expression profiling of human mesenchymal stem cells derived from bone marrow during expansion and osteoblast differentiation. BMC Genomics 8:70

Mareschi K, Biasin E, Piacibello W, Aglietta M, Madon E, Fagioli F (2001) Isolation of human mesenchymal stem cells: bone marrow versus umbilical cord blood. Haematologica 86:1099–1100

Oka C, Tsujimoto R, Kajikawa M et al (2004) HtrA1 serine protease inhibits signaling mediated by Tgf beta family proteins. Development 131:1041–1053

Pattappa G, Heywood HK, De Bruijn JD et al (2011) The metabolism of human mesenchymal stem cells during proliferation and differentiation. J Cell Physiol 226:2562–2570

Panepucci RA, Siufi JL, Silva WA et al (2004) Comparison of gene expression of umbilical cord vein and bone marrow-derived mesenchymal stem cells. Stem Cells 22:1263–1278

Rajewsky N. (2011) microRNAs and the Operon paper (2011). J Mol Biol 409:70–75

Sarugaser R, Lickorish D, Baksh D et al (2005) Human umbilical cord perivascular (HUCPV) cells: a source of mesenchymal progenitors. Stem Cells 23:220–229

Schilling T, Küffner R, Klein-Hitpass L et al (2008) Microarray analyses of transdifferentiated mesenchymal stem cells. J Cell Biochem 103:413–433

Secco M, Moreira YB, Zucconi E et al (2009) Gene expression profile of mesenchymal stem cells from paired umbilical cord units: cord is different from blood. Stem Cell Rev 5:387–401

Shekaran A, Garcia AJ (2011) Extracellular matrix-mimetic adhesive biomaterials for bone repair. J Biomed Mater Res A 96:261–272

Shi S, Robey PG, Gronthos S et al (2001) Comparison of human dental pulp and bone marrow stromal stem cells by cDNA microarray analysis. Bone 29:532–539

Stadtfeld M, Hochedlinger K (2010) Induced pluripotency: history, mechanisms, and applications. Genes Dev 24:2239–2263

Tanabe S (2014) Role of mesenchymal stem cells in cell life and their signaling. World J Stem Cells 26:24–32

Tormos KV, Anso E, Hamanaka RB et al (2011) Mitochondrial complex III ROS regulate adipocyte differentiation. Cell Metab 14:537–544

Ulloa-Montoya F, Kidder BL, Pauwelyn KA et al (2007) Comparative transcriptome analysis of embryonic and adult stem cells with extended and limited differentiation capacity. Genome Biol 8:R163

Wagner W, Wein F, Seckinger A et al (2005) Comparative characteristics of mesenchymal stem cells from human bone marrow, adipose tissue, and umbilical cord blood. Exp Hematol 33:1402–1416

Wang H S, Hung SC, Peng ST et al (2004) Mesenchymal stem cells in the Wharton's jelly of the human umbilical cord. Stem Cells 22:1330–1337

Welsh IC, Hagge-Greenberg A, O'Brien TP et al (2007) A dosage-dependent role for Spry2 in growth and patterning during palate development. Mech Dev 124:746–761

Wexler SA, Donladson C, Denning-Kendall P et al (2003) Adult bone marrow is a rich source of human mesenchymal 'stem' cells but umbilical cord and mobilized adult blood are not. Br J Haematol 121:368–374

Part II
Transcriptome in Disease

Chapter 7
Thymus Gene Coexpression Networks: A Comparative Study in Children with and Without Down Syndrome

Carlos Alberto Moreira-Filho, Silvia Yumi Bando, Fernanda Bernardi Bertonha, Filipi Nascimento Silva, Luciano da Fontoura Costa and Magda Carneiro-Sampaio

Abstract In this chapter we characterized trisomy 21-driven transcriptional alterations in human thymus through gene coexpression network (GCN) analysis. We used whole thymic tissue (corticomedullar sections)—obtained at heart surgery from Down syndrome (DS) and karyotipically normal individuals (CT)—and a network-based approach for GCN analysis allowing the study of interactions between all the system's constituents based on community detection. Changes in the degree of connections observed for hierarchically important hubs in DS and CT gene networks corresponded to sub-network changes, i.e. module (communities) changes. Distinct communities of highly interconnected gene sets were topologically identified for DS and CT networks. Trisomy 21 gene dysregulation in thymus may therefore be viewed as the breakdown and altered reorganization of functional modules.

Electronic supplementary material The online version of this chapter (doi: 10.1007/978-3-319-11985-4_7) contains supplementary material, which is available to authorized users. Videos can also be accessed at http://www.springerimages.com/videos/978-3-319-11984-7.

C. A. Moreira-Filho (✉)
Departamento de Pediatria, Faculdade de Medicina da Universidade de São Paulo,
Av. Dr. Arnaldo, 455 CEP 01246-903, São Paulo, São Paulo, Brazil
e-mail: carlos.moreira@hc.fm.usp.br

S. Y. Bando · F. B. Bertonha · M. Carneiro-Sampaio
Departamento de Pediatria, Faculdade de Medicina da Universidade de São Paulo,
São Paulo, São Paulo, Brazil
e-mail: silvia.bando@hc.fm.usp.br

F. B. Bertonha
e-mail: bertonhafb@gmail.com

M. Carneiro-Sampaio
e-mail: magdacs@usp.br

F. N. Silva · L. d. F. Costa
Instituto de Física de São Carlos, Universidade de São Paulo, São Carlos, São Paulo, Brazil
e-mail: filipinascimento@gmail.com

L. d. F. Costa
e-mail: ldfcosta@gmail.com

© Springer International Publishing Switzerland 2014
G. A. Passos (ed.), *Transcriptomics in Health and Disease,*
DOI 10.1007/978-3-319-11985-4_7

123

7.1 Introduction

Thymus provides the specialized microenvironment for the proliferation, differentiation, T-cell antigen receptor (TCR) gene rearrangement and T-cell repertoire selection (Anderson and Takahama 2012). The thymic microenvironment encompasses
thymic epithelial cells (TEC), fibroblasts, thymic myoid cells, and bone marrow-
derived accessory cells such as B lymphocytes, macrophages and dendritic cells
(Jablonska-Mestanova et al. 2013). Therefore, T-cell selection involves cellular processes driven by coordinate changes in the expression of hundreds of genes in the
thymus (Macedo et al. 2009; Abramson et al. 2010; Mingueneau et al. 2013).

In Down syndrome (Mégarbané et al. 2009) gene expression dysregulation caused
by trisomy 21—specifically by gene imbalance dosage involving the Hsa21 region
in chromosome 21 (Aït Yahia-Graison et al. 2007; Prandini et al. 2007; Korbel et al.
2009)—causes thymic structural and functional abnormalities. DS patients present
abnormal thymuses, characterized by lymphocyte depletion, cortical atrophy, and
loss of corticomedullary demarcation. These long time recognized DS thymic abnormalities (Levin et al. 1979; Larocca et al. 1990) are not related to DS precocious
senescence: DS immune system is intrinsically deficient from the very beginning
(Kusters et al. 2010). This was recently confirmed by imaging studies. Sonographic
thymic measurements showed that the majority of DS fetuses have smaller thymuses
than control (De Leon-Luis 2011). Thymic-toraxic ratio (TT-ratio) evaluations obtained through ultrasound examinations showed that fetuses with trisomy 21 have a
small thymus, suggesting accelerated thymic involution in utero (Karl 2012).

Measuring the total number of signal joint TCR excision circles per ml blood,
Bloemers et al. (2011) found out that DS thymus has a decreased thymic output,
concluding that "reduced thymic output, but not reduced peripheral generation nor
increased loss of naive T cells, results in the low naive T cell numbers found in DS".
Studying the Ts65DN mouse model of DS, Lorenzo et al. (2013) showed that immature thymocyte defects underlie immune dysfunction in DS and that increased oxidative stress and reduced cytokine signaling impair T-cell development. Since DS
autoimmune diseases are more represented in DS, Pellegrini et al. (2012) investigated phenotypic and functional alterations of natural T regulatory cells (nTreg) in DS
people and found an over-expressed peripheral nTreg population with a defective
inhibitory activity, what may be correlated with autoimmunity in DS. In the same
line of research, our group investigated the expression of the autoimmune regulator
(AIRE) gene in the thymuses of DS children and we found, by immunohistochemistry, a reduced expression of AIRE protein in DS thymic tissues (Lima et al. 2011).
On the other hand, Xu et al. (2013), studying global changes and chromosome distribution characteristics of miRNAs expression in lymphocytes from DS children by
high-throughput sequencing technology, discovered that most of the overexpressed
miRNAs in DS were not Hsa21-derived. Therefore, miRNA abnormal expression in
DS should be probably associated with the dysregulation of disomic genes caused
by trisomy 21. This result clearly evidenced the importance of performing global
transcriptome analysis in DS thymuses in order to characterize GCN changes that
could better explain the mechanisms involved in DS thymic hypofunction.

7.2   Gene Expression Analysis in Thymic Tissue

Our previous study showing AIRE decreased expression and global thymic hypofunction in DS (Lima et al. 2011) was limited to a few hundred differentially
expressed genes and network transcriptional analysis was accomplished without
considering distinct levels of node hierarchy (i.e. without considering concentric
node degrees; see Chap. 3). Here we employed a network-based analytical strategy
that allows the study of GCNs encompassing differentially expressed as well as all
valid GO annotated transcripts from trisomic and kariotypically normal thymuses,
leading to the identification of modular transcriptional repertoires distinctive for
each condition.

7.2.1   Experimental Approach

The mRNA samples used in DNA microarray experiments were extracted from
fresh corticomedullar sections of thymic tissue obtained at heart surgery from DS
and karyotipically normal individuals. In this analysis we compared ten Down syndrome (DS group) patients with ten patients without DS (control group—CT). All
patients were gender and age matched (age ranging from 2 to 18 months). The use
of whole tissue coupled with community structure analysis of gene interaction networks (see Sect. 8.3.2) is a strategy that may be adopted for circumventing tissue
microdissection (Chaussabel and Baldwin 2014).

A total of 12,989 valid GO annotated genes (CO) were obtained after exclusion
of up to three outliers per gene for each group and comparative quality control of
featured intensities for all patients by boxplot analysis. Hereafter all valid GO genes
were uploaded to MeV software version 4.8.1 for statistical analysis. A total of 538
differentially expressed GO annotated genes (DE) were identified using the Significance Analysis of Microarrays (SAM) procedure, all up-regulated in the DS group.
Coexpression gene networks (GCNs) were obtained for differentially expressed
genes (DE networks) and for the complete set of GO annotated valid transcripts
(CO networks).

7.3   Gene Coexpression Networks (GNs): Visualization,
Analysis and Community Detection

GCNs were inferred for DS and CT groups using DE or CO subsets of genes after
gene-gene Pearson's correlation method. A 0.965 link-strength cut-off was adopted
for DE networks. The resulting networks had 251 genes and 425 links for DS group
(Fig. 7.1a) or 256 genes and 1218 links for CT group (Fig. 7.1c). We adopted a
higher link-strength cut-off (0.990) to finalize the CO networks, which had 8753
genes and 24,744 links for DS group or 8205 genes and 38,925 links for CT group.
All these networks had scale-free node degree distribution (see Chap. 3).

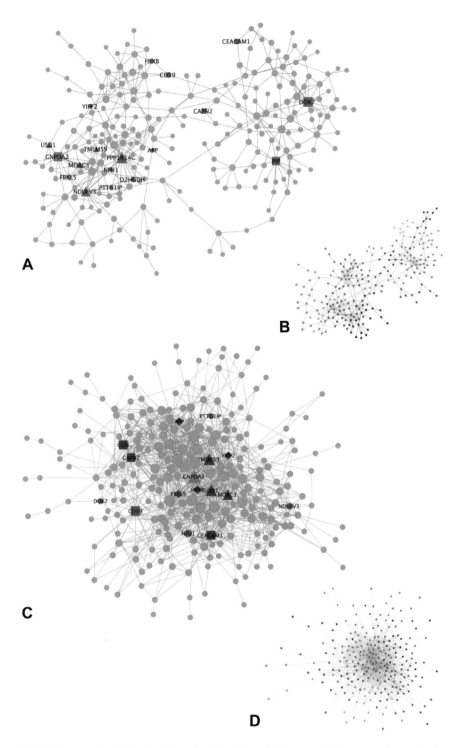

Fig. 7.1 Comparative DE networks analysis for DS and CT groups. DE coexpression networks for DS (**a**) and CT (**c**) groups. Hubs, VIPs and high-hubs are indicated by *rectangles*, *diamonds*

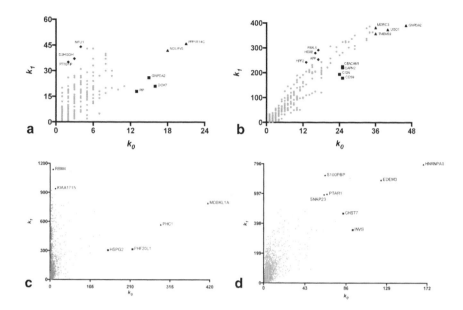

Fig. 7.2 Scatter plot of node degree (k_0) *vs* concentric node degree (k_1) measures for DS and CT groups. DE networks appear in **a** (DS group) and **b** (CT group). CO networks appear in **c** (DS) and **d** (CT). Hubs, VIPs and high-hubs are indicated by *rectangles*, *diamonds* and *triangles*, respectively

We developed a methodology for GCN visualization (3D) and analysis (see Chap. 3 and Bando et al., 2013) that allows the categorization of network nodes according to distinct hierarchical levels of gene-gene connections: hubs are highly connected nodes, VIPs have low node degree but connect only with hubs, and high-hubs have VIP status and high overall number of connections. We classified network nodes as VIPs, hubs or high-hubs by obtaining the node degree, k_0, and the first level concentric node degree, k_1, which takes into account all node connections leaving from its immediate neighborhood, then projecting all node values in a k_0 vs k_1 graphic (Fig. 7.2a–7.2d).

7.3.1 Connectivity

Network connectivity k for non-directed networks was calculated by k=2 L/N, where L stands for the number of edges and N for the number of nodes.

and *triangles*, respectively, in both networks. Hubs, VIPs and high-hubs are *red-colored* in DS-DE network and *blue-colored* in CT-DE network. *Green-colored* symbol indicates a common node. *Circular-shaped blue nodes* in DS-DE indicate nodes displaying high-hierarchy in CT-DE. *Circular-shaped red nodes* in CT-DE indicate nodes displaying high-hierarchy in DS-DE. **b** and **d**: Community analysis for DE networks (DS in **b** and CT in **d**). The clusters (communities) are indicated by different colors

7.3.2 Community Detection

Community detection in complex networks is usually accomplished by discovering the network modular structure that optimizes the modularity measurement. Modularity takes into account the relationship between the number of links inside a community compared to connections between nodes in distinct communities (Newman and Girvan 2004). A diverse range of optimization techniques exist to optimize the modularity. Here we applied the method proposed by Blondel et al. (2008) which attains good modularity values and at same time presents excellent performance.

Communities can be defined as partition sets of nodes $C = c_1, c_2, c_3 \ldots$ in a network that share more links between those inside them than to those in other partitions. Because connectivity is a relative measurement, which depends on the nature of each network, Newman and Girvan (Newman and Girvan 2004) introduced the concept of modularity by comparing the connectivity inside each community between the network and a random realization with the same degree distribution. Formally, the modularity Q_C is defined by equation 7.1.

$$Q_C = \frac{1}{2m} \sum_{c \in C} \sum_{v,w \in c} \left[A_{vw} - \frac{k_v k_w}{2m} \right] \qquad ((7.1))$$

The sum in Q_C is taken over all pair of nodes (v, w) that are in each community c, where A_{vw} is the adjacency matrix of the network ($A_{vw} = 1$ if v and w are connected, otherwise $A_{vw} = 0$), k_v is the node degree (number of connections) of node v, and m is the total number of edges of the network. This can be written in more general terms of the network by equation 7.2:

$$Q_C = \frac{1}{2m} \sum_{c \in C} (e_c - a_c^2) \qquad ((7.2))$$

where

$$e_c = \sum_{v,w \in c} \frac{A_{vw}}{2m} \qquad ((7.3))$$

and

$$a_c = \sum_{v \in c} \frac{k_v}{2m} \qquad ((7.4))$$

In summary, Q_C is the fraction of connections restricted to the same community, e_c, compared to the total fraction of edges connected to nodes in the community a_c. Higher modularity values imply better quality of the community organization for a partition set C in a network.

Most of the methods to detect community generate hierarchical structures. The Newman-Girvan method uses the edge betweenness centrality measurement as a criterion for removing edges and obtaining connected components that correspond to each network partition. This builds a tree of communities with branches occurring every time a component is divided in two. Agglomerative methods start from a set of communities, where each node corresponds to a different community, which are progressively merged together according to a similarity criterion or to directly maximize the change of modularity (Clauset et al. 2004). In both cases, a dendrogram of the partition hierarchy is obtained. The optimal set of communities is then obtained by a cut for the highest value of modularity.

Figure 7.3 illustrates the hierarchical community structure of a small network. Starting from non existent community structure the network is split into two partitions [AB, C] and next into tree by splitting AB into [A, B], therefore resulting in the set [A, B, C]. When the modularity of both configurations is calculated, it reveals that the one with only two communities present much more community-like structure than when considering the other, more granular, partition set.

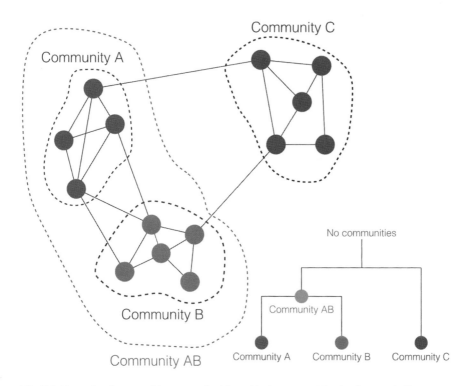

Fig. 7.3 Example of a network's community hierarchical structure. The dendrogram indicates two realizations of network partitions. One made of communities AB and C, and another where community AB is split into A and B, thus forming the partition set [A, B, C]

Community structure is present in many real systems, such as social networks, knowledge networks, the Internet, etc. This is no different for gene coexpression networks, in which communities can encompass complex mechanisms that work together to maintain the cellular processes across different conditions. For example, community structure analysis of gene coexpression networks obtained from skeletal muscle cells' transcriptome revealed different biological pathways for Duchenne muscular dystrophy patients comparatively to normal individuals (Narayanan and Subramaniam 2013). The same modular approach has successfully been used for investigating immune response to infections (Banchareau et al. 2012) and vaccines (Obermoser et al. 2013) using whole blood transcriptome data sets (reviewed in Chaussabel and Baldwin 2014).

7.4   Results

Hubs, VIPs and high-hubs showed network centrality in DS and CT networks, what is consistent with the network disease model, where a group of nodes whose perturbation (e.g. 21 trisomy) leads to a disease phenotype occupies a central position in the network (Barabási et al. 2011; Bando et al. 2013). Moreover, network connectivity is lower in DS networks (DS-DE = 3.386; CT-DE = 9.516; DS-CO = 2.83; CT-CO = 4.74). Gene hierarchy changes are evident in the GCNs presented in Fig. 7.1a and 7.1c and also in the scatter plots presented in Fig. 7.2a–7.2d (DE and CO data). While both complete networks (CO) present good quality of community structure, with modularity around 0.60, the DE networks present significant discrepancy in theirs modularity values. The DS-DE is much more modular than the CT-DE network, with modularity values of 0.75 and 0.41 respectively.

7.4.1   Gene Communities

An overall picture of DE gene communities (modules) is depicted in Fig. 7.1b for DS-DE and in Fig. 7.1d for CT-DE networks. DS-CO and CT-CO networks and respective communities can be properly visualized only in 3D (see videos 7.1 and 7.2 http://www.springerimages.com/videos/978-3-319-11984-7). Different node colors identify the distinct gene communities. The lower connectivity of DS-DE network—also observed for DS-CO in comparison to CT-CO—contrasts with its higher modularity. DS modules are more sparsely connected to each other than CT modules, what may reflect some dysregulation in cell's functional organization (Barabási et al. 2004; Sahni et al. 2013). A full analysis of these gene communities surpasses the scope of this chapter and will be published elsewhere.

7.4.2 DE Networks

In this section we briefly discuss the biological functions of some selected hubs, VIPs and high-hubs found in DE networks. In the DS-DE network (Fig. 7.1a) two high-hubs are clearly identified. The first, PPP1R14C (aliase KEPI) is capable to activate the MEK-ERK pathway and therefore to interfere in thymocyte development (Gallo et al. 2007), whereas the second, NDUVF3, is a gene overexpressed in DS (Pereira et al. 2009) which codes for a protein harboring a peptide sequence that can induce phenotypic and functional differentiation of CD8 thymocytes (Sasada 2001). Two of the VIPs in this network—NFU1 and D2HGDH—are mitochondrial enzymes and the third, PTTG1IP, is a gene overexpressed in DS with a function in T-lymphocyte activation (Stoika and Melmed 2002). One of the hubs, PIP, codes for the prolactin-induced protein and plays a role on the proliferation of thymic epithelial and dendritic cells (Barnard et al. 2008), as well as in the suppression of glucocorticoid-induced thymocyte apoptosis (Krishnan et al. 2003). The other two hubs here considered are: GNPDA2, a regulator of energy metabolism (Yang et al. 2009) and DOK7, a member of the DOK (downstream of kinase) protein family that was not previously found to be expressed in thymic tissues. Interestingly, two members of DOK family, DOK1 and DOK2, are involved in T cell homeostasis maintenance (Guittard et al. 2009) and DOK4 is a negative modulator of T-cell activation (Gérard et al. 2009).

In CT-DE (Fig. 7.1b) the high-hubs are GNPDA2, a gene also found as a hub in DS-DE, USO1, involved in ER-to-Golgi transport (Kim et al. 2012), MORC3, a gene that regulates cell senescence via p53 (Takahashi et al. 2007) and TMEM59, which controls autophagy via LC3 (Boada-Romero et al. 2013). Autophagy is a very important process in the immune system: in the thymus, autophagy can modulate the selection of CD4$^+$ T-cell clones (Puleston and Simon 2014). The VIPs in this network contain relevant genes for thymic function: APP, which, equally to TMEM59, controls autophagy via LC3 (Tian et al. 2013), YIPF2, involved in ER-to-Golgi transport (Tanimoto et al. 2011) HEXB, a gene that causes thymic involution when mutated (Kanzaki et al. 2010), and FBXL5, a regulator of oxidative stress (Ruiz and Bruick 2014). All the hubs in this network are also important for thymus functioning. Firstly, CGN, which codes for cingulin, a protein that controls claudin-3 expression (Gillemot et al. 2013) and, therefore, exerts an important role in AIRE (the autoimmune regulator gene, Mathis and Benoist 2009) expression in thymic epithelial cells (Hamazaki et al. 2007; Hollaender 2007). The other hubs are CEACAM1, a potent regulator of T-cell stimulation (Kammerer et al. 2001), CAPN2, which codes for calpain, an inducer of T-cell apoptosis (Ishihara et al. 2013), and CD59, strongly expressed in Hassall's corpuscles (Berthelot et al. 2010) and involved in T cell polarization (Izsepi et al. 2013).

7.4.3   CO Networks

The first high-hub in DS-CO network (Fig. 7.2c), MOBKL1A, is a regulator of T-cell proliferation (Zhou et al. 2008) via MST1/MST2, two kinases that control the activation of rho family GTPases and thymic egress of mature thymocytes (Mou et al. 2012). The other high-hub is PHC1, a human orthologue of the Drosophila polyhomeotic member of polycomb group (PcG), which acts on chromatin remodeling and regulation of HOX genes during development (Awad et al. 2013). One of the two main VIPs is RBM4, a gene which codes for a multifunctional RNA-binding protein, involved in alternative splicing of pre-mRNA, translation, and micro-RNA-mediated gene regulation, and found to be hypoexpressed in DS brain tissue (Markus and Morris 2009). The other VIP is KIA1715, a gene that integrates the HOXD gene cluster and has multiple roles in embryonic development (Spitz et al. 2003). It is tempting to suppose functional interactions between this gene and the high-hub PHC1. The two main hubs in this network are PHF20L1, involved in regulating DNMT1 activity and DNA methylation in cells (Estéve et al. 2014) and HSPG2, which codes for the pelercan protein, an integrant of the lamimin-5 contaning conduits in human thymus, a system responsible for the transport of small molecules (Drumea-Mirancea et al. 2006).

The CT-CO network (Fig. 7.2d) contains two main high-hubs, HNRNPA0, which codes for a RNA-binding protein with a role in determining hematopoietic cell fate (Young et al. 2014) and EDEM-3, which enhances glycoprotein endoplasmic reticulum-associated degradation (Hirao et al. 2006).

The three VIPs in this network are involved basic biological functions required for thymus functioning, as follows. Firstly comes S100BP, a member of the S100 gene family, the largest subfamily of calcium binding proteins of EF-hand type (Chen et al. 2014) and well expressed in human newborn thymus dendritic cells (Jablonska-Mestanova et al. 2013). The next is PTAR1, involved in ubiquitination (Kim et al. 2011); a cell process required for the selection of natural regulatory T cells (Oh et al. 2013). The third VIP is SNAP23, a gene involved in both the determination of cell polarity (Low et al. 2006) and phagosome formation and maturation, presumably by mediating SNARE-based membrane traffic (Sakurai et al. 2012). Finally we have two CT-CO hubs, also playing important roles. CHST7 is a gene involved in apoptosis control of lymphoid cells (Nakayama et al. 2013) and INVS codes for inversin, a protein that interacts with calmodulins (Morgan et al., 2002) and thus is involved in the control of apoptosis in thymocyte development (Liang et al. 2007) and directional cell migration processes (Veland et al. 2013).

7.5   Discussion and Conclusions

The data presented here show that thymus global hypofunction in Down syndrome correlates with distinctive GCN topology and node hierarchy. The comparative genetic and topological analysis of DE networks disclose how significant is the impact of trisomy 21 on thymus functioning. The high modularity of DS-DE network contrasts with its reduced connectivity, thus indicating a certain degree of disorganization of modular interactions caused by gene dosage imbalance. Since the CO networks are more similar in their connectivity and modularity (although not in hubs' hierarchy), the DS-DE network may represent the "ground zero", where trisomy-driven gene dysregulation strongly impacted the normal gene-gene interactions, with the derived "shock-waves" reflecting on the DS-CO network.

In conclusion, alterations observed in DS networks regarding connectivity, modularity, and communities' structure (this latter aspect not fully discussed here) reflect chromosome 21 dysregulation and its consequences. As a whole, the results indicate that GCNs' functional and topological modules correspond, and that trisomy 21 may be interpreted as the breakdown and altered reorganization of functional modules. This mechanism is coherent with the network-based model of human disease (Barabási et al. 2011; Sahani et al. 2013) and has been experimentally confirmed both for chronic non-communicable diseases (Bando et al. 2013; Narayanan and Subramaniam 2013) and for infectious diseases (Chaussabel and Baldwin 2014).

Financial Support This work was funded by Fundação de Amparo à Pesquisa do Estado de São Paulo (FAPESP) research grant 2009/53443-1 and Conselho Nacional de Desenvolvimento Científico e Tecnológico (CNPq) grant 305635/2009-3 to CAM-F. MC-S was funded by FAPESP grant 2008/58238-4. L da FC was funded by FAPESP grants 2005/00587-5 and 2011/50761-2 and CNPq grants 301303/06-1 and 73583/2008-0. CAM-F and L da FC are funded by FAPESP-MCT/CNPq/PRONEX grant 2011/50761-2. FNS is the recipient of a CAPES fellowship.

Supporting Information (Videos)

Video 7.1 and Video 7.2—Community analysis for CO networks (DS in video 7.1 and CT in 7.2). The clusters (communities) are indicated by different colors.

References

Abramson J, Giraud M, Benoist C et al (2010) Aire's partners in the molecular control of immunological tolerance. Cell 140:123–135
Aït Yahya-Graison E, Aubert J, Dauphinot L et al (2007) Classification of human chromosome 21 gene-expression variations in Down syndrome: impact on disease phenotypes. Am J Hum Genet 81:475–91
Anderson G, Takahama Y (2012) Thymic epithelial cells: working class heroes for T cell development and repertoire selection. Trends Immunol 33:256–263
Awad S, Al-Dosari MS, Al-Yacoub N et al (2013) Mutation in PHC1 implicates chromatin remodeling in primary microcephaly pathogenesis. Hum Mol Genet 22:2200–2213
Banchereau R, Jordan-Villegas A, Ardura M et al (2012) Host immune transcriptional profiles reflect the variability in clinical disease manifestations in patients with *Staphylococcus aureus* infections. PLoS ONE 7(4):e34390

Bando SY, Silva FN, Costa Lda F et al (2013) Complex network analysis of CA3 transcriptome reveals pathogenic and compensatory pathways in refractory temporal lobe epilepsy. PLoS ONE 8:e79913

Barabási AL, Oltvai ZN (2004) Network biology: understanding the cell's functional organization. Nat Rev Genet 5:101–113

Barabási AL, Gulbahce N, Loscalzo J (2011) Network medicine: a network based approach to human disease. Nat Rev Genet 13:56–68

Barnard A, Layton D, Hince M et al (2008) Impact of the neuroendocrine system on thymus and bone marrow function. Neuroimmunomodulation 15:7–18

Berthelot JM, le Goff B, Maugars Y (2010) Thymic Hassall's corpuscles, regulatory T-cells, and rheumatoid arthritis. Semin Arthritis Rheum 39:347–355

Bloemers BL, Bont L, de Weger RA et al (2011) Decreased thymic output accounts for decreased naive T cell numbers in children with Down syndrome. J Immunol 186:4500–4507

Blondel VD, Guillaume JL, Lambiotte R et al (2008) Fast unfolding of communities in large networks. J Stat Mech. doi:10.1088/1742-5468/2008/10/P10008

Boada-Romero E, Letek M, Fleischer A et al (2013) TMEM59 defines a novel ATG16L1-binding motif that promotes local activation of LC3. EMBO J 32:566–582

Chaussabel D, Baldwin N (2014) Democratizing systems immunology with modular transcriptional repertoire analyses. Nat Rev Immunol 14:271–280

Chen H, Xu C, Jin Q et al (2014) S100 protein family in human cancer. Am J Cancer Res 4:89–115

Clauset A, Newman MEJ, Moore C (2004) Finding community structure in very large networks. Phys Rev E 70:066111

De Leon-Luis J, Santolaya J, Gamez F et al (2011) Sonographic thymic measurements in Down syndrome fetuses. Prenat Diagn 31:841–845

Drumea-Mirancea M, Wessels JT, Müller CA et al (2006) Characterization of a conduit system containing laminin-5 in the human thymus: a potential transport system for small molecules. J Cell Sci 119:1396–1405

Estève PO, Terragni J, Deepti K et al (2014) Methyllysine reader plant homeodomain (PHD) finger protein 20-like 1 (PHF20L1) antagonizes DNA (Cytosine-5) methyltransferase 1 (DNMT1) proteasomal degradation. J Biol Chem 289:8277–8287

Gallo EM, Winslow MM, Canté-Barrett K et al (2007) Calcineurin sets the bandwidth for discrimination of signals during thymocyte development. Nature 450:731–735

Gérard A, Ghiotto M, Fos C et al (2009) Dok-4 is a novel negative regulator of T cell activation. J Immunol 182:7681–7689

Guillemot L, Spadaro D, Citi S (2013) The junctional proteins cingulin and paracingulin modulate the expression of tight junction protein genes through GATA-4. PLoS ONE 8:e55873

Guittard G, Gérard A, Dupuis-Coronas S et al (2009) Cutting edge: Dok-1 and Dok-2 adaptor molecules are regulated by phosphatidylinositol 5-phosphate production in T cells. J Immunol 182:3974–3978

Hamazaki Y, Fujita H, Kobayashi T et al (2007) Medullary thymic epithelial cells expressing Aire represent a unique lineagederived from cells expressing claudin. Nat Immunol 8:304–311

Hirao K, Natsuka Y, Tamura T et al (2006) EDEM3, a soluble EDEM homolog, enhances glycoprotein endoplasmic reticulum-associated degradation and mannosetrimming. J Biol Chem 281:9650–9658

Holländer GA (2007) Claudins provide a breath of fresh Aire. Nat Immunol 8:234–236. PubMed PMID: 17304232

Ishihara M, Araya N, Sato T et al (2013) Preapoptotic protease calpain-2 is frequently suppressed in adult T-cell leukemia. Blood 121:4340–4347

Izsepi E, Himer L, Szilagyi O et al (2013) Membrane microdomain organization, calcium signal, and NFAT activation as an important axis in polarized Th cell function. Cytometry A 83:185–196

Jablonska-Mestanova V, Sisovsky V, Danisovic L et al (2013) The normal human newborns thymus. Bratisl Lek Listy 114:402–408

Kammerer R, Stober D, Singer BB et al (2001) Carcinoembryonic antigen-related cell adhesion molecule 1 on murine dendritic cells is a potent regulator of T cell stimulation. J Immunol 166:6537–6544

Kanzaki S, Yamaguchi A, Yamaguchi K et al (2010) Thymic alterations in GM2 Gangliosidoses model mice. PLoS ONE 5(8):e12105

Karl K, Heling KS, Sarut Lopez A et al (2012) Thymic-thoracic ratio in fetuses with trisomy 21, 18 or 13. Ultrasound Obstet Gynecol 40:412–417

Kim W, Bennett EJ, Huttlin EL et al (2011) Systematic and quantitative assessment of the ubiquitin-modified proteome. Mol Cell 44:325–340

Kim S, Hill A, Warman ML et al (2012) Golgi disruption and early embryonic lethality in mice lacking USO1. PLoS ONE 7(11):e50530

Korbel JO, Tirosh-Wagner T, Urban AE et al (2009) The genetic architecture of Down syndrome phenotypes revealed by high-resolution analysis of human segmental trisomies. Proc Natl Acad Sci U S A 106:12031–12036

Krishnan N, Thellin O, Buckley DJ et al (2003) Prolactin suppresses glucocorticoid-induced thymocyte apoptosis in vivo. Endocrinology 144:2102–2110

Kusters MA, Gemen EF, Verstegen RH et al (2010) Both normal memory counts and decreased naive cells favor intrinsic defect over early senescence of Down syndrome T lymphocytes. Pediatr Res 67:557–562

Larocca LM, Lauriola L, Ranelletti FO et al (1990) Morphological and immunohistochemical study of Down syndrome thymus. J Med Genet Suppl. 7:225–230

Levin SM, Schlesinger Z, Handzel T et al (1979) Thymic deficiency in Down's syndrome. Pediatrics 63:80–87

Liang H, Coles AH, Zhu Z et al (2007) Noncanonical Wnt signaling promotes apoptosis in thymocyte development. J Exp Med 204:3077–3084

Lima FA, Moreira-Filho CA, Ramos PL et al (2011) Decreased AIRE expression and global thymic hypofunction in Down syndrome. J Immunol 187:3422–3430

Lorenzo LP, Shatynski KE, Clark S et al (2013) Defective thymic progenitor development and mature T-cell responses in a mouse model for Down syndrome. Immunology 139:447–458

Low SH, Vasanji A, Nanduri J et al (2006) Syntaxins 3 and 4 are concentrated in separate clusters on the plasma membranebefore the establishment of cell polarity. Mol Biol Cell 17:977–989

Macedo C, Evangelista AF, Magalhães DA et al (2009) Evidence for a network transcriptional control of promiscuous gene expression in medullary thymic epithelial cells. Mol Immunol 46:3240–3244

Markus MA, Morris BJ (2009) RBM4: a multifunctional RNA-binding protein. Int J Biochem Cell Biol 41:740–743

Mathis D, Benoist C (2009) Aire. Annu Rev Immunol 27:287–312

Mégarbané A, Ravel A, Mircher C et al (2009) The 50th anniversary of the discovery of trisomy 21: the past, present, and future of research and treatment of Down syndrome. Genet Med 11:611–616

Mingueneau M, Kreslavsky T, Gray D et al (2013) The transcriptional landscape of $\alpha\beta$ T cell differentiation. Nat Immunol 14:619–632

Morgan D, Goodship J, Essner JJ et al (2002) The left-right determinant inversin has highly conserved ankyrin repeat and IQ domains and interacts with calmodulin. Hum Genet 110:377–384

Mou F, Praskova M, Xia F et al (2012) The Mst1 and Mst2 kinases control activation of rho family GTPases and thymic egress of mature thymocytes. J Exp Med 209:741–759

Nakayama F, Umeda S, Ichimiya T et al (2013) Sulfation of keratan sulfate proteoglycan reduces radiation-induced apoptosis in human Burkitt's lymphoma cell lines. FEBS Lett 587:231–237

Narayanan T, Subramaniam S (2013) Community structure analysis of gene interaction networks in Duchenne Muscular Dystrophy. PLoS ONE 8(6):e67237

Newman MEJ, Girvan M (2004) Finding and evaluating community structure in networks. Phys Rev E 69:026113

Obermoser G, Presnell S, Domico K et al (2013) Systems scale interactive exploration reveals quantitative and qualitative differences in response to influenza and pneumococcal vaccines. Immunity 38:831–844

Oh J, Wu N, Baravalle G et al (2013) MARCH1-mediated MHCII ubiquitination promotes dendritic cell selection of natural regulatory T cells. J Exp Med 210:1069–1077

Pellegrini FP, Marinoni M, Frangione V et al (2012) Down syndrome, autoimmunity and T regulatory cells. Clin Exp Immunol 169:238–243

Pereira PL, Magnol L, Sahún I et al (2009) A new mouse model for the trisomy of the Abcg1-U2af1 region reveals the complexity of the combinatorial genetic code of Down syndrome. Hum Mol Genet 18:4756–6926

Prandini P, Deutsch S, Lyle R et al (2007) Natural gene-expression variation in Down syndrome modulates the outcome of gene-dosage imbalance. Am J Hum Genet 81:252–263

Puleston DJ, Simon AK (2014) Autophagy in the immune system. Immunology 141:1–8

Ruiz JC, Bruick RK (2014) F-box and leucine-rich repeat protein 5 (FBXL5): sensing intracellular iron and oxygen. J Inorg Biochem 133:73–77

Sahni N, Yi S, Zhong Q et al (2013) Edgotype: a fundamental link between genotype and phenotype. Curr Opin Genet Dev 23:649–657

Sakurai C, Hashimoto H, Nakanishi H et al (2012) SNAP-23 regulates phagosome formation and maturation in macrophages. Mol Biol Cell 23:4849–4863

Sasada T, Ghendler Y, Neveu JM et al (2001) A naturally processed mitochondrial self-peptide in complex with thymic MHC molecules functions as a selecting ligand for a viral-specific T cell receptor. J Exp Med 194:883–892

Spitz F, Gonzalez F, Duboule D (2003) A global control region defines a chromosomal regulatory landscape containing the HoxD cluster. Cell 113:405–417

Stoika R, Melmed S (2002) Expression and function of pituitary tumour transforming gene for T-lymphocyte activation. Br J Haematol 119:1070–1074

Takahashi K, Yoshida N, Murakami N et al (2007) Dynamic regulation of p53 subnuclearlocalization and senescence by MORC3. Mol Biol Cell 18:1701–1709

Tanimoto K, Suzuki K, Jokitalo E et al (2011) Characterization of YIPF3 and YIPF4, cis-Golgi Localizing Yip domain family proteins. Cell Struct Funct 36:171–185

Tian Y, Chang JC, Fan EY et al (2013) Adaptor complexAP2/PICALM, through interaction with LC3, targets Alzheimer's APP-CTF forterminal degradation via autophagy. Proc Natl Acad Sci U S A 110:17071–17076

Veland IR, Montjean R, Eley L et al (2013) Inversin/Nephrocystin-2 is required for fibroblast polarity and directional cell migration. PLoS ONE 8:e60193

Xu Y, Li W, Liu X et al (2013) Identification of dysregulated microRNAs in lymphocytes from children with Down syndrome. Gene 530:278–286

Yang H, Youm YH, Vandanmagsar B et al (2009) Obesity accelerates thymic aging. Blood 114:3803–3812

Young DJ, Stoddart A, Nakitandwe J et al (2014) Knockdown of Hnrnpa0, a del(5q) gene, alters myeloid cell fate in murine cells through regulation of AU-rich transcripts. Haematologica 99(6):1032–1040 PubMed PMID: 24532040

Zhou D, Medoff BD, Chen L et al (2008) The Nore1B/Mst1 complex restrains antigen receptor-induced proliferation of naïve T cells. Proc Natl Acad Sci U S A 105:20321–20326

Chapter 8
Transcriptome Profiling in Autoimmune Diseases

Cristhianna V. A. Collares and Eduardo A. Donadi

Abstract Autoimmune diseases are a group of different inflammatory disorders characterized by systemic or localized inflammation, affecting approximately 0.1–1 % of the general population. Several studies suggest that genetic risk loci are shared between different autoimmune diseases and pathogenic mechanisms may also be shared. The strategy of performing differential gene expression profiles in autoimmune disorders has unveiled new transcripts that may be shared among these disorders. Microarray technology and bioinformatics offer the most comprehensive molecular evaluations and it is widely used to understand the changes in gene expression in specific organs or in peripheral blood cells. The major goal of transcriptome studies is the identification of specific biomarkers for different diseases. It is believed that such knowledge will contribute to the development of new drugs, new strategies for early diagnosis, avoiding tissue autoimmune destruction, or even preventing the development of autoimmune disease. In this review, we primarily focused on the transcription profiles of three typical autoimmune disorders, including type 1 diabetes mellitus (destruction of pancreatic islet beta cells), systemic lupus erythematosus (immune complex systemic disorder affecting several organs and tissues) and multiple sclerosis (inflammatory and demyelinating disease of the central nervous system).

8.1 Autimmune diseases have been prominent in the medical-scientific scenario Have been Prominent in the Medical-Scientific Scenario

The main function of immune system is the protection of the organism against invasion of pathogens and restore tissue integrity. Autoimmune diseases are a group of diverse inflammatory disorders characterized by systemic or localized inflammation, usually leading to ischemia and tissue destruction. Therefore, systemic

E. A. Donadi (✉) · C. V. A. Collares
Department of Medicine, Division of Clinical Immunology, Ribeirão Preto Medical School,
University of São Paulo, 14049–900 Ribeirão Preto, SP, Brazil
e-mail: eadonadi@fmrp.usp.br

© Springer International Publishing Switzerland 2014
G. A. Passos (ed.), *Transcriptomics in Health and Disease,*
DOI 10.1007/978-3-319-11985-4_8

autoimmune disorders encompass diseases caused by a fault in the immune system, losing its ability to identify particular self-antigens. In this process, a chronic over-reactivity of B and T cells may arise producing unsafe signals released for cells or tissues when they are in dangerous and have abnormal cell death (Kamradt and Mitchison 2001; Matzinger 2002). Other changes occur in the immune system resulting from this failure, such as loss of inflammation control resulting in continuous immune activation without any infection.

The knowledge on the pathogenesis of autoimmune diseases has not yet been fully elucidated. Some studies suggest that genetic risk susceptibility factors are shared between different systemic autoimmune diseases, and the pathogenic mechanisms may be similar (Zhernakova et al. 2009). However, different loci have been identified as disease specific, suggesting the presence of many immunopathogenic pathways (Cho and Gregersenet 2011).

Although many genetic loci for autoimmune diseases were described, additional elements have been identified and associated with development of autoimmunity. Approximately 0.1–1 % of the general population develops autoimmune diseases during life. Considering first-degree relatives, this incidence increase five times, and in monozygotic twins this rate is five-fold increased. Thus, the risk is increased with increasing genetic similarity to an affected individual. However, the highest autoimmune disease susceptibility rate among monozygotic twins is not higher than 20–30 %, showing that there are additional elements and/or multiple factors involved in these processes, ranging from genetic to environmental factors (Mackay 2009). Together, these factors are influencing this rate and may play an important role on clinical manifestations in a genetically predisposed individual. In addition, some autoimmune diseases are influenced by hormonal factors and may affected more women than men, i.e., systemic lupus erythematosus, that affects 80–90 % more women than men, and the peak of its incidence occurs during childbearing ages (Straub 2007).

There are many autoimmune diseases and in this revision we will devote special attention to three of them: type 1 diabetes mellitus, systemic lupus erythematosus and multiple sclerosis. These three diseases have been prominent in the medical-scientific scenario, because recent studies reveal important information about their transcriptome, and this knowledge may contribute for a better understanding of disease features, as well as for the development of new drugs, new diagnostic methods and discovery of biomarkers for early diagnosis.

8.1.1 Type 1 Diabetes Mellitus

Diabetes mellitus is one of the most studied diseases and a great amount of information is available in public databases regarding genetic association, meta-analysis of diabetes and associated complications. Type 1 diabetes (T1D) accounts for approximately 10 % of all cases of diabetes, affecting individuals under the age of 30, but can also manifested later (Geenen et al. 2010). T1D is caused by an autoimmune mechanism against pancreatic beta cells, resulting in decrease and/or disruption of insulin production. It is estimated that 80–95 % of pancreatic beta cells are destroyed when T1D is diagnosed. Until now, T1D affects more than 300 million of adults in worldwide populations, and this rate will be increase to 440 million by 2030 (Shaw et al. 2010). Thus, T1D fits as an example of an autoimmune disease, in which target organ destruction is allied to the presence of disease-specific target organ autoantigens and autoantibodies, and the disease is considered to be a T cell-mediated autoimmune disorder (Battaglia 2014). The knowledge of the T1D still has significant shortcomings, particularly about asymptomatic early stages of auto-immune attack. This occurs because the difficult accessibility to the pancreas. The role of islet autoantibodies is still unknown, although they have important function as serological marker for the disease (Achenbach et al. 2004). Actually, autoanti-bodies for islet cells antigens (ICAs), glutamic acid decarboxylase (GAD), islet antigens insulin (IAA) and protein tyrosine phosphatase-like protein IA-2 (IA-2 A) are used for the prediction and progression of T1D (Achenbach et al. 2004). In recent years, other biomarkers have been used like as autoantibodies against the zinc efflux transporter ZnT8 (Herold et al. 2009; Wenzlau et al. 2007). However, it is necessary to discovery new biomarkers capable to diagnosis T1D before the complete islet cell destruction. Nowadays, it is known that diabetes is a polygenic disease resulting of a highly selective autoimmune response that causes inflammation (insulitis) followed by destruction of insulin-secreting pancreatic beta cells, leading to deregulation of glucose metabolism and insulin deficiency (Erlich et al. 2008; Dib et al. 2008; Geenen et al. 2010).

Despite the identification of several risk factors for T1D, its etiology is still unknown. The early prediction is still missing due to insufficient predictive power of the individual risk factors (Purohit and She 2008). Moreover, Jayaraman et al. (2013) showed that chromatin remodeling resulted in simultaneous down-regulation of several inflammatory genes and up-regulation of many genes responsible for a set of cellular functions, including glucose homeostasis. Thus, a complex disease like T1D involves so many genes that contribute in a different manner in multiple signaling and metabolic pathways.

Furthermore, several genes and gene regions distributed throughout the genome have been reported in association with T1D in population, family and linkage studies. The strongest genetic susceptibility contribution comes from the human leucocyte antigen complex (HLA) region at chromosome 6p21 (IDDM1), responsible for up to 40–50 % of T1DM susceptibility, and from the insulin gene (*INS*) region (IDDM2) (Pugliese and Miceli 2002). Moreover, three principal cellular types, i.e., macrophages, dendritic cells and lymphocytes are involved on T1D pathogenesis, acting in a complex way contributing to the loss of tolerance against pancreatic autoantigens, including: (i) decreased or very poor expression of insulin in thymus; (ii) molecules encoded by the *HLA-DQA1**05:01, *HLA-DQB1**03:02 and *02:01 and *HLA-DRB1**03/*04 alleles that may mediate the presentation of autoantigens, contributing to the development of anti-GAD, insulin (IAA), islet antigen 2 (IA2A) and ISS autoantibodies.; (iii) deficient immunoregulation, mediated by specific surface and intracellular molecules, including IL-2, IL-2RA, IL-2RB, CTLA-4, PTPN-2, PTPN-22; iii) decreased quantity of regulatory T cells; (iv) function-modified molecules involved on innate immune response, (v) regulation by invariant natural killer-like T (iNKT) cells (Cipolleta et al. 2005; Sia 2006; Li et al. 2007; Chentoufi et al. 2008; Knip and Siljander 2008; McDevitt and Unanue 2008; Tisch and Wang 2008; Karumuthil-Melethil et al. 2008; Todd 2010; Pociot et al. 2010; Buschard 2011; Novak and Lehuen 2011).

As the course of insulitis, the progression of diabetes involves interactions between environmental factors (nutrition, viral infections, etc) and genetic background of the patient. Therefore, it is important to identify the molecular mechanisms that are involved on the survival of pancreatic beta cells and production of inflammatory mediators such as chemokines and cytokines. In this context, several genes were identified by genome wide association studies (GWAS) (Barrett et al. 2009; Nerup et al. 2009; Plagnol et al. 2011) and by transcriptome analyses in animal models (Grinberg-Bleyer et al. 2010; Fornari et al. 2011), in peripheral blood mononuclear cells (Rassi et al. 2008; Han et al. 2011; Collares et al. 2013a, b), in pancreatic beta cells (Planas et al. 2010) and in whole peripheral blood cells (Reynier et al. 2010), and using these approaches several other genes have been associated with T1D.

In 2009, Barrett et al., using GWAS showed more than 40 loci containing probable contributors to T1D (McDevitt and Unanue 2008). Generally, it is assumed that most if not all of the candidate genes to T1D modulate the immune system (Concannon et al. 2009). Following this hypothesis, the beta cells undergo a process that starts in the pancreas, but is regulated elsewhere, even though 61 % of the genes responsible for T1D are expressed in human pancreatic islets (Eizirik et al. 2012). Another interesting observation is that the expression of these genes is modified after exposure to proinflammatory cytokines or double-stranded sRNA (by-product resulting from viral infection) that can contribute for T1D (Eizirik et al. 2009; Moore et al. 2009; Colli et al. 2010; Eizirik et al. 2012). Two genes deserve special mention as they play an important role on the production of cytokines/chemokines and apoptosis of pancreatic beta cells: IFIH-1 (Colli et al. 2010) and PTPN-2 (Moore et al. 2009; Colli et al. 2010; Santin et al. 2011).

At least 3000 genes associated with inflammation, innate immune response and apoptosis have their expression controlled by cytokines. In human islets, some very relevant cytokine and chemokine genes are induced, such as *CCL2, CCL5, CCL3, CXCL9, CXCL10, CXCL11, IL-6* and *IL-8* (Eizirik et al. 2012). Among them, the CCL2 and CXCL10 molecules attract macrophages and may be involved on the recruitment of immune cells at the beginning of insulitis.

The gene expression profiles have unveiled new transcriptional alterations in several disorders. Microarray technology and bioinformatics offer the most comprehensive molecular evaluations and it is widely used to understand the changes in gene expression in specifics organs, including in pancreas during the course of T1D in NOD mice (Vukkadapu et al. 2005). The great advantage of this technology is that it enables the achievement of a set of data from different experiments; these data can be combined into a single database, allowing the comparison of gene expression profiles between different samples. These analyses have led to define new transcriptional changes associated with many autoimmune diseases, including T1D. The major goal of transcriptome studies is the identification of specific biomarkers for different diseases. Using microarray technologies, Collares et al. (2013a) evaluated the transcriptome of diabetic patients, including T1D, type 2 diabetes (T2D) and gestational diabetes (GDM) patients, to see specific expression profile signatures of each type of diabetes. The results revealed that the overall gene expression profile is characteristic for each group of diabetic patients and that gene expression profile of GDM was closer to T1D than to T2D. An *in silico* analysis showed that the similarities observed in the transcriptional profile of GDM and T1D were due to the role of genes associated with inflammation (Evangelista et al. 2014). The higher expression of these genes in some T1D and GDM patients seems to influence the global gene expression pattern of diabetic patients. Indeed, several important molecular mechanisms identified in this cluster account for an intricate array of inflammation pathways. Using the DAVID database (http://david.abcc.ncifcrf.gov/) it is possible to obtain functionality of these genes, and observed the involvement of modulated genes in different biological functions. Our group observed (data not published), in a meta-analysis study of T1D, T2D and GDM that induced genes were grouped into five major groups of biological function: (i) development of multicellular organism (20.3%), (ii) signal transduction (17.9%), (iii) stress response (12.2%) (iv) cell differentiation (10.7%), (v) processes the immune system (6.8%). About repressed genes, we showed that these genes were clustered into three main biological processes: (i) regulation of metabolic processes (30%), (ii) biosynthetic processes (26.9%), (iii) transcriptional processes (22%) (Fig. 8.1).

Performing a more restrictive analysis, we considered just modulated genes that showed fold change≥ 2 for all comparisons of the three groups of diabetes, and we observed 10 most significant genes, of which seven were induced in GDM and T1D and repressed in T2D and three genes were repressed in GDM and T1D and induced in T2D. Taking into account these 10 genes, Fig. 8.2 shows the location of each gene.

In terms of gene expression regulation, the non-coding RNAs have been identified as important or even major regulators of gene expression and include small microRNA (miRNA) and long noncoding RNA (lncRNA) (Mattick and Makunin 2006; Ponting et al. 2009). Some miRNAs have been shown as key players involved in pancreatic development and homeostasis. Among them, miR-375 has been considered a key miRNA that regulates B-cell insulin secretion and, hence, overall glucose homeostasis of the body (Dumortier and Van Obberghen 2012). MiRNAs play regulatory roles in many biological processes associated with diabetes, including adipocyte differentiation, metabolic integration, insulin resistance and appetite regulation (Krützfeldt and Stoffel 2006). The role of miRNAs in diabetes has been associated with several pathogenic features. For example, miR-410, miR-200a and miR-130a regulate secretion of insulin in response to stimulatory levels of glucose, and overexpression of miR-410 enhances the levels of glucose-stimulated insulin secretion (Hennessy et al. 2010). MiR-30d is upregulated in pancreatic beta-cells and collaborates for increasing insulin gene expression (Tang et al. 2009a) and miR-9 acts in the fine-tuning of glucose metabolism (Plaisance et al. 2006). An important miR responsible for insulin gene expression and secretion is miR-375 (El Ouaamari et al. 2008; Poy et al. 2004).

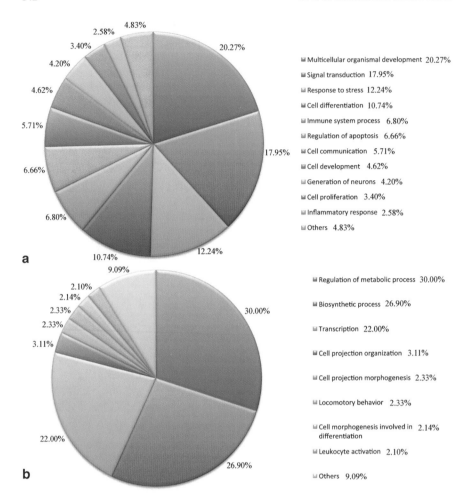

Fig. 8.1 Biological function of the significant and differentially expressed genes (3747 transcripts), which were modulated after comparing GDM, T1DM and T2DM. In *panel A* are represented upregulated transcripts (from GDM to T1D to T2D), which were clustered into five groups according to their biological functions: (i) development of multicellular organism (20.3%), (ii) signal transduction (17.9%), (iii) stress response (12.2%) (iv) cell differentiation (10.7%), v) immune system processes (6, 8%). *Panel B* shows the downregulated transcripts, which clustered into three main groups: (i) regulation of metabolic processes (30%), (ii) biosynthetic processes (26.9%) (iii) transcriptional processes (22%)

In a recent study, the analysis of mRNA/miRNA signatures encompassing T1D, T2D and GDM patients, pinpointed some miRNAs shared among the three types of diabetes; selected miRNAs specific for each type of diabetes; and identified non-described miRNAs associated with each type of diabetes (Collares et al.

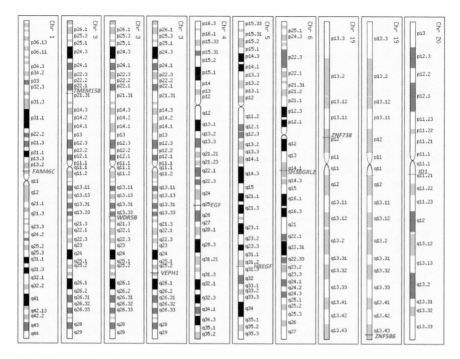

Fig. 8.2. Location of the modulated genes obtained from meta-analysis of the gene profiles obtained for GDM, T1D and T2D patients, showing those upregulated (*red*) and those down-regulated (*green*) genes

2013b). The authors showed 9 miRNAs shared among the three types of diabetes, including hsa-miR-126, hsa-miR-144, hsa-miR-27a, hsa-miR-29b, hsa-miR-1307, hsa-miR-142-3p, hsa-miR-142-5p, hsa-miR-199a-5p, and hsa-miR-342-3p, and suggested that these miRNAs are associated with diabetes *per se*. For T1D, they pinpointed some miRNAs as candidate to be involved in T1D and the only miRNA linked to glucose metabolism was let-7f, which was previously suggested as a potential therapy for T2D (Frost and Olson 2011).

Therefore, there are many studies on the transcriptome on autoimmune diabetes; however, knowledge is still at the beginning of a long journey. Although linkage studies have been associated with transcriptome and microRNAs and, most recently proteomes, many issues are still studied to unveil the intricate mechanisms associated with the development of diabetes. It is believed that such knowledge will contribute to the development of new drugs, new strategies for diabetes care, including early diagnosis (before there is destruction of most of the pancreatic beta cells), avoiding autoimmune destruction of pancreatic beta-cells, or even preventing the development of T1D.

8.1.2 Systemic Lupus Erythematous

Systemic lupus erythematosus (SLE) is a systemic autoimmune disorder characterized by the presence of high amounts of circulating immune complexes, leading to accumulation of organ damage over time as a result of persistent tissue inflammation. SLE involves the activation of the innate and adaptive immune response and usually is considered a severe and potentially life-threatening disease, which may represent a therapeutic challenge because of its heterogeneous organ manifestations. The central features of SLE encompass T- and B-cells abnormalities that could lead to autoantibody production. Innate immune cells produce type 1 interferon (IFN) that has a central role in systemic autoimmunity and in the activation of B and T cells. Autoantibodies produced by B-cells stimulate dendritic cell IFN production, encompassing the role of innate and adaptive systems. On the other hand, infectious agents combined with genetic factors may serve as an initial trigger for autoimmunity hyperactivity and development of SLE (Crow 2008). Patients may present a variety of clinical manifestations and immunological features. Thus, it may cause many symptoms as inflammation of nervous system, nephritis, leukopenia, arthritis and skin rashes (Petri 1995). SLE pathogenesis involves an increased production of immune complexes and an excess of innate immune activation involving Toll-like receptors and type I interferon, which in concert yield an abnormal activation of lymphocytes. A special attention has been devoted to SLE as a consequence of elevated mortality rate that has decreased over the past 30 years, but is still high worldwide (Bertsias et al. 2010). Among SLE patients, women are nine times more affected than men, suggesting that sex-related factors are crucial in the development of the disease (Schwartzman-Morris and Putterman 2012; Weckerle and Niewold 2011).

Genome-wide association and linkage studies have revealed over 40 genes associated with SLE and loci linked to pathways of immune system regulation, tissue response to injury, endothelial function and other yet undefined functions (Moser et al. 2009; Deng and Tsao 2010; Guerra et al. 2012). In addition, as in other autoimmune diseases, genetic risk factors for SLE include alleles of the human leukocyte antigen (HLA) region, and other genes including *IRF5, STAT4, BLK, TNFAIP3, TNIP1, FCGR2B,* and *TNFSF13* (Koga et al. 2011). These SLE susceptibility genes are also involved in other autoimmune diseases and the genetic polymorphism of cytokines may deregulate lymphocyte activity.

Despite numerous studies, the etiology of SLE remains uncertain and disease pathogenesis has been associated to the interaction of genetic, epigenetic and environmental factors (Liu and Davidson 2012). Additionally, exposure to viruses and bacterial infections, and also ultra-violet radiation are known to trigger SLE (Doria et al. 2008). In terms of the genetic risk for developing SLE, only 10–20% of the cases can be explained by heritability, and genetic variability of individuals have a smaller contribution. The great challenge of the studies on lupus is the identification of these variants that are, in 90% of the cases, in non-coding, intronic or intergenic regions (Moser et al. 2009; Deng and Tsao 2010; Guerra et al. 2012; Costa et al. 2013; Kilpinen and Dermitzakis 2012). Moreover, in monozygotic twins, the concordance rate is 24–69% and in dizygotic twins is 2–5% (Jarvinen et al. 1992;

Deapen et al. 1992), evidencing the key role of environmental factors in the development of SLE. Thus, the role of the environment on SLE risk is overt by the high disagreement rate among monozygotic twins. More specifically, there were no genomic differences between monozygotic twins, but exist some epigenomic and gene expression variations that may provide evidences as how the candidate genes exert their roles in the pathogenesis of SLE (Furukawa et al. 2013). Environmental factors mediate epigenetic effects, i.e., DNA/RNA methylation and histone modification, that may influence the expression of miRNAs, added to microRNA effects, regulate gene expression. Maybe this variation on epigenetic factors should explain part of the missing heritability (Kilpinen and Dermitzakis 2012; Costa et al. 2013).

Gene expression studies cover genetic and epigenetic effects and are important to identify deregulation of molecular and cellular pathways in SLE, which are responsible for its effects on phenotype. The term epigenetics refers to changes in the DNA or surrounding chromatin that may alter gene expression, but without modification on the genetic composition (Mattick et al. 2010). In this context, peripheral blood mononuclear cells (PBMC) have been used in large scale. Studies with PBMC have shown interesting results when healthy individuals were compared to SLE patients. For instance, the signature of interferon (IFN) is very characteristic and more prominent in patients with more active and severe form of SLE. Any abnormalities of B cells, autoantibody production, complement activation and production of type 1 interferon (IFN) are very important in the SLE pathogenesis (Crow and Kirou 2004). In special, the production of IFN-alfa may be detected in serum (Bengtsson et al. 2000) in specific phenotypes of active disease, i.e., several SLE patients have high levels of type I IFN in the circulation (Weckerle et al. 2011). Microarray analysis showed overexpression of IFN-alfa-regulated genes in SLE patients, including the type 1 IFN signature (Crow and Wohlgemuth 2003), and many SLE families with high levels of IFN-alfa have been clustered (Niewold et al. 2007). Moreover, it has shown that viral infection treated with IFN-alfa may contribute to de novo SLE development, which disappears when treatment is discontinued (Niewold and Swedler 2005; Ronnblom et al. 1990).

Even when considered the subpopulations of purified cells of peripheral blood mononuclear cells, there is a unique pattern of expression common among these subpopulations, which is the signature of interferon expression. A specific signature has been identified in SLE T CD4+ cells, involving IFN transcripts, and most differentially expressed genes in these cells had promoters with binding sites for interferon regulatory factor (IRF) -3 and -7 (Lyons et al. 2010; Li et al. 2010). The involvement of IFN and IFN-induced genes also appear when evaluating target organ or tissue. For example, comparing SLE patients with rheumatoid arthritis or osteoarthritis, is observed, in the transcriptome of the synovial membrane, induction of expression of IFN induced genes and repression of genes involved in extracellular matrix homeostasis (Toukap et al. 2007).

However, considering SLE, the evaluation of bone marrow is more informative than PBMC, since bone marrow is a central lymphoid organ with hematopoietic and immunoregulatory function and exhibits a variety of histopathological abnormalities in SLE (Voulgarelis et al. 2006). In 2008, Nakou et al. showed the microarray analysis of the bone marrow, in which it was possible to visualize the clear differentiation between active from inactive SLE. Using bioinformatics and evaluating bone marrow cells, pathways related with cellular growth, cell survival and immune reactions, were identified playing important role in the pathogenesis of SLE (Nakou et al. 2010). In addition, some studies have been done in platelets and microarray using platelets from SLE patients versus health individual revealed increased expression of genes encoding cytokines, chemokines and proteins involved in apoptosis, and overexpression of type I IFN-regulated genes (Lood et al. 2010). These studies confirm the type I INF-related genes expression profile in platelets from SLE patients, as well as its related proteins. Regardless of the biological material studied, several lines of evidence show that IFN is extremely important for SLE and its expression is induced in tissues from SLE patients. Approximately 60 % of patients exhibit increased expression of genes induced by type 1 IFN, which is directly associated with the disease activity (Baechler et al. 2003; Bennett et al. 2003; Han et al. 2003; Kirou et al. 2004; Kirou et al. 2005; Feng et al. 2006). In addition, the signaling pathways induced by type 1 IFN are also more activated in SLE (Yao et al. 2009). Grammatikos et al. (2014) showed also three important overexpressed genes in SLE patients: (i) *IL10* (interleukin 10) that is involved on B cells maturation and antibody production; (ii) *CD70*, which is expressed on activated T cells and is involved in T cell proliferation; iii) *OAS2* (2′–5′-oligoadenylate synthetase 2), which is an interferon-inducible gene. Other important gene that has been studied in SLE patients is interferon regulatory factor 5 (*IRF5*), in which polymorphisms have been shown to confer risk or protective effect in SLE. Recent data have shown characteristic and differential gene expression signatures between SLE patients and health individuals, including the identification of IRF5-SLE risk haplotype and defined the four most abundant haplotypes in SLE patients (Stone et al. 2013). Furthermore, the gene expression profile studies include induction of specific transcripts of granulocytes, repression of genes related to DNA repair, differential expression of genes involved in cell apoptosis and motility, as well as autoimmune gene signature differentiating SLE patients and unaffected first degree relatives (Baechler et al. 2003; Han et al. 2003; Rus et al. 2004; Maas et al. 2005; Lee et al. 2011).

Many transcription factors were shown to be crucial for immune system and their differences in the expression and activity may imply in discovering novel biomarkers in some diseases, including SLE. Comparing levels of transcription factors in PBMC of SLE patients, Sui et al. (2012) found 92 differentially expressed transcription factors, and indicated activator protein-1 (AP-1), Pbx1 and myocyte enhancer factor-2 (MEF-2) as candidates involved in pathogenesis of SLE and new diagnosis biomarker for this disease. The transcription factor FOXO1 was also related to SLE, which was downregulated in PBMCs from SLE and rheumatoid arthritis patients (Kuo and Lin 2007). The transcript family FOXO involves transcription factors that play an important role in controlling lymphocyte activation and proliferation. A member of nuclear factor (NF)-kB/Rel family of transcription factors, c-Rel, was found in higher levels in PBMCs from SLE patients (Burgos et al. 2000). Since cytokines are produced by T-help cell 1 (Th1) and 2 (Th2), probably transcription factors related to T-help cells must have an important role in SLE (Foster and

Kelly 1999). The principal transcription factors for differentiation of Th1 and Th2 are T-bet and GATA-3, that were found overexpressed and underexpressed in SLE patients, respectively (Chan et al. 2006; Lit et al. 2007). Other transcription factors, including AP-1, NF-kB, and IRF5 increase STAT-4 expression, which risk haplotype for SLE is overexpressed (Remoli et al. 2007) and is important for type 1 IFN receptor signaling. Moreover, the IRF5 is mediator of Toll-like receptor-triggered expression of proinflammatory cytokines such as type 1 IFN and TNF-alfa (Kawai and Akira 2006).

Regarding the control of gene expression within the context of SLE, the levels of RNA may be controlled by epigenetic mechanism including microRNAs, which usually acts by degradation of target mRNA or inhibiting its translation. Many studies have reported microRNAs deregulation in SLE and more than 42 differentially expressed microRNAs were detected in PBMCs from SLE patients, and some of them were pinpointed as biomarker candidates. It has been demonstrated that microRNAs deregulation are implicated in different systemic autoimmune diseases (Stagakis et al. 2011). MiR-21 acts partly through inhibition of *PDCD4* (selective protein translation inhibitor of genes involved in immune responses) and it was found upregulated in T and B-cells (Stagakis et al. 2011) and in CD4T cells (Pan et al. 2010) of SLE patients comparing to control group, suggesting it as future biomarker for SLE.

An increased expression of miR-224 (Lu et al. 2013), miR-148a (Pan et al. 2010), miR-15 (Yuan et al. 2012), miR-142-3p and miR-181a (Carlsen et al. 2013), miR-189, miR-61, miR-78, miR-21, miR-142-3p, miR-342, miR-299-3p, miR-198 and miR-298 (Dai et al. 2007) have been described in patients and animal models. However, many studies have shown downregulation of different microRNAs in SLE patients, including: miR-146a (Tang et al. 2009), miR-145 in T cells (Lu et al. 2013), miR-155 in serum and urine from SLE patients (Wang et al. 2010), miR-181a in pediatric patients (Lashine et al. 2011), miR-19b and miR-20a in monocytes (Teruel et al. 2011), miR-125a (Zhao et al. 2010), miR-17, miR-20a, miR-106a, miR-92a, and miR-203 in the circulation (Carlsen et al. 2013), miR-196a, miR-17-5p, miR-409-3p, miR-141, miR-383, miR-112 and miR-184 in PBMCs (Dai et al. 2007).

In the context of SLE, some specific knowledge about some microRNAs and their target genes may help to develop new drugs and diagnostics. For example, the decreased expression of miR-145 and induction of its target protein activator of transcription-1 (STAT-1), seems to be associated with lupus nephritis, and may contribute to the immunopathogenesis of SLE (Lu et al. 2013). MiR-146a, which targets STAT1 and IRF5 in innate immune cells and is negative regulator of type 1 IFN and TLR7 signaling pathways, was described as repressed compared to controls (Tang et al. 2009b). Some reports show microRNAs, such as miR-148a, miR-126, miR-21, and miR-29b downregulating, directly or indirectly, DNA methyltransferase-1 expression and, thus, contributing to global hypomethylation observed in SLE (Deng et al. 2001; Pan et al. 2010; Zhao et al. 2011; Layer et al. 2003; Qin et al. 2013). MiR-125a is involved in inflammatory chemokine pathway and contributes to higher expression of RANTES, an inflammatory chemokine, indicating that this miR can be used as a novel target for SLE treatment (Zhao et al. 2010). Thus, it is believed that many of these microRNAs and other yet to be described may be used as biomarkers of SLE.

8.1.3 Multiple Sclerosis

Multiple sclerosis (MS) is a common, severe, chronic inflammatory autoimmune and demyelinating disease of the central nervous system (CNS), due to immune reaction against myelin proteins, mainly affecting white matter in which autoreactive T cells attack the myelin-oligodendrocyte complex (Noseworthy et al. 2000). Generally, it begins at third or fourth decade of life, affects more women (60% of MS cases) than men (Weinshenker 1994), and approximately 80% of MS patients have relapsing-remitting MS form (Lublin and Reingold 1996). The role of autoimmunity becomes clear by the presence of autoreactive T cells for myelin components of CNS and peripheral blood of MS patients. The etiology of MS is still unknown; however, evidence indicates a multifactorial and complex nature, where genetic and environmental factors may influence their onset (Noseworthy et al. 2000). In addition, evidence pinpoints for polygenic susceptibility and multiple environmental triggering factors (Poo 2001).

It is believed that T lymphocytes are activated at lymph nodes in the periphery and bind to receptors on endothelial cell, continuing to cross the blood brain barrier into the interstitial matrix (Karpuj et al. 1997). Activation of T cells induce the release of cytokines the further opening access to the CNS through the blood-brain barrier, stimulating chemotaxis, resulting in a second recruitment of inflammatory cells, and leakage of plasma proteins into the CNS, triggering a series of mechanisms responsible for myelin damage. The main pathological feature in MS is the plaque, a well-demarcated white matter injury histologically characterized by inflammation, T cells and macrophages, demyelination and gliosis, and axonal loss (Lucchinetti et al. 1998).

Current knowledge of MS allows the formation of the concept of circulating T-cell receptor-selected T cells in MS and that CD8+ T cells may be essential in the pathophysiology of the disease. The study of abnormalities of blood T cells in MS may contribute to better understanding of the disease and the discovery of new drugs against MS (Laplaud et al. 2004). Studying monozygotic twins and observing the differences in the manifestation of MS, it has been interpreted that this difference was due to the influence of the environment, which is considered of great importance in this autoimmune disease (Mumford et al. 1994; Sadovnick et al. 1996; Willer et al. 2003; Nielsen et al. 2005; Islam et al. 2006; Chitnis 2007; Oksenberg et al. 2008). However, the genetic influence is very important for disease development.

In genetically determined diseases, the genetic component is valued at higher relative risk of siblings of affected individual presenting the same disease, and there is also a higher concordance rate in monozygotic than in dizygotic twins (Sadovnick and Ebers 1993; Willer et al. 2003). Considering MS, the most striking susceptibility genes are encoded at the major histocompatibility complex MHC, especially class II alleles (Dyment et al. 2004). The HLA-DR2 phenotype (*DRB1*15:01-DQB1*06:02*) has been described in different populations (Sadovnick and Ebers 1993; Epplen et al. 1997; Barcellos et al. 2003). It is believed that there are specific standards of ethnic, environmental or both association patterns, wherein *HLA-DRB1* alleles may have different behavior in different environmental contexts (Brum et al. 2007). Furthermore, the allele group *DRB1*15* was suggested as significant factor MS susceptibility and development (Kaimen-Maciel et al. 2009), as previously demonstrated in the Brazilian Caucasian population (Brum et al. 2007).

Many research studies of gene expression on MS have been performed using brain tissue from patients, and gene profile may be altered in acute, chronic or silent lesions or normal tissue. The most important discoveries of this period was a set of 16 genes related to autoimmunity, with seven of them associated with SLE and two associated with T1D (Tajouri et al. 2007). However, MS PBMCs are also widely used for obtaining mRNA for gene expression studies.

Approximately 300 differentially expressed genes were detected in a study done in PBMCs of MS patients (Bomprezzi et al. 2003). Among them, overexpression of (i) platelet activating factor acetyl hydrolase (*PAFAH1B1*), a gene associated with brain development and chemoattraction during inflammation and allergy; (ii) tumor necrosis factor receptor (*TNFR* or *CD27*), which is co-stimulator for T cell activation and fundamental for immune response development; (iii) T cell receptor (*TCR*), crucial for T cell mediated immune response and it was associated with MS susceptibility (Beall et al. 1993); (iv) zeta chain associated protein kinase (*ZAP70*), gene responsible for TCR induced T cell activation (Chan et al. 1992); (v) interleukin 7 receptor (*IL7R*), involved in B and T cells activation. In the same study, several genes were repressed, such as: tissue inhibitor of metalloproteinase 1 (*TIMP1*), plasminogen activator inhibitor 1 (*SERPINE 1*), histone coding genes, and heat shock protein 70 (*HSP70*). Additionally, the evaluation of T cells from MS patients stressed the importance of transcriptional regulation of NF-kB, which is responsible for regulating gene expression during MS relapse; deregulation of NF-kB on T cell transcriptome may be used as a molecular biomarker for clinical disease activity (Satoh et al. 2008).

Alternative study of the transcription profile of MS patients is the use of cerebrospinal fluid. Brynedal et al. (2010) investigated gene expression profile in leukocytes of CSF from MS patients and found *AIF1, MGC29506, POU2AF1, PLAUR* and *TNFRSF17* as differentially expressed. Recently, a comparative study between MS patients at relapse and healthy controls was performed (Jernas et al. 2013), showing the overexpression of genes involved in T and NK cell process, genes belonging to pathways involved in T-cell co-stimulation, activated T-cell proliferation, regulation of cell surface receptors and NK-cell activation. The authors also showed a decreased expression of genes associated with innate immunity, B-cell activation and immunoglobulin secretion and T helper 2 responses in leukocytes of CSF, highlighting the *HMOX1* gene. The deletion of this gene was associated to enhance demyelination (Chora et al. 2007). The induced genes were: i) *EDN1,* associated with integrity of blood-brain-barrier; ii) *CXCL11* that is important for recruitment of T-cells to the CNS when disease activity is higher; iii) *CXCL13*, which may be important for the T-helper cell recruitment during relapses. Furthermore, certain *CXCL13* polymorphic sites associated with high levels of the chemokine are more frequent in patients with MS (Linden et al. 2013). Cerebrospinal fluid was also used for the study of the hypothalamus–pituitary–adrenal (HPA) axis activity in MS, because of its association with disease progression and comorbid mood disorders. The activity of axis was determined by measuring cortisol in cerebrospinal fluid and the results reveled, in MS patients, low HPA axis activity and associated it with increased disease severity (Melief et al. 2013).

Additionally, in a comparative study between three neurodegenerative disorders (Associated Neurocognitive Disorders, Alzheimer's disease and Multiple Sclerosis), was observed the common overexpression of *BACE2*, gene previously associated to Alzheimer's disease (Holler et al. 2012) that codes for an amyloid-beta peptide (Borjabad and Volsky 2012). In the same study, among the repressed genes were the *GABRG2* (GABA receptor 2), impairing GABAergic neuron signal transmission and memory (Melzer et al. 2012). Observing T cell genes in whole blood of MS patients, Ganghi et al. (2010) showed overexpressed genes in MS patients comparing to control subjects, and most of them was expressed on cells from antigen presenting cell, suggesting that excessive T cell activity as a hallmark of disease.

Brain derived neurotrophic factor (BDNF) was suggested as neuroprotective factor for MS (Frota et al. 2009) and the overexpression of anti-inflammatory pathway, BNDF related neuroprotection, showed by overexpression of BDNF, BDNF upstream activator-TNK and BDNF receptor NTRK3, was demonstrated during acute relapse (Gurevich and Achiron 2012). In addition, some transcript factors were described influencing MS disease, in special the YY1, which is related to processes that affect myelin protein generation (Berndt et al. 2001), immune response process (Guo et al. 2001; Guo et al. 2008) and viral replication (Oh and Broyles 2005) and are involved in differential gene expression in MS patients (Riveros et al. 2010).

Recently, many studies are shown de epigenetic role in the pathophysiology of MS disease. The epigenetic mechanisms can alter gene expression and can also modulate the response to environmental factors, affecting de MS susceptibility. Three principal epigenetic mechanisms include: DNA methylation, histone modifications, and micro-RNA-mediated genetic silencing. Among them, microRNAs have been extensively evaluated for their influence on the manifestation of various autoimmune diseases, including MS. Some of microRNAs were found to be induced in different studies, such as: miR-17-5p, that can act on lipid kinases and regulate the development of lymphocyte (Lindberg et al. 2010); miR-326, which is associated with disease severity in MS patients (Du et al. 2009), miR-214 and miR-23a, that were present in active and inactive MS lesions and in oligodendrocyte differentiation, suggesting their involvement in remyelination (Junker et al. 2009); miR-23a was also overexpressed in PBMCs from relapsing-remitting MS patients (Ridolfi et al. 2013); miR-338, miR-491 and miR-155 (also referred as miR-155-5p) in patients with more advanced stage of MS (Noorbakhsh et al. 2011). The last one, miR-155, was found overexpressed in circulating monocytes from MS patients and in myeloid cells from MS brain lesions (Moore et al. 2013). Still on induced microRNAs in MS patients, miR-145, miR-660 and miR-939 were detected upregulated in PBMCs from patients (Sondergaard et al. 2013); miR-223 from PBMCs (Ridolfi et al. 2013), blood (Keller et al. 2009; Cox et al. 2010) and regulatory T cells (De Santis et al. 2010)

Considering the repressed microRNA described, there are: miR-219 and miR-338-5p in inactive lesions of MS patients, that have target genes responsible for maintain integrity of myelin (Junker et al. 2009); free circulating miR-15b and miR-223, that could implicated in induction of their target genes involved in disease (Fenoglio et al. 2013); members of the mir-29 family in PBMCs from relapsing-remitting MS patients, that were associated with apoptotic processes and IFN feedback loops (Hecker et al. 2013); miR-20a-5p in whole blood of patients (Keller et al. 2014), that target *CDKN1A* gene, which collaborates in T cell activation and has been associated with systemic autoimmunity (Santiago-Raber et al. 2001); miR-17 and miR-20a, which are related with control of immune function, are involved in T cell activation and are implicated in development of MS (Cox et al. 2010).

In conclusion, in this revision we highlightened three representative autoimmune diseases: T1D, SLE and MS. In all of them, the non-genetic factors may have important role in development of the disorders. The discovery of which genes and microRNAs are involved in the development of each one of these pathologies is one of the great challenges for better understanding of their respective causes and course of developments. Thus, new drugs and new diagnostic methods may be determined. Prevention also becomes possible when biomarkers will be known and used frequently, because the environmental factors associated with the development of these diseases can be avoided, or even eliminated. In concert, these approaches may be used to decrease morbidity of individuals with genetic propensity for the development of various autoimmune diseases.

References

Achenbach P, Warncke K, Reiter J et al (2004) Stratification of type 1 diabetes risk on the basis of islet autoantibody characteristics. Diabetes 53:384–392

Baechler EC, Batliwalla FM, Karypis G et al (2003) Interferon-inducible gene expression signature in peripheral blood cells of patients with severe lupus. Poc Natl Acad Sci U S A 100:2610–2615

Barcellos LF, Oksenberg JR, Begovich AB et al (2003) HLA-DR2 dose effect on susceptibility to multiple sclerosis and influence on disease course. Am J Hum Genet 72:710–716

Barrett JC, Clayton DG, Concannon P et al (2009) Genome-wide association study and meta-analysis find that over 40 loci affect risk of type 1 diabetes. Nat Genet 41:703–707

Battaglia M (2014) Neutrophils and type 1 autoimmune diabetes. Curr Opin Hematol 21:8–15

Beall SS, Biddison WE, McFarlin DE et al (1993) Susceptibility for multiple sclerosis is determined, in part, by inheritance of a 175-kb region of the TcR V beta chain locus and HLA class II genes. J Neuroimmunol 45:53–60

Bengtsson AA, Sturfelt G, Truedsson L et al (2000) Activation of type I interferon system in systemic lupus erythematosus correlates with disease activity but not with antiretroviral antibodies. Lupus 9:664–671

Bennett L, Palucka AK, Arce E et al (2003) Interferon and granulopoiesis signatures in systemic lupus erythematosus blood. J Exp Med 197:711–723

Berndt JA, Kim JG, Tosic M et al (2001) The transcriptional regulator Yin Yang 1 activates the myelin PLP gene. J Neurochem 77:935–942

Bertsias GK, Salmon JE, Boumpas DT (2010) Therapeutic opportunities in systemic lupus erythematosus: state of the art and prospects for the new decade. Ann Rheum Dis 69:1603–1611

Bomprezzi R, Ringner M, Kim S et al (2003) Gene expression profile in multiple sclerosis patients and healthy controls: identifying pathways relevant to disease. Hum Mol Genet 12:2191–2199

Borjabad A, Volsky DJ (2012) Common transcriptional signatures in brain tissue from patients with HIV-associated neurocognitive disorders, alzheimer's disease, and multiple sclerosis. J Neuroimmune Pharmacol 7:914–926

Brum DG, Barreira AA, Louzada-Junior P et al (2007) Association of the HLA-DRB1*15 allele group and the DRB1*1501 and DRB1*1503 alleles with multiple sclerosis in White and Mulatto samples from Brazil. J Neuroimmunol 189:118–124

Brynedal B, Khademi M, Wallstrom E et al (2010) Gene expression profiling in multiple sclerosis: a disease of the central nervous system, but with relapses triggered in the periphery? Neurobiol Dis 37:613–621

Burgos P, Metz C, Bull P et al (2000) Increased expression of c-rel, from the NF-KB/Rel family, in T cells from patients with systemic lupus erythematosus. J Rheumatol 27:116–127

Buschard K (2011) What causes type 1 diabetes? Lessons from animal models. APMIS Suppl 119(132):1–19

Carlsen AL, Schetter AJ, Nielsen CT et al (2013) Circulating microRNA expression profiles associated with systemic lupus erythematosus. Arthritis Rheum 65:1324–1334

Chan AC, Iwashima M, Turck CW et al (1992) ZAP-70: a 70 kd protein-tyrosine kinase that associates with the TCR zeta chain. Cell 71:649–662

Chan RW, Lai FM, Li EK et al (2006) Imbalance of Th1/Th2 transcription factors in patients with lupus nephritis. Rheumatology (Oxford) 45:951–957

Chentoufi AA, Binder NR, Berka N et al (2008) Advances in type I diabetes associated tolerance mechanisms. Scand J Immunol 68:1–11

Chitnis T (2007) The role of CD4 T cells in the pathogenesis of multiple sclerosis. Int Rev Neurobiol 79:43–72

Cho JH, Gregersen PK (2011) Genomics and the multifactorial nature of human autoimmune disease. N Engl J Med 365:1612–1623

Chora AA, Fontoura P, Cunha A et al (2007) Heme oxygenase-1 and carbon monoxide suppress autoimmune neuroinflammation. J Clin Invest 117:438–447

Cipolletta C, Ryan KE, Hanna EV et al (2005) Activation of peripheral blood CD14$^+$ monocytes occurs in diabetes. Diabetes 54:2779–2786

Collares CV, Evangelista AF, Xavier DJ et al (2013a) Transcriptome meta-analysis of peripheral lymphomononuclear cells indicates that gestational diabetes is closer to type 1 diabetes than to type 2 diabetes mellitus. Mol Biol Rep 40:5351–5358

Collares CV, Evangelista AF, Xavier DJ et al (2013b) Identifying common and specific microRNAs expressed in peripheral blood mononuclear cell of type 1, type 2, and gestational diabetes mellitus patients. BMC Res Notes 6:491

Colli ML, Moore F, Gurzov EN et al (2010) MDA5 and PTPN2, two candidate genes for type 1 diabetes, modify pancreatic b-cell responses to the viral by-product double-stranded RNA. Hum Mol Genet 19:135–146

Concannon P, Rich SS, Nepom GT (2009) Genetics of type 1A diabetes. N Engl J Med 360:1646–1654

Costa V, Aprile M, Esposito R et al (2013) RNA-Seq and human complex diseases: recent accomplishments and future perspectives. Eur J Hum Genet 21:134–142

Cox MB, Cairns MJ, Gandhi KS et al (2010) MicroRNAs miR-17 and miR-20a inhibit T cell activation genes and are under-expressed in MS whole blood. PLoS One 5:e12132

Crow MK (2008) Collaboration, genetic associations, and lupus erythematosus. N Engl J Med 358:956–961

Crow MK, Kirou KA (2004) Interferon-alpha in systemic lupus erythematosus. Curr Opin Rheumatol 16:541–547

Crow MK, Wohlgemuth J (2003) Microarray analysis of gene expression in lupus. Arthritis Res Ther 5:279–287

Dai Y, Huang YS, Tang M et al (2007) Microarray analysis of microRNA expression in peripheral blood cells of systemic lupus erythematosus patients. Lupus 16:939–946

De Santis G, Ferracin M, Biondani A et al (2010) Altered miRNA expression in T regulatory cells in course of multiple sclerosis. J Neuroimmunol 226:165–171

Deapen D, Escalante A, Weinrib L et al (1992) A revised estimate of twin concordance in systemic lupus erythematosus. Arthritis Rheum 35:311–318

Deng Y, Tsao BP (2010) Genetic susceptibility to systemic lupus erythematosus in the genomic era. Nat Rev Rheumatol 6:683–692

Deng C, Kaplan MJ, Yang J et al (2001) Decreased Ras-mitogen-activated protein kinase signaling may cause DNA hypomethylation in T lymphocytes from lupus patients. Arthritis Rheum 44:397–407

Dib SA (2008) Heterogeneity of type 1 diabetes mellitus. Arq Bras Endocrinol Metabol 52:205–218

Doria A, Canova M, Tonon M et al (2008) Infections as triggers and complications of systemic lupus erythematosus. Autoimmun Rev 8:24–28

Du C, Liu C, Kang J et al (2009) MicroRNA miR-326 regulates TH-17 differentiation and is associated with the pathogenesis of multiple sclerosis. Nat Immunol 10:1252–1259

Dumortier O, Van Obberghen E (2012) MicroRNAs in pancreas development. Diabetes Obes Metab 14(Suppl 3):22–28

Dyment DA, Ebers GC, Sadovnick AD (2004) Genetics of multiple sclerosis. Lancet Neurol 3:104–110

Eizirik DL, Colli ML, Ortis F (2009) The role of inflammation in insulitis and b-cell loss in type 1 diabetes. Nat Rev Endocrinol 5:219–226

Eizirik DL, Sammeth M, Bouckenooghe T et al (2012) The human pancreatic islet transcriptome: expression of candidate genes for type 1 diabetes and the impact of pro-inflammatory cytokines. PLoS Genet 8:e1002552

El Ouaamari A, Baroukh N, Martens GA et al (2008) MiR-375 targets 3'-phosphoinositide-dependent protein kinase-1 and regulates glucose-induced biological responses in pancreatic beta-cells. Diabetes 57:2708–2017

Epplen C, Jackel S, Santos EJ et al (1997) Genetic predisposition to multiple sclerosis as revealed by immunoprinting. Ann Neurol 41:341–352

Erlich H, Valdes AM, Noble J et al (2008) HLA DR-DQ haplotypes and genotypes and type 1 diabetes risk: analysis of the type 1 diabetes genetics consortium families. (Type 1 Diabetes Genetics Consortium). Diabetes 57:1084–1092

Evangelista AF, Collares CVA, Xavier DJ et al (2014) Integrative analysis of the transcriptome profiles observed in type 1, type 2 and gestational diabetes mellitus reveals the role of inflammation. BMC Med Genomics 7:28

Feng X, Wu H, Grossman JM et al (2006) Association of increased interferon-inducible gene expression with disease activity and lupus nephritis in patients with systemic lupus erythematosus. Arthritis Rheum 54:2951–2962

Fenoglio C, Ridolfi E, Cantoni C et al (2013) Decreased circulating miRNA levels in patients with primary progressive multiple sclerosis. Mult Scler 19:1938–1942

Fornari TA, Donate PB, Macedo C et al (2011) Development of type 1 diabetes mellitus in non-obese diabetic mice follows changes in thymocyte and peripheral T lymphocyte transcriptional activity. Clin Dev Immunol 2011:158735

Foster MH, Kelley VR (1999) Lupus nephritis: update on pathogenesis and disease mechanisms. Semin Nephrol 19:173–181

Frost RJ, Olson EN (2011) Control of glucose homeostasis and insulin sensitivity by the Let-7 family of microRNAs. Proc Natl Acad Sci U S A 108:21075–21080

Frota ER, Rodrigues DH, Donadi EA et al (2009) Increased plasma levels of brain derived neurotrophic factor (BDNF) after multiple sclerosis relapse. Neurosci Lett 460:130–132

Furukawa H, Oka S, Matsui T et al (2013) Genome, epigenome and transcriptome analyses of a pair of monozygotic twins discordant for systemic lupus erythematosus. Hum Immunol 74:170–175

Geenen V, Mottet M, Dardenne O et al (2010) Thymic self-antigens for the design of a negative/tolerogenic self-vaccination against type 1 diabetes. Curr Opin Pharmacol 10:461–472

Grammatikos AP, Kyttaris VC, Kis-Toth K et al (2014) A T cell gene expression panel for the diagnosis and monitoring of disease activity in patients with systemic lupus erythematosus. Clin Immunol 150:192–200

Grinberg-Bleyer Y, Baeyens A, You S et al (2010) IL-2 reverses established type 1 diabetes in NOD mice by a local effect on pancreatic regulatory T cells. J Exp Med 207:1871–1878

Guerra SG, Vyse TJ, Cunninghame Graham DS (2012) The genetics of lupus: a functional perspective. Arthritis Res Ther 14:211

Guo J, Casolaro V, Seto E et al (2001) Yin-Yang 1 activates interleukin-4 gene expression in T cells. J Biol Chem 276:48871–48878

Guo J, Lin X, Williams MA et al (2008) Yin-Yang 1 regulates effector cytokine gene expression and T(H)2 immune responses. J Allergy Clin Immunol 122:195–201

Gurevich M, Achiron A (2012) The switch between relapse and remission in multiple sclerosis: continuous inflammatory response balanced by Th1 suppression and neurotrophic factors. J Neuroimmunol 252:83–88

Han GM, Chen SL, Shen N et al (2003) Analysis of gene expression profiles in human systemic lupus erythematosus using oligonucleotide microarray. Genes Immun 4:177–186

Han D, Leyva CA, Matheson D et al (2011) Immune profiling by multiple gene expression analysis in patients at-risk and with type 1 diabetes. Clin Immunol 139:290–301

Hecker M, Thamilarasan M, Koczan D et al (2013) MicroRNA expression changes during interferon-beta treatment in the peripheral blood of multiple sclerosis patients. Int J Mol Sci 14:16087–16110

Hennessy E, Clynes M, Jeppesen PB et al (2010) Identification of microRNAs with a role in glucose stimulated insulin secretion by expression profiling of MIN6 cells. Biochem Biophys Res Commun 396:457–462

Herold KC, Brooks-Worrell B, Palmer J et al (2009) Validity and reproducibility of measurement of islet autoreactivity by T-cell assays in subjects with early type 1 diabetes. Diabetes 58:2588–2595

Holler CJ, Webb RL, Laux AL et al (2012) BACE2 expression increases in human neurodegenerative disease. Am J Pathol 180:337–350

Islam T, Gauderman WJ, Cozen W et al (2006) Differential twin concordance for multiple sclerosis by latitude of birthplace. Ann Neurol 60:56–64

Jarvinen P, Kaprio J, Makitalo R et al (1992) Systemic lupus erythematosus and related systemic diseases in a nationwide twin cohort: an increased prevalence of disease in MZ twins and concordance of disease features. J Intern Med 231:67–72

Jayaraman S, Patel A, Jayaraman A et al (2013) Transcriptome analysis of epigenetically modulated genome indicates signature genes in manifestation of type 1 diabetes and its prevention in NOD mice. PLoS One 8:e55074

Jernas M, Malmeström C, Axelsson M et al (2013) MS risk genes are transcriptionally regulated in CSF leukocytes at relapse. Mult Scler 19:403–410

Junker A, Krumbholz M, Eisele S et al (2009) MicroRNA profiling of multiple sclerosis lesions identifies modulators of the regulatory protein CD47. Brain 132:3342–3352

Kaimen-Maciel DR, Reiche EM, Borelli SD et al (2009) HLA-DRB1* allele-associated genetic susceptibility and protection against multiple sclerosis in Brazilian patients. Mol Med Rep 2:993–998

Kamradt T, Mitchison NA (2001) Tolerance and autoimmunity. N Engl J Med 344:655–664

Karpuj MV, Steinman L, Oksenberg JR (1997) Multiple sclerosis: a polygenic disease involving epistatic interactions, germline rearrangements and environmental effects. Neurogenetics 1:21–28

Karumuthil-Melethil S, Perez N, Li R et al (2008) Induction of innate immune response through TLR2 and dectin 1 prevents type 1 diabetes. J Immunol 181:8323–8334

Kawai T, Akira S (2006) TLR signaling. Cell Death Differ 13:816–825

Keller A, Leidinger P, Lange J et al (2009) Multiple sclerosis: microRNA expression profiles accurately differentiate patients with relapsing remitting disease from healthy controls. PLoS One 4:e7440

Keller A, Leidinger P, Steinmeyer F et al (2014) Comprehensive analysis of microRNA profiles in multiple sclerosis including next-generation sequencing. Mult Scler 20:295–303

Kilpinen H, Dermitzakis ET (2012) Genetic and epigenetic contribution to complex traits. Hum Mol Genet 21:R24–R28

Kirou KA, Lee C, George S et al (2004) Coordinate overexpression of interferon-alpha-induced genes in systemic lupus erythematosus. Arthritis Rheum 50:3958–3967

Kirou KA, Lee C, George S et al (2005) Activation of the interferon-alpha pathway identifies a subgroup of systemic lupus erythematosus patients with distinct serologic features and active disease. Arthritis Rheum 52:1491–1503

Knip M, Siljander H (2008) Autoimmune mechanisms in type 1 diabetes. Autoimmun Rev 7:550–557

Koga M, Kawasaki A, Ito I et al (2011) Cumulative association of eight susceptibility genes with systemic lupus erythematosus in a Japanese female population. J Hum Genet 56(7):503–507

Krützfeldt J, Stoffel M (2006) MicroRNAs: a new class of regulatory genes affecting metabolism. Cell Metab 4:9–12

Kuo CC, Lin SC (2007) Altered FOXO1 transcript levels in peripheral blood mononuclear cells of systemic lupus erythematosus and rheumatoid arthritis patients. Mol Med 13:561–566

Laplaud DA, Ruiz C, Wiertlewski S et al (2004) Blood T-cell receptor beta chain transcriptome in multiple sclerosis. Characterization of the T cells with altered CDR3 length distribution. Brain 127:981–995

Lashine YA, Seoudi AM, Salah S et al (2011) Expression signature of microRNA-181-a reveals its crucial role in the pathogenesis of paediatric systemic lupus erythematosus. Clin Exp Rheumatol 29:351–357

Layer K, Lin G, Nencioni A et al (2003) Autoimmunity as the consequence of a spontaneous mutation in Rasgrp1. Immunity 19:243–255

Lee HM, Sugino H, Aoki C et al (2011) Underexpression of mitochondrial-DNA encoded ATP synthesis-related genes and DNA repair genes in systemic lupus erythematosus. Arthritis Res Ther 13:R63

Li R, Perez N, Karumuthil-Melethil S et al (2007) Bone marrow is a preferential homing site for autoreactive T-cells in type 1 diabetes. Diabetes 56:2251–2259

Li QZ, Zhou J, Lian Y et al (2010) Interferon signature gene expression is correlated with autoantibody profiles in patients with incomplete lupus syndromes. Clin Exp Immunol 159:281–291

Lindberg RL, Hoffmann F, Mehling M et al (2010) Altered expression of miR-17–5p in CD4+ lymphocytes of relapsing-remitting multiple sclerosis patients. Eur J Immunol 40:888–898

Lindén M, Khademi M, Lima Bomfim I et al (2013) Multiple sclerosis risk genotypes correlate with an elevated cerebrospinal fluid level of the suggested prognostic marker CXCL13. Mult Scler 19:863–870

Lit LC, Wong CK, Li EK et al (2007) Elevated gene expression of Th1/Th2 associated transcription factors is correlated with disease activity in patients with systemic lupus erythematosus. J Rheumatol 34:89–96

Liu Z, Davidson A (2012) Taming lupus—a new understanding of pathogenesis is leading to clinical advances. Nat Med 18:871–882

Lood C, Amisten S, Gullstrand B et al (2010) Platelet transcriptional profile and protein expression in patients with systemic lupus erythematosus: up-regulation of the type I interferon system is strongly associated with vascular disease. Blood 116:1951–1957

Lu MC, Lai NS, Chen HC et al (2013) Decreased microRNA(miR)-145 and increased miR-224 expression in T cells from patients with systemic lupus erythematosus involved in lupus immunopathogenesis. Clin Exp Immunol 171:91–99

Lublin FD, Reingold SC (1996) Defining the clinical course of multiple sclerosis: results of an international survey. Neurology 46:907–911

Lucchinetti CF, Brueck W, Rodriguez M et al (1998) Multiple sclerosis: lessons from neuropathology. Semin Neurol 18:337–349

Lyons PA, McKinney EF, Rayner TF et al (2010) Novel expression signatures identified by transcriptional analysis of separated leukocyte subsets in systemic lupus erythematosus and vasculitis. Ann Rheum Dis 69:1208–1213

Maas K, Chen H, Shyr Yu et al (2005) Shared gene expression profiles in individuals with autoimmune disease and unaffected first-degree relatives of individuals with autoimmune disease. Hum Mol Genet 14:1305–1314

Mackay IR (2009) Clustering and commonalities among autoimmune diseases. J Autoimmun 33:170–177

Mattick JS, Makunin IV (2006) Non-coding RNA. Hum Mol Genet 15(Spec No 1):R17–R29

Mattick JS, Taft RJ, Faulkner GJ (2010) A global view of genomic information–moving beyond the gene and the master regulator. Trends Genet 26:21–28

Matzinger P (2002) The danger model: a renewed sense of self. Science 296:301–305

McDevitt HO, Unanue ER (2008) Autoimmune diabetes mellitus–much progress, but many challenges. Adv Immunol 100:1–12

Melief J, de Wit SJ, Van Eden CG et al (2013) HPA axis activity in multiple sclerosis correlates with disease severity, lesion type and gene expression in normal-appearing white matter. Acta Neuropathol 126:237–249

Melzer S, Michael M, Caputi A et al (2012) Long-range-projecting GABAergic neurons modulate inhibition in hippocampus and entorhinal cortex. Science 335:1506–1510

Moore F, Colli ML, Cnop M et al (2009) PTPN2, a candidate gene for type 1 diabetes, modulates interferon-gamma-induced pancreatic b-cell apoptosis. Diabetes 58:1283–1291

Moore CS, Rao VT, Durafourt BA et al (2013) miR-155 as a multiple sclerosis-relevant regulator of myeloid cell polarization. Ann Neurol 74:709–720

Moser KL, Kelly JA, Lessard CJ et al (2009) Recent insights into the genetic basis of systemic lupus erythematosus. Genes Immun 10:373–379

Mumford CJ, Wood NW, Kellar-Wood H et al (1994) The British Isles survey of multiple sclerosis in twins. Neurology 44:11–15

Nakou M, Knowlton N, Frank MB et al (2008) Gene expression in systemic lupus erythematosus: bone marrow analysis differentiates active from inactive disease and reveals apoptosis and granulopoiesis signatures. Arthritis Rheum 58(11):3541–3549

Nakou M, Bertsias G, Stagakis I et al (2010) Gene network analysis of bone marrow mononuclear cells reveals activation of multiple kinase pathways in human systemic lupus erythematosus. PLoS One 5:e13351

Nerup J, Nierras C, Plagnol V et al (2009) Genome-wide association study and meta-analysis find that over 40 loci affect risk of type 1 diabetes. Nat Genet 41:703–707

Nielsen NM, Westergaard T, Rostgaard K et al (2005) Familial risk of multiple sclerosis: a nationwide cohort study. Am J Epidemiol 162:774–778

Niewold TB, Swedler WI (2005) Systemic lupus erythematosus arising during interferon-alpha therapy for cryoglobulinemic vasculitis associated with hepatitis C. Clin Rheumatol 24:178–181

Niewold TB, Hua J, Lehman TJ et al (2007) High serum IFN-alpha activity is a heritable risk factor for systemic lupus erythematosus. Genes Immun 8:492–502

Noorbakhsh F, Ellestad KK, Maingat F et al (2011) Impaired neurosteroid synthesis in multiple sclerosis. Brain 134:2703–2721

Noseworthy JH, Lucchinetti C, Rodriguez M et al (2000) Multiple sclerosis. N Engl J Med 343:938–952

Novak J, Lehuen A (2011) Mechanism of regulation of autoimmunity by iNKT cells. Cytokine 53:263–270

Oh J, Broyles SS (2005) Host cell nuclear proteins are recruited to cytoplasmic vaccinia virus replication complexes. J Virol 79:12852–12860

Oksenberg JR, Baranzini SE, Sawcer S et al (2008) The genetics of multiple sclerosis: SNPs to pathways to pathogenesis. Nature Rev Genet 9:516–526

Pan W, Zhu S, Yuan M et al (2010) MicroRNA-21 and microRNA-148a contribute to DNA hypomethylation in lupus CD4$^+$ T cells by directly and indirectly targeting DNA methyltransferase 1. J Immunol 184:6773–6781

Petri M (1995) Clinical features of systemic lupus erythematosus. Curr Opin Rheumatol 7:395–401

Plagnol V, Howson JM, Smyth DJ et al (2011) Genome-wide association analysis of autoantibody positivity in type 1 diabetes cases. PLoS Genet 7:e1002216

Plaisance V, Abderrahmani A, Perret-Menoud V et al (2006) MicroRNA-9 controls the expression of granuphilin/Slp4 and the secretory response of insulin-producing cells. J Biol Chem 281:26932–26942

Planas R, Pujol-Borrell R, Vives-Pi M (2010) Global gene expression changes in type 1 diabetes: insights into autoimmune response in the target organ and in the periphery. Immunol Lett 133:55–61

Pociot F, Akolkar B, Concannon P et al (2010) Genetics of type 1 diabetes: what's next? Diabetes 59:1561–1571

Ponting CP, Oliver PL, Reik W (2009) Evolution and functions of long noncoding RNAs. Cell 136:629–641

Poo MM (2001) Neurotrophins as synaptic modulators. Nat Rev Neurosci 2:24–32

Poy MN, Eliasson L, Krutzfeldt J et al (2004) A pancreatic islet-specific microRNA regulates insulin secretion. Nature 432:226–230

Pugliese A, Miceli D (2002) The insulin gene in diabetes. Diabetes Metab Res Rev 18:13–25

Purohit S, She JX (2008) Biomarkers for type 1 diabetes. Int J Clin Exp Med 1:98–116

Qin H, Zhu X, Liang J et al (2013) MicroRNA-29b contributes to DNA hypomethylation of CD4$^+$ T cells in systemic lupus erythematosus by indirectly targeting DNA methyltransferase 1. J Dermatol Sci 69:61–67

Rassi DM, Junta CM, Fachin AL et al (2008) Gene expression profiles stratified according to type 1 diabetes mellitus susceptibility regions. Ann N Y Acad Sci 1150:282–289

Remoli ME, Ragimbeau J, Giacomini E et al (2007) NF-{kappa}B is required for STAT-4 expression during dendritic cell maturation. J Leukoc Biol 81:355–363

Reynier F, Pachot A, Paye M et al (2010) Specific gene expression signature associated with development of autoimmune type-I diabetes using whole-blood microarray analysis. Genes Immun 11:269–278

Ridolfi E, Fenoglio C, Cantoni C et al (2013) Expression and genetic analysis of MicroRNAs involved in multiple sclerosis. Int J Mol Sci 14:4375–4384

Riveros C, Mellor D, Gandhi KS et al (2010) A transcription factor map as revealed by a genome-wide gene expression analysis of whole-blood mRNA transcriptome in multiple sclerosis. PLoS One 5:e14176

Ronnblom LE, Alm GV, Oberg KE (1990) Possible induction of systemic lupus erythematosus by interferon-alpha treatment in a patient with a malignant carcinoid tumour. J Intern Med 227:207–210

Rus V, Chen H, Zernetkina V et al (2004) Gene expression profiling in peripheral blood mononuclear cells from lupus patients with active and inactive disease. Clin Immunol 112:231–234

Sadovnick AD, Ebers GC (1993) Epidemiology of multiple sclerosis: a critical overview. Can J Neurol Sci 20:17–29

Sadovnick AD, Ebers GC, Dyment DA et al (1996) Evidence for genetic basis of multiple sclerosis. The Canadian Collaborative Study Group. Lancet 347:1728–1730

Santiago-Raber ML, Lawson BR, Dummer W et al (2001) Role of cyclin kinase inhibitor p21 in systemic autoimmunity. J Immunol 167:4067–4074

Santin I, Moore F, Colli ML et al (2011) PTPN2, a candidate gene for type 1 diabetes, modulates pancreatic β-cell apoptosis via regulation of the BH3-only protein Bim. Diabetes 60:3279–3288

Satoh J, Misawa T, Tabunoki H et al (2008) Molecular network analysis of T-cell transcriptome suggests aberrant regulation of gene expression by NF-kappaB as a biomarker for relapse of multiple sclerosis. Dis Markers 25:27–35

Schwartzman-Morris J, Putterman C (2012) Gender differences in the pathogenesis and outcome of lupus and of lupus nephritis. Clin Dev Immunol 2012:604892

Shaw JE, Sicree RA, Zimmet PZ (2010) Global estimates of the prevalence of diabetes for 2010 and 2030. Diabetes Res Clin Pract 87:4–14

Sia C (2006) Replenishing peripheral CD4$^+$ regulatory T cells: a possible immune-intervention strategy in type 1 diabetes? Rev Diabet Stud 3:102–107

Sondergaard HB, Hesse D, Krakauer M et al (2013) Differential microRNA expression in blood in multiple sclerosis. Mult Scler 19:1849–1857

Stagakis E, Bertsias G, Verginis P et al (2011) Identification of novel microRNA signatures linked to human lupus disease activity and pathogenesis: miR-21 regulates aberrant T cell responses through regulation of PDCD4 expression. Ann Rheum Dis 70:1496–1506

Stone RC, Du P, Feng D et al (2013) RNA-Seq for enrichment and analysis of IRF5 transcript expression in SLE. PLoS One 8:e54487

Straub RH (2007) The complex role of estrogens in inflammation. Endocr Rev 28:521–574

Sui WG, Lin H, Chen JJ et al (2012) Comprehensive analysis of transcription factor expression patterns in peripheral blood mononuclear cell of systemic lupus erythematosus. Int J Rheum Dis 15:212–219

Tajouri L, Fernandez F, Griffiths LR (2007) Gene expression studies in multiple sclerosis. Curr Genomics 8:181–189

Tang X, Muniappan L, Tang G et al (2009a) Identification of glucose-regulated miRNAs from pancreatic {beta} cells reveals a role for miR-30d in insulin transcription. RNA 15:287–293

Tang Y, Luo X, Cui H et al (2009b) MicroRNA-146A contributes to abnormal activation of the type I interferon pathway in human lupus by targeting the key signaling proteins. Arthritis Rheum 60:1065–1075

Teruel R, Corral J, Pérez-Andreu V et al (2011) Potential role of miRNAs in developmental haemostasis. PLoS One 6:e17648

Tisch R, Wang B (2008) Dysrulation of T cell peripheral tolerance in type 1 diabetes. Adv Immunol 100:125–149

Todd JA (2010) Etiology of type 1 diabetes. Immunity 32:457–467

Toukap AN, Galant C, Theate I et al (2007) Identification of distinct gene expression profiles in the synovium of patients with systemic lupus erythematosus. Arthritis Rheum 56:1579–1588

Voulgarelis M, Giannouli S, Tasidou A et al (2006) Bone marrow histological findings in systemic lupus erythematosus with hematological abnormalities: a clinicopathological study. Am J Hematol 81:590–597

Vukkadapu SS, Belli JM, Ishii K et al (2005) Dynamic. interaction between T cell-mediated beta-cell damage and beta-cell repair in the. run up to autoimmune diabetes of the NOD mouse. Physiol Genomics 21:201–211

Wang G, Tam LS, Li EK et al (2010) Serum and urinary cell-free MiR-146a and MiR-155 in patients with systemic lupus erythematosus. J Rheumatol 37:2516–2522

Weckerle CE, Niewold TB (2011) The unexplained female predominance of systemic lupus erythematosus: clues from genetic and cytokine studies. Clin Rev Allergy Immunol 40:42–49

Weckerle CE, Franek BS, Kelly JA et al (2011) Network analysis of associations between serum interferon-alpha activity, autoantibodies, and clinical features in systemic lupus erythematosus. Arthritis Rheum 63:1044–1053

Weinshenker BG (1994) Natural history of multiple sclerosis. Ann Neurol 36:S6–S11

Wenzlau JM, Juhl K, Yu L et al (2007) The cation efflux transporter ZnT8 (Slc30A8) is a major autoantigen in human type 1 diabetes. Proc Natl Acad Sci U S A 104:17040–17045

Willer CJ, Dyment DA, Risch NJ et al (2003) Twin concordance and sibling recurrence rates in multiple sclerosis. Proc Natl Acad Sci U S A 100:12877–12882

Yao Y, Higgs BW, Morehouse C et al (2009) Development of potential pharmacodynamic and diagnostic markers for anti-IFN-α monoclonal antibody trials in systemic lupus erythematosus. Hum Genomics Proteomics pii: 374312

Yuan Y, Kasar S, Underbayev C et al (2012) Role of microRNA-15a in autoantibody production in interferon-augmented murine model of lupus. Mol Immunol 52:61–70

Zhao X, Tang Y, Qu B et al (2010) MicroRNA-125a contributes to elevated inflammatory chemokine RANTES levels via targeting KLF13 in systemic lupus erythematosus. Arthritis Rheum 62:3425–3435

Zhao S, Wang Y, Liang Y et al (2011) MicroRNA-126 regulates DNA methylation in CD4+ T cells and contributes to systemic lupus erythematosus by targeting DNA methyl-transferase 1. Arthritis Rheum 63:1376–1386

Zhernakova A, van Diemen CC, Wijmenga C (2009) Detecting shared pathogenesis from the shared genetics of immune-related diseases. Nat Rev Genet 10:43–55

Chapter 9
Expression of DNA Repair and Response to Oxidative Stress Genes in Diabetes Mellitus

Paula Takahashi, Danilo J. Xavier and Elza T. Sakamoto-Hojo

Abstract Diabetes Mellitus (DM) is a group of chronic metabolic diseases that arises from a deficiency in insulin secretion and/or action, resulting in hyperglycemia. The two main categories of DM are type 1 diabetes mellitus (T1DM) and type 2 diabetes mellitus (T2DM). An interplay between oxidative stress and both T1DM and T2DM has been observed, with evidence indicating that oxidative stress can be the cause and also a consequence of both types of DM. In fact, a number of studies has detected elevated levels of oxidative stress markers and DNA damage (a consequence of oxidative stress), as well as an impaired antioxidant system in patients suffering from T1DM or T2DM. Accordingly, several works have identified differentially expressed genes that are associated with responses to oxidative stress and DNA damage in T1DM as well as in T2DM patients. In addition, a set of microRNAs that has been previously shown to clearly distinguish T1DM patients from healthy subjects potentially targets a plethora of genes involved in DNA repair and response to oxidative stress. Collectively, these studies indicate that patients with DM present changes in the gene expression profiles as a response to the insults to which they are subjected as part of the development and/or as a consequence of the disease.

9.1 Introduction

9.1.1 Reactive Oxygen Species: Definition and Consequences

Reactive oxygen species (ROS) are reactive small molecules that contain an oxygen atom in their structure (Lenzen 2008, Halliwell and Gutteridge 2007). These small

E. T. Sakamoto-Hojo (✉) · P. Takahashi · D. J. Xavier
Department of Genetics, Ribeirão Preto Medical School,
University of São Paulo, Ribeirão Preto, São Paulo, Brazil
e-mail: etshojo@usp.br

E. T. Sakamoto-Hojo
Department of Biology, Faculty of Philosophy, Sciences and Letters of Ribeirão Preto,
University of São Paulo, Ribeirão Preto, São Paulo, Brazil

© Springer International Publishing Switzerland 2014
G. A. Passos (ed.), *Transcriptomics in Health and Disease*,
DOI 10.1007/978-3-319-11985-4_9

molecules can be free radicals with an unpaired electron, such as superoxide radical ($O_2^{\cdot-}$) and hydroxyl radical ($OH^{\cdot}$); non-radicals, such as hydrogen peroxide (H_2O_2); anions, including superoxide (O_2^-) and peroxynitrite ($ONOO^-$); non-ions, including H_2O_2 and $OH^{\cdot}$. The reactivity of all these different reactive species varies, with $OH^{\cdot}$ being the most reactive oxygen radical (Lenzen 2008). Besides exogenous sources, including smoke, air pollutants, ultraviolet radiation, γ-irradiation, and many drugs, ROS can also be derived from several endogenous sources, such as NADPH oxidases (NOXs), the mitochondrial respiratory chain, xanthine oxidase, lipoxygenases, cyclooxygenases, cytochrome P450 enzymes, nitric oxide synthases, among others (Nathan and Cunningham-Bussel 2013, Jiang et al. 2011).

In normal conditions, ROS play a role in physiological processes that trigger adequate cellular responses (Rains and Jain 2011). ROS provide protection to the host by killing invading pathogens, as well as act as cellular messengers in a network of intra and intercellular communication pathways (Kalyanaraman 2013, Edeas et al. 2010). In contrast, an excessive amount of ROS can be detrimental to the cells. To counteract ROS, cells are equipped with a number of enzymatic and non-enzymatic mechanisms as well as with an adaptive mechanism that results in the expression of antioxidant genes. Concerning the enzymatic machinery, superoxide dismutases (SODs) are involved in the normal dismutation of superoxide and there are at least three types of SODs: cytosolic copper/zinc SOD, mitochondrial manganese SOD, and extracellular SOD (Kalyanaraman 2013). H_2O_2 is detoxified by catalase (CAT) mainly when its levels are low and by glutathione peroxidase enzyme (GPx) when its levels are higher. GPx also metabolizes other lipid peroxides (LOOH) (Kalyanaraman 2013). Regarding the non-enzymatic detoxification mechanisms, the small molecular weight antioxidants include ascorbic acid (vitamin C), α-tocopherol (vitamin E), reduced glutathione (GSH), β-carotene, among others. While vitamin C reacts rapidly with several ROS, such as superoxide and hydroxyl radical, vitamin E can halt lipid peroxidation (Kalyanaraman 2013). The constant exposure to a mild amount of oxidants triggers the elevated production of antioxidant enzymes and this intrinsic mechanism ultimately leads to the re-establishment of the cellular oxidant/antioxidant homeostasis. This adaptive mechanism consists of the nuclear factor erythroid 2-related factor 2 (Nrf2) binding to DNA sequences present in antioxidant response elements (ARE) and inducing the transcription of antioxidant genes, including, but not limited to SOD, GPx, and CAT (Kalyanaraman 2013).

However, a state known as oxidative stress, in which the levels of prooxidants overcome those of antioxidants, can occur. Oxidative stress can be a consequence of reduced concentrations of antioxidants or antioxidant enzymes or impaired adaptive mechanism, and it can also be a consequence of elevated generation of ROS (Kalyanaraman 2013). Oxidative stress can cause oxidative damage to DNA (the focus of this chapter); it is important to note that proteins, lipids, and carbohydrates are also subjected to oxidative damage (Storr et al. 2013, Kalyanaraman 2013).

ROS can react with both purines and pyrimidines of DNA, generating a number of DNA base products, and because guanine exhibits a low redox potential, it is preferentially oxidized, with 7,8-dihydro-8-oxoguanine (8-oxoguanine; 8-oxoG) being the most extensively studied DNA lesion (Dizdaroglu 2012, Storr et al. 2013). ROS can also cause DNA intrastrand and interstrand crosslinks, DNA-protein crosslinks, DNA single and double strand breaks (SSB and DSB, respectively), as well as damage to the sugar moiety of DNA (Dizdaroglu 2012, Storr et al. 2013). DNA damage can be mutagenic, hence, perturbing maintenance of genomic stability. Thus, repair pathways have evolved to remove oxidative DNA lesions and restore DNA structure, protecting cells from such a harmful condition (Dizdaroglu 2012, Storr et al. 2013). Base-excision repair (BER) and nucleotide-excision repair (NER) are the two key pathways in repairing oxidative DNA damage (Dizdaroglu 2012, Friedberg et al. 2006). However, there exist other mechanisms: mismatch repair (MMR), a repair pathway in the cellular nucleotide pool, homologous recombination (HR), and non-homologous end-joining (NHEJ) (Dizdaroglu 2012).

The ROS pathway, from the exogenous and endogenous sources to the repair pathways involved in removing the DNA lesions caused by ROS, is better depicted in Fig. 9.1

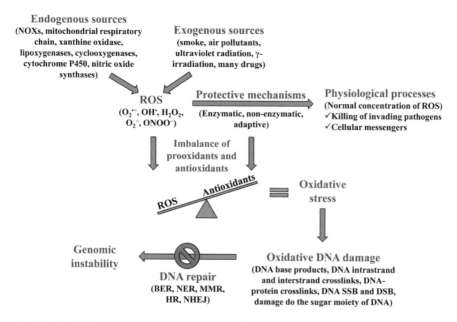

Fig. 9.1 The ROS pathway. ROS can be generated by both endogenous and exogenous sources. Protective mechanisms exist to maintain ROS homeostasis. However, oxidative stress, which is the imbalance between the levels of ROS and those of antioxidants, favoring the former, may occur. Oxidative stress can damage DNA, generating several types of lesions, which, in turn, can lead to genomic instability. Because of that, cells are equipped with many DNA repair mechanisms. *NOXs*, NADPH oxidases; *ROS* reactive oxygen species; $O_2^{\bullet-}$ superoxide radical; OH$^\bullet$ hydroxyl radical; H_2O_2 hydrogen peroxide; O_2^- superoxide; ONOO$^-$ peroxynitrite; *SSB* single strand break; *DSB* double strand break; *BER* base-excision repair; *NER* nucleotide-excision repair; *MMR* mismatch repair; *HR* homologous recombination; *NHEJ* non-homologous end-joining

9.1.2 Diabetes Mellitus and ROS

Diabetes Mellitus (DM) is a group of chronic metabolic diseases that arises from a deficiency in insulin secretion and/or action, which, in turn, leads to chronic high blood glucose levels or hyperglycemia. The latter has been implicated in long-term complications involving a variety of organs, including kidneys, eyes, heart, nerves, and blood vessels (ADA 2013). Individuals are diagnosed with diabetes when displaying one of the following: glycated hemoglobin levels (HbA1C)$\geq$6.5%, fasting plasma glucose levels (FPG)$\geq$126 mg/dL (7.0 mmol/L), 2-h plasma glucose levels after 75 g glucose load$\geq$200 mg/dL (11.1 mmol/L) (in the absence of unequivocal hyperglycemia, these three parameters should be confirmed by retaking the test), or for individuals with classic hyperglycemic symptoms/hyperglycemic crisis, casual plasma glucose levels$\geq$200 mg/dL (11.1 mmol/L) (ADA 2013). According to the International Diabetes Federation, there were approximately 382 million people between the ages of 20 and 79 years with diabetes worldwide in 2013 (http://www.idf.org/diabetesatlas) (IDF 2013). Hence, DM represents a relevant public health issue.

The two major forms of DM are type 1 diabetes mellitus (T1DM) and type 2 diabetes mellitus (T2DM). There is evidence of an association between oxidative stress and both types of DM: oxidative stress can be a consequence of these disorders due to hyperglycemia, but it can also be a contributing factor to the pathogenesis of both diseases because reactive molecules play a crucial role in pancreatic β-cell damage. Hyperglycemia may lead to increased oxidative stress by the direct production of ROS or by changes in the redox homeostasis through the disruption of a variety of mechanisms: elevated polyol pathway flux, higher intracellular production of advanced glycation end-products (AGEs), activation of protein kinase C, or even increased generation of superoxide by the mitochondrial electron transport chain (Rains and Jain 2011; Brownlee 2001; Ahmad et al. 2005). On the other hand, in the case of T1DM, the invading immune cells release pro-inflammatory cytokines into the target β-cells and those cytokines, in turn, increase the production of reactive species, which leads to β-cell destruction (Lenzen 2008). Regarding T2DM, β-cell failure is the major contributing factor to its pathogenesis. In this type of diabetes, glucotoxicity (which includes ROS generation) and lipotoxicity are involved in the β-cell dysfunction as these cells are subjected to longstanding exposure to elevated levels of glucose and free fatty acids (Drews et al. 2010). In addition, according to some studies, low levels of GPx or CAT protein and activity have been observed in human islets (Drews et al. 2010; Tonooka et al. 2007, Robertson and Harmon 2007), rendering these cells more vulnerable to reactive species.

Taken together, these studies have indicated a crosstalk between the two major forms of DM and oxidative stress (which is better depicted in Fig. 9.2), which, in turn, can trigger DNA damage. Thus, it is reasonable that in order to cope with the disease and its consequences, patients suffering from DM are responding to all these insults in several ways, including via alterations in the gene expression profiles.

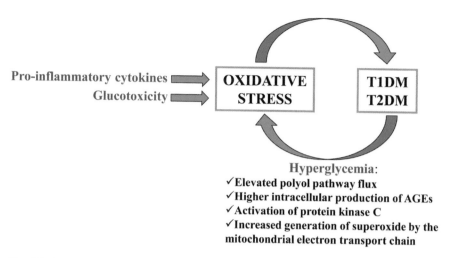

Fig. 9.2 Crosstalk between oxidative stress and the two main categories of Diabetes Mellitus. Oxidative stress may be a contributing factor to diabetes, but it can also be a consequence of the latter. T1DM, Type 1 Diabetes Mellitus; T2DM, Type 2 Diabetes Mellitus; AGEs, advanced glycation end-products

9.2   Type 1 Diabetes Mellitus

T1DM is a consequence of the autoimmune elimination of the insulin-producing pancreatic β-cells, which eventually ceases insulin production and hence the absorption of glucose by the tissues of the body (ADA 2013; Wållberg and Cooke 2013). Approximately 5–10 % of all diabetic patients present T1DM, which can occur even in the elderly, although it usually arises during childhood and adolescence (ADA 2013). For the development of T1DM, an immune response with strong pro-inflammatory features against the pancreatic β-cell antigens must arise and the control of the autoimmune responses must be impaired so those responses can become chronic, leading to the elimination of β-cells (Wållberg and Cooke 2013).

The exact cause of this form of diabetes has not been elucidated, but it has been suggested that several susceptibility genes combined with environmental insults contribute to the development of this disease (ADA 2013; Wållberg and Cooke 2013). Regarding the genetic background, the HLA genes located on chromosome 6p21 confer the highest genetic risk for T1DM; other genes including those encoding insulin (*INS*) on 11p15, cytotoxic T-lymphocyte-associated protein 4 (*CTLA4*) on 2q33, protein tyrosine phosphatase, non-receptor type 22 (*PTPN22*) on 1p13, and interleukin 2 receptor alpha (*IL2RA*) on 10p15, also present strong associations with the disorder (Burren et al. 2011, Barrett et al. 2009, Bell et al. 1984, Nisticò et al. 1996, Bottini et al. 2004, Lowe et al. 2007). With respect to environmental factors, both the presence and the absence of infections, as well as climate and diet have been suggested to contribute to T1DM onset (Wållberg and Cooke 2013, von Herrath 2009, Zaccone and Cooke 2011, Cooper et al. 2011, Zipitis and Akobeng 2008).

9.2.1 Oxidative Stress and DNA Damage in T1DM

Several studies have investigated the levels of antioxidants, markers of oxidative stress, and DNA damage in patients suffering from T1DM relative to healthy subjects. Codoñer-Franch and colleagues (2010) found significantly elevated levels of the three oxidative stress markers (circulating levels of lipoperoxides (LPO) and malondialdehyde (MDA) and plasma concentration of carbonyl groups (CG)), a slightly decreased erythrocyte GPx activity, and a significant decrease in α-tocopherol/total cholesterol ratio in T1DM children with good glycemic control *versus* age-matched control subjects. On the contrary, the same authors did not observe differences in the erythrocyte concentration of GPx's cofactor GSH or in the serum levels of β-carotene. Moreover, another work has demonstrated reduced SOD and GPx activities in leukocytes from men and women with T1DM in comparison with their corresponding controls (Dinçer et al. 2003). In the same work, the authors reported that strand breakage and formamidopyrimidine DNA glycosylase (Fpg)-sensitive sites (oxidised DNA damage detected by the comet assay with the DNA repair enzyme Fpg) were elevated in leukocytes of the two groups of patients. Furthermore, according to Goodarzi et al. (2010), plasma MDA and glycated serum protein (GSP) levels were significantly increased in T1DM patients relative to healthy subjects. In agreement with Hata et al. (2006), the previously mentioned study also observed that urinary concentrations of the oxidative DNA damage marker 8-hydroxydeoxyguanosine (8-OHdG) were significantly higher in T1DM patients when compared with the control group. Likewise, significantly elevated levels of both nuclear DNA fragmentation and concentrations of 8-OHdG have also been reported in spermatozoa from T1DM patients relative to non-diabetic fertile men (Agbaje et al. 2008). Regarding basal levels of DNA damage, significantly elevated rates were found in neutrophils from T1DM patients presenting acceptable glycemic control relative to age- and sex-matched controls (Hannon-Fletcher et al. 2000). Collectively, these studies suggest the impairment of the antioxidant defense system and an increase in oxidative stress and DNA damage in T1DM patients, even in subjects presenting a satisfactory glycemic control.

Results from another study indicated that serum copper-to-zinc ratio (reduced levels of zinc and elevated levels of copper indicate the presence of oxidative stress), serum SOD activity, blood and urinary MDA, as well as urinary 8-OHdG were significantly higher in young T1DM patients (in particular in poorly-controlled patients ($HbA1c \geq 9\%$) compared with the control group (Lin et al. 2014). It is important to note that the findings regarding SOD activity contradict the aforementioned results from the work of Dinçer et al. (2003). A plausible explanation given by Lin et al. (2014) for the increased SOD activity in T1DM patients observed in their work would be that it might compensate for the oxidative stress in these individuals. Interestingly, the levels of serum copper, serum copper-to-zinc ratio, urinary MDA, and urinary 8-OHdG were significantly elevated in the poorly-controlled patients relative to the optimal-and-suboptimal-glycemic-control patients ($HbA1c < 9\%$) (Lin et al. 2014). Taken together, these results corroborate the previously cited works (except for the

SOD activity findings) and also indicate that poor glycemic control is associated with augmented oxidative stress and DNA damage in T1DM patients.

9.2.2   Transcriptional Expression of Protein-Coding Genes and MicroRNAs Related to Oxidative Stress and DNA Repair in T1DM

A number of large-scale transcriptional profiling studies has been performed to compare gene expression displayed by T1DM patients relative to healthy subjects by carrying out microarray experiments. A study investigated gene expression profiles of endothelial progenitor cells (EPC), which were *in vitro* differentiated from peripheral blood mononuclear cells (PBMCs), from T1DM patients pre- and post-supplementation with folic acid (FA, a B-vitamin with antioxidant properties) and non-diabetic individuals (van Oostrom et al. 2009). The 1591 genes found to be differentially expressed between T1DM patients pre-FA treatment and the control group were classified into Gene Ontology (GO) terms, including response to stress' and 'response to hypoxia'. Among the up-regulated genes (related to these two terms) detected in EPC from T1DM patients were dual oxidase 2 (*DUOX2*), a NADPH oxidase that can produce superoxide, nitric oxide synthase 2A (*NOS2A*) that is capable of generating nitric oxide, thioredoxin reductase 2 (*TXNRD2*), a major enzyme involved in the control of the intracellular redox balance, lactoperoxidase (*LPO*) and NADPH oxidase organizer 1 (*NOXO1*), which is associated with the generation of ROS. Importantly, FA treatment notably affected the EPC transcriptome, resulting in normalization of gene expression in diabetic EPC to levels similar to those exhibited by healthy individuals. In fact, 513 of the 1591 genes found differentially expressed in the patients *versus* the control group returned to control levels after FA supplementation. As expected, FA altered the expression of oxidative stress-associated genes in EPC, with four (*DUOX2, NOS2A, NOXO1* and *LPO*) being included among the 513 normalized genes. In addition, another differentially expressed gene (down-regulated) in T1DM patients that was normalized by FA treatment was the transcription factor (TF) V-maf musculoaponeurotic fibrosarcoma oncogene homolog F (*MAFF*) (van Oostrom et al. 2009). This TF can bind to another TF, Nrf2, which, in turn, induces the expression of ARE-dependent genes (Katsuoka et al. 2005, Blank 2008).

Irvine et al. (2012) investigated whether there were differences in gene expression of purified peripheral blood CD14[+]monocytes between recently diagnosed T1DM children and adult healthy controls by whole-genome microarrays, followed by validation of a group of genes by quantitative polymerase chain reaction (qPCR). Results indicated that the monocyte expression profiles exhibited by the patients clustered into two subgroups, with one of them (group B) clustering separate from the other patient subgroup and the healthy controls. At diagnosis, both subgroups of patients were clinically identical, however, group B presented increased levels of HbA1c 3 and 6 months after diagnosis and needed significantly higher insulin doses during the first year of the disease. Expression profiles in monocytes from patients belonging to group B indicated cellular activation through stress, including the unfolded protein response (UPR), which results from endoplasmic reticulum (ER) stress (*IRE1*, *GRP78*, *DDIT3*, *XBP1*). Furthermore, *HIF1A*, a major mediator of oxidative stress, and several of its targets (*DDIT4*, *PFKFB3*, and *ADM*) (Ruiz-García et al. 2011; Geiger et al. 2011) were up-regulated in group B monocytes, while genes that play a role in mitochondrial oxidative phosphorylation (*PDHB*, *MDH1*, *IDH1*, *SDHC*, *ACLY*) were found repressed. Moreover, mitochondrion was the most significantly enriched cellular component term for the down-regulated genes in group B. In addition, repression of many genes related to cellular anti-oxidant pathways (*CAT*, *G6PD*, *OXR1*, *PRDX1*, *PRDX3*) were observed in group B monocytes, indicating perturbation of protective systems (Irvine et al. 2012). The biological processes oxidative and ER stresses are closely associated. Oxidative stress can promote ER stress, and in response to that, ER activates the UPR transcriptional program (Martinon and Glimcher 2011). Failure of the UPR results in prolonged ER stress, which, in turn, triggers apoptosis and inflammation. Accordingly, genes controlling apoptosis were enriched in monocytes from group B patients. Hence, collectively, these findings imply that the group B monocytes are intrinsically susceptible to stress or exist in a stressful environment, as well as indicate the persistence of ER stress (Irvine et al. 2012).

Intriguingly, Stechova and co-workers (2011) compared gene expression profiles of freshly isolated PMBCs from patients with T1DM, their first-degree relatives with higher genetic risk of developing the disease, and non-diabetic individuals by the microarray technology. They observed a clear difference between the expression profiles of relatives of patients (in particular the autoantibody-negative ones) and healthy controls. Moreover, the highest number of differentially activated cell signalling processes (99 pathways) was reported in the comparison between the relatives, regardless of autoantibody status, and the control group. Interestingly, DNA damage and oxidative stress were among those pathways (Stechova et al. 2011). Thus, these findings showed that non-diabetic relatives of T1DM patients also present alterations in the expression of genes.

Regarding the target tissue, whole-genome transcript expression for four T1DM pancreases (collected at different T1DM stages) and for purified islets from two of them was compared with that of the control group by carrying out microarray experiments, followed by qPCR validation of a group of genes (Planas et al. 2010). In agreement with the aforementioned studies, Planas and colleagues (2010) observed changes in the expression of oxidative stress genes (all up-regulated), including metallothioneins, such as *MT1M* (in the four cases and in purified islets) and *SOD2*, ceruloplasmin and thioredoxin interacting protein (case 1, pancreas and islets). Moreover, islets shared some alterations in the expression of apoptosis-related genes, such as repression of pro-apoptotic genes (*MLLT11*, *PRUNE2*, and *NLRP1*).

With respect to the expression of non-coding protein genes, microRNAs (miRNAs) have been indicated both as potential biomarkers for the earlier diagnosis of diabetes and as therapeutic targets for the treatment of this disorder (Mao et al. 2013). miRNAs are endogenous non-coding RNA molecules of approximately 22 nucleotides that are involved in the post-transcriptional regulation of protein-coding gene expression (Bartel 2004) by base-pairing to specific sites in the 5' untranslated regions (UTR) (Grey et al. 2010, Helwak et al. 2013), coding sequences (Hafner et al. 2010, Helwak et al. 2013, Reczko et al. 2012), and 3' UTRs of the mRNA targets; in this way, miRNAs lead to the degradation and/or translational repression of their targets (Bartel 2009, Krol et al. 2010, Lee et al. 1993, Lim et al. 2005b, Wightman et al. 1993). Recently, we compared the miRNA expression profiles displayed by PBMCs from T1DM patients with those exhibited by PBMCs from healthy non-diabetic controls by performing microarray experiments (Takahashi et al. 2014). In this study, we were able to identify a set of 44 differentially expressed miRNAs (35 induced and nine repressed) that clearly stratified patients with T1DM from the healthy subjects. After target prediction, results pointed to 10,827 and 6636 potential targets of the up- and down-regulated miRNAs, respectively; of note, a total of 85 and 75 genes implicated in DNA repair and response to oxidative stress, respectively, are potential targets of the 44 differentially modulated miRNAs in T1DM (unpublished data).

Taken together, these works on the whole-transcript expression in T1DM patients are consistent with the aforementioned studies that have detected elevated oxidative stress and DNA damage levels, as well as a perturbation in the antioxidant defense mechanisms in these individuals (Sect. 9.2.1). Moreover, genes involved in DNA repair and response to oxidative stress are putative targets of a set of miRNAs that clearly distinguished T1DM patients from healthy individuals. Finally, altogether, these works support the hypothesis that patients with T1DM respond to the increased oxidative stress and DNA lesions by changing their gene expression profiles.

9.3   Type 2 Diabetes Mellitus

T2DM represents approximately 90 % of all diagnosed cases of diabetes (ADA 2011). The disease is mainly characterized by resistance to insulin action and also by deficiency in the secretion of this hormone, presenting great correlation with aging, obesity, and lack of physical activity (Golay and Ybarra 2005). It is not known whether during the development of the disease it is the insulin resistance or the secretion deficiency that occurs first; however, insulin resistance appears as a crucial factor, especially when related to obesity, given the fact that about 60–90 % of all T2DM patients are or were obese (Gerich 1999, Golay and Ybarra 2005, Stumvoll et al. 2005).

Obesity causes a chronic inflammatory response in the adipose tissue, characterized by an abnormal production of cytokines, which include mostly molecules playing roles in stress response processes (Sethi and Hotamisligil 1999, Lebrun and Van Obberghen 2008). One of these cytokines is the Tumor Necrosis Factor-alpha (TNF-α), which is released at large amounts by adipocytes and acts on the insulin receptor, inhibiting its tyrosine kinase activity, which culminates in insulin resistance (Sethi and Hotamisligil 1999, Kohn et al. 2005). In addition, excess body fat leads to an increase in the number of fatty acid molecules in the blood. Consequently, a preferential use of lipids as an energy source occurs, especially by muscles, which prevents glucose utilization and glycogen synthesis, leading to hyperglycemia. Furthermore, there is an increase in insulin secretion to compensate for the insulin receptor resistance, and this condition gradually leads to the development of the disease (Lam et al. 2003, Golay and Ybarra 2005).

The biochemical mechanisms and physiological processes that characterize T2DM are not well understood. Some susceptibility genes have been identified, including genes related to cellular metabolism, such as PPAR gamma (*PPARG*) (Barroso et al. 2006), *KCNJ* (Schwanstecher and Schwanstecher 2002), and *CAPN10* (Cox et al. 2004), as well as the transcription factors *HNF4A* (Damcott et al. 2004, Hara et al. 2006) and *TCF7L2* (Florez et al. 2003, Barroso et al. 2006). Other genes, such as *ENPP1*, *RBP4*, and *SIRT1* are strong T2DM candidate genes, although they still need to be validated (Freeman and Cox 2006). The protein encoded by the *ENPP1* gene acts both on insulin resistance and on obesity development (Meyre et al. 2005). High concentrations of RBP4 protein promote systemic insulin resistance and when present at low levels, RBP4 promotes an increased sensitivity to the hormone (Yang et al. 2005). The overexpression of SIRT1 protein increases the secretion of insulin by β-cells, presumably due to increased efficiency in ATP production by the oxidative phosphorylation process (Moynihan et al. 2005).

Currently, the treatment for patients with T2DM is limited to drug therapies that improve disease conditions, such as insulin resistance. However, these drugs do not aim to restore the normal glucose metabolism, an event that exposes patients to the risk of disease complications. Instead, treatment for T2DM aims to reduce hyperglycemia by two main mechanisms: increased secretion of insulin by the pancreas or decreased production of glucose by the liver. Metformin is one of the most used drugs to treat T2DM patients. PPAR gamma antagonists, which act by increasing the sensitivity of insulin receptor, have also been used (Moller 2001). Still, sulphonylureas, thiazolidinediones, and insulin are also among the medications indicated for the treatment and glycemic control of T2DM patients (Ismail-Beigi 2012). However, treatment is complicated, mainly due to the lack of control of insulin secretion performed by the β-cells, which accurately adjust the amount of insulin secreted in accordance with the needs of the organism. Therefore, patients often have episodes of hypoglycemia and hyperglycemia, both with serious consequences for the diabetic patient, because the former can lead to coma, while the latter can lead to blindness, kidney failure, and vascular diseases (Taylor 1999, Kruger et al. 2006).

9.3.1   Oxidative Stress and DNA Damage in T2DM

In T2DM, hyperglycemia contributes significantly to the production of free radicals. Moreover, the antioxidant defense mechanisms are not able to compensate or neutralize the amount of radicals formed, since there is evidence of a reduced activity of antioxidant enzymes in these patients, such as SOD, CAT, and glutathione reductase (GSR) (Seghrouchni et al. 2002). Thus, an increase in ROS generation and hence increased oxidative damage occur (Nishikawa et al. 2000, Blasiak et al. 2004). Additionally, as previously mentioned, there is the contribution of other processes, such as formation of AGEs that initiate cascades related to oxidative damage and are toxic to the cells, including pancreatic β-cells (Piperi et al. 2012).

As a consequence of oxidative stress, T2DM patients also show increased levels of lipid peroxidation products when compared with healthy individuals (Slatter et al. 2000). Accordingly, Abou-Seif and Youssef (2004) performed a study evaluating a series of markers in diabetic patients compared with healthy individuals, including lipid peroxidation products (MDA), antioxidants (GSH, SOD, and CAT) as well as oxidation protein products and AGEs markers. As a result, the authors found in the diabetic patient group decreased levels of the three antioxidants and increased levels of MDA, oxidation protein products, and AGEs, providing further support to the information aforementioned.

Hereupon, glycemic control has beneficial effects as it reduces the harmful effects of hyperglycemia in diabetic patients (Stolar et al. 2008, Ismail-Beigi 2012). Concerning the importance of glycemic control, Cakatay (2005) verified protein oxidation using different markers comparing hyperglycemic and non-hyperglycemic T2DM patients, both groups without any comorbidity. The author found higher levels of protein oxidation in hyperglycemic T2DM patients. As patients had no comorbidities, high protein oxidation in these patients indicates that oxidative stress may not be a result of complications of the disease, Lodovici et al. (2008) also conducted a comparative study between hyperglycemic and non- hyperglycemic T2DM patients, but focusing on the antioxidant defense status. The authors observed that patients with elevated glucose levels exhibited decreased antioxidant condition when compared with non-hyperglycemic patients. Therefore, these studies suggest that oxidative stress resulting from hyperglycemia is an important factor involved in the decline of antioxidant defense and in the increase in oxidative damage, being necessary a proper control of both, blood glucose levels and free radical production, thus, avoiding the action of the latter on different macromolecules, such as lipids and proteins.

Regarding the generation of DNA damage, it is well known that the exposure of cells to oxidative stress as a consequence of the increased generation of ROS induces higher DNA damage levels in PBMCs from diabetic patients when compared with healthy individuals (Bonnefont-Rousselot et al. 2000, Lee and Wei 2007, Song et al. 2007, da Silva et al. 2013). Furthermore, Binici et al. (2013) verified an increased genomic instability in T2DM patients. In addition, the efficiency of DNA repair was compared between a group of poorly controlled T2DM patients and a group of healthy subjects, by measuring DNA damage levels (comet assay) caused by hydrogen peroxide and by doxorubicin (Blasiak et al. 2004); their results demonstrated that besides having higher baseline DNA damage than healthy subjects, when exposed to mutagens, T2DM patients also showed a lower efficiency of DNA repair, although both groups were able to repair the damage.

Even in T2DM patients with good glycemic control, the lack of proper physiological adjustment of insulin secretion, as present in healthy individuals (Kruger et al. 2006), may lead these patients to experience periods of hyperglycemia, triggering oxidative stress as a consequence. Even though this might not be enough to cause significant damage to the nuclear DNA, there is still the possibility of generation of damage in the mitochondrial DNA (mtDNA), since cellular respiration in mitochondria makes this organelle the site of increased production of free radicals inside the cell (Fernandez-Sanchez et al. 2011). Some studies suggest that DNA damage can be repaired in the mitochondria less efficiently than in the nucleus (Arnheim and Cortopassi 1992, Lim et al. 2005a). Santos et al. (2003) found that fibroblast strains exposed to hydrogen peroxide also showed differences in nuclear DNA and mtDNA efficiency of repair. Under certain conditions, damage in the nuclear DNA was fully repaired after a certain period of time, while damage in mtDNA was not significantly repaired, increasing apoptosis; these findings suggest the existence of a threshold in the mitochondrial repair and when it is crossed, mechanisms of cell death are triggered.

9.3.2 Alterations of Transcriptional Expression Profiles in T2DM

Since the initial work of Schena et al. (1995), the microarray technique became a common and important tool in medical and biological research. Thus, in the last years, many studies used this technology to analyze the gene expression profiles exhibited by T2DM patients in order to reveal new genes and pathways involved in the disease. Takamura et al. (2007) compared the gene expression profiles of PBMCs from T2DM patients with those from non-diabetic subjects by analyzing the modulation of specific gene set categories. The authors found that genes related to the c-Jun N-terminal kinase (JNK) and mitochondrial oxidative phosphorylation (OXPHOS) pathways (possibly related to stress response) were differentially modulated in T2DM patients. Palsgaard et al. (2009) compared the gene expression profiles of muscle biopsies from T2DM patients with those from their healthy first degree relatives in an attempt to investigate if the latter have an increased risk of developing the disorder. Their findings indicated that the up-regulated genes in the group of T2DM relatives were involved in insulin signaling, which, according to the author, could be a compensatory mechanism for a reduced insulin signaling activity. Bikopoulos et al. (2008) analyzed the gene expression profiles of human pancreatic islets under chronic exposure to free fatty acids; aside from showing a significantly reduced glucose-stimulated insulin secretion and increased ROS generation, the pancreatic islets chronically exposed to oleate also presented altered expression of 40 genes; these genes were related to the oleate metabolism inflammation, and also to antioxidant defense (which were up-regulated), highlighting the importance of free fatty acids as risk factors for the development of T2DM.

Our group also conducted a study comparing the transcriptional expression patterns exhibited by PBMCs from T2DM patients with those from healthy subjects, focusing on pathways such as oxidative stress response, DNA repair, response to hypoxia, inflammation, fatty acid processing, and immune response (Manoel-Caetano et al. 2012). We obtained a list of 92 differentially expressed genes (52 up-regulated and 40 down-regulated) in diabetic patients compared to the control group, and these genes were associated with the six aforementioned biological processes; among them, genes related to oxidative stress responses and hypoxia (*OXR1*, *SMG1*, and *UCP3*) were highly up-regulated, possibly in an attempt to deal with increased oxidative stress. Concerning the down-regulated genes, many were involved in inflammation, immune response and DNA repair (including *SUMO1*, *ATRX*, and *MORF4L2*). The down-regulation of several DNA repair genes is in agreement with the decreased efficiency of DNA repair verified in T2DM patients (Blasiak et al. 2004, Pacal et al. 2011). In another study conducted by Marselli et al. (2010), genes related to glucotoxicity, oxidative stress (up-regulated), cell cycle, apoptosis, or ER stress were found differentially expressed in β-cell-enriched samples obtained from T2DM patients relative to control individuals.

Lately, the role of ER stress has been studied in T2DM regarding both the pathogenesis of the disease and its consequences. The ER is the major organelle responsible for regulating not only the synthesis of proteins but also its folding, maturation, and transport, representing the most significant sensor of nutrients in the cell and a major coordinator of metabolic responses. The ER maintains a controlled balance between the synthesis and proper protein folding (Ron and Walter 2007, Piperi et al. 2012). However, several conditions can break this homeostatic balance, such as excess of nutrients, insulin resistance, increased levels of ROS, and inflammation related to obesity (Scheuner et al. 2005). The disturbance of this homeostasis leads to an accumulation of misfolded proteins in the organelle, either by an increased rate of protein synthesis, or by alterations in the ER milieu, compromising the efficiency of protein folding. Regardless the case, the UPR response is triggered to restore protein homeostasis (Sano and Reed 2013, Biden et al. 2014). Mainly three proteins are responsible for the activation of UPR response: inositol-requiring protein-1α (IRE1α), protein kinase RNA (PKR)-like ER kinase (PERK), and activating transcription factor 6 (ATF6). Under normal conditions, the Binding immunoglobulin Protein (BiP) chaperone is bound to the luminal regions of the PERK and ATF6 proteins, maintaining an inactive conformation. During ER stress, as a response to the accumulation of misfolded proteins, BiP is released from PERK and ATF6, in order to assist with the proper protein folding (Gardner and Walter 2011, Sano and Reed 2013). Differently, IRE1α seems to become active when bound to misfolded proteins. The activation of these three proteins leads to signaling pathways which diminish the accumulation of proteins in the ER. This is accomplished by the translation inhibition of a series of mRNAs (which restricts the protein influx into the ER) and by promoting the transport of misfolded proteins outside the ER, where they will be ubiquitinated and directed to degradation (Sano and Reed 2013).

UPR has been related to the carbohydrate metabolism, according to studies in T2DM. ER stress can inhibit the suppression of gluconeogenic enzymes; as a consequence, there is an increase in the production of glucose by the liver, leading to obesity and the development of diabetes (Kimura et al. 2012). Wang et al. (2009) also showed that acute ER stress promoted the activation and nuclear entry of CRTC2, which, in turn, induced the expression of ATF6 and activated gluconeogenesis. Insulin resistance mechanism was also related to ER stress as a consequence of a high fat diet. The XBP-1 protein binds to promoters of genes related to UPR and to the ER assisted degradation, in order to restore protein homeostasis in the ER (Sano and Reed 2013). According to Hage Hassan et al. (2012), silencing XBP-1 in C2C12 cells and human myotubes increased the sensitivity to ER stress and insulin resistance, as a consequence of the insulin receptor substrate-1 (IRS-1) degradation. In addition, ER stress provokes IRS-1 serine phosphorylation, disassembling it from the insulin receptor and increasing insulin resistance (Bailly-Maitre et al. 2010). Likewise, pancreatic β-cell death was related to ER stress and UPR response, due to the high insulin demand, which causes an increased dependence of ER functioning, to ensure proper synthesis and insulin folding (Sano and Reed 2013). For instance, Back et al. (2009) showed that the absence of eIF2α phosphorylation in mice β-cells caused dysregulated proinsulin translation, increased oxidative damage, and defective ER trafficking of proteins and apoptosis.

There is evidence that hyperglycemia is also associated with ER stress. Excess glucose may react with other molecules such as lipids and proteins, leading to AGEs formation, thereby resulting in alterations in the ER homeostasis (Inagi 2011, Piperi et al. 2012). The effects of ongoing stress to ER have also been associated with the development of inflammation and also with various T2DM complications (Hayashi et al. 2005, Piperi et al. 2012). As such, Oslowski et al. (2012) reported that ER stress induced thioredoxin interacting protein (TXNIP) through IRE1α and PERK action in mice and human pancreatic β-cells. TXNIP, in turn, promoted IL-1β production in these cells. Some other studies have demonstrated the importance of ER stress to T2DM. Casas et al. (2007) evaluated the influence of amyloid polypeptide, which is toxic to pancreatic β-cells, on the expression profiles of MIN6 cells and primary cultures of human pancreatic islets. The authors observed not only the influence of amyloid polypeptide aggregation on the induction of genes related to ER stress, but also an impairment of the proteasome function, which contributed to apoptosis.

Komura et al. (2010) studied the immune-mediated response by comparing the transcriptional expression profiles of PBMCs from T2DM patients *versus* healthy individuals. The authors detected an elevated expression of markers of ER stress in the T2DM group. Finally, Iwasaki et al. (2014) verified the influence of ATF4, a transcription factor activated after metabolic stresses (including ER stress), on the inflammation mediated by free fatty acids. Using macrophages, the authors provided evidence that the ATF4 pathway was activated by free fatty acids, linking metabolic stress to inflammation process in these cells.

Altogether, the information in the literature regarding transcriptional expression profiles highlights not only altered pathways in T2DM, such as inflammation, oxidative stress, immune response, and ER stress, but also establishes a link between biological processes. In addition, the use of microarrays and other techniques for whole-genome profiling brings forth a large amount of data, whose analysis may reveal new altered pathways in T2DM, increasing our knowledge about the disease and also providing new therapy possibilities.

9.4 Conclusions

Diabetes mellitus is a group of chronic metabolic diseases that has a great impact on public health. A number of studies have demonstrated a link between the two main types of diabetes, T1DM and T2DM, and oxidative stress, since the latter may ultimately result in β-cell damage as well as can be a consequence of diabetes itself due to hyperglycemia. In agreement with that, increased levels of oxidative stress markers and DNA damage and impaired antioxidant system have been reported in patients suffering from T1DM or T2DM. Thus, it is expected that these patients respond to those insults in many ways, including via alterations in the expression of genes, which in fact have been described in this chapter.

References

Abou-Seif MA, Youssef AA (2004) Evaluation of some biochemical changes in diabetic patients. Clin Chim Acta 346:161–170

ADA (American Diabetes Association) (2011) Standards of medical care in diabetes–2011. Diabetes Care 34(Suppl 1):11–61

ADA (American Diabetes Association) (2013) Diagnosis and classification of diabetes mellitus. Diabetes Care 36(Suppl 1):S67–S74

Agbaje IM, McVicar CM, Schock BC et al (2008) Increased concentrations of the oxidative DNA adduct 7,8-dihydro-8-oxo-2-deoxyguanosine in the germ-line of men with type 1 diabetes. Reprod Biomed Online 16:401–409

Ahmad FK, He Z, King GL (2005) Molecular targets of diabetic cardiovascular complications. Curr Drug Targets 6:487–494

Arnheim N, Cortopassi G (1992) Deleterious mitochondrial DNA mutations accumulate in aging human tissues. Mutat Res 275:157–167

Back SH, Scheuner D, Han J et al (2009) Translation attenuation through eIF2alpha phosphorylation prevents oxidative stress and maintains the differentiated state in beta cells. Cell Metab 10:13–26

Bailly-Maitre B, Belgardt BF, Jordan SD et al (2010) Hepatic Bax inhibitor-1 inhibits IRE1alpha and protects from obesity-associated insulin resistance and glucose intolerance. J Biol Chem 285:6198–6207

Barrett JC, Clayton DG, Concannon P et al (2009) Genome-wide association study and meta-analysis find that over 40 loci affect risk of type 1 diabetes. Nat Genet 41:703–707

Barroso I, Luan J, Sandhu MS et al (2006) Meta-analysis of the Gly482Ser variant in PPARGC1A in type 2 diabetes and related phenotypes. Diabetologia 49:501–505

Bartel DP (2004) MicroRNAs: genomics, biogenesis, mechanism, and function. Cell 116:281–297

Bartel DP (2009) MicroRNAs: target recognition and regulatory functions. Cell 136:215–233

Bell GI, Horita S, Karam JH (1984) A polymorphic locus near the human insulin gene is associated with insulin-dependent diabetes mellitus. Diabetes 33:176–183

Biden TJ, Boslem E, Chu KY, Sue N (2014) Lipotoxic endoplasmic reticulum stress, beta cell failure, and type 2 diabetes mellitus. Trends Endocrinol Metab. Advance online publication. doi:10.1016/j.tem.2014.02.003

Bikopoulos G, da Silva PA, Lee SC et al (2008) Ex vivo transcriptional profiling of human pancreatic islets following chronic exposure to monounsaturated fatty acids. J Endocrinol 196:455–464

Binici DN, Karaman A, Coskun M et al (2013) Genomic damage in patients with type-2 diabetes mellitus. Genet Couns 24:149–156

Blank V (2008) Small Maf proteins in mammalian gene control: mere dimerization partners or dynamic transcriptional regulators? J Mol Biol 376:913–925

Blasiak J, Arabski M, Krupa R et al (2004) DNA damage and repair in type 2 diabetes mellitus. Mutat Res 554:297–304

Bonnefont-Rousselot D, Bastard JP, Jaudon MC, Delattre J (2000) Consequences of the diabetic status on the oxidant/antioxidant balance. Diabetes Metab 26:163–176

Bottini N, Musumeci L, Alonso A et al (2004) A functional variant of lymphoid tyrosine phosphatase is associated with type I diabetes. Nat Genet 36:337–338

Brownlee M (2001) Biochemistry and molecular cell biology of diabetic complications. Nature 414:813–820

Burren OS, Adlem EC, Achuthan P et al (2011) T1DBase: update 2011, organization and presentation of large-scale data sets for type 1 diabetes research. Nucleic Acids Res 39:D997–D1001

Cakatay U (2005) Protein oxidation parameters in type 2 diabetic patients with good and poor glycaemic control. Diabetes Metab 31:551–557

Casas S, Gomis R, Gribble FM et al (2007) Impairment of the ubiquitin-proteasome pathway is a downstream endoplasmic reticulum stress response induced by extracellular human islet amyloid polypeptide and contributes to pancreatic beta-cell apoptosis. Diabetes 56:2284–2294

Codoñer-Franch P, Pons-Morales S, Boix-García L, Valls-Bellés V (2010) Oxidant/antioxidant status in obese children compared to pediatric patients with type 1 diabetes mellitus. Pediatr Diabetes 11:251–257

Cooper JD, Smyth DJ, Walker NM et al (2011) Inherited variation in vitamin D genes is associated with predisposition to autoimmune disease type 1 diabetes. Diabetes 60:1624–1631

Cox NJ, Hayes MG, Roe CA et al. (2004) Linkage of calpain 10 to type 2 diabetes: the biological rationale. Diabetes 53(Suppl 1):19–25

da Silva BS, Rovaris DL, Bonotto RM et al (2013) The influence on DNA damage of glycaemic parameters, oral antidiabetic drugs and polymorphisms of genes involved in the DNA repair system. Mutagenesis 28:525–530.

Damcott CM, Hoppman N, Ott SH et al (2004) Polymorphisms in both promoters of hepatocyte nuclear factor 4-alpha are associated with type 2 diabetes in the Amish. Diabetes 53:3337–3341

Dinçer Y, Akçay T, Ilkova H et al (2003) DNA damage and antioxidant defense in peripheral leukocytes of patients with Type I diabetes mellitus. Mutat Res 527:49–55

Dizdaroglu M (2012) Oxidatively induced DNA damage: mechanisms, repair and disease. Cancer Lett 327:26–47

Drews G, Krippeit-Drews P, Düfer M (2010) Oxidative stress and beta-cell dysfunction. Pflugers Arch 460:703–718

Edeas M, Attaf D, Mailfert AS et al (2010) Maillard reaction, mitochondria and oxidative stress: potential role of antioxidants. Pathol Biol (Paris) 58:220–225

Fernandez-Sanchez A, Madrigal-Santillan E, Bautista M et al (2011) Inflammation, oxidative stress, and obesity. Int J Mol Sci 12:3117–3132

Florez JC, Hirschhorn J, Altshuler D (2003) The inherited basis of diabetes mellitus: implications for the genetic analysis of complex traits. Annu Rev Genomics Hum Genet 4:257–291

Freeman H, Cox RD (2006) Type-2 diabetes: a cocktail of genetic discovery. Hum Mol Genet 15 Spec No 2:R202–R209

Friedberg EC, Walker GC, Siede W et al (2006) DNA Repair and Mutagenesis. ASM, Washington, DC

Gardner BM, Walter P (2011) Unfolded proteins are Ire1-activating ligands that directly induce the unfolded protein response. Science 333:1891–1894

Geiger K, Leiherer A, Muendlein A et al (2011) Identification of hypoxia-induced genes in human SGBS adipocytes by microarray analysis. PLoS One 6:e26465

Gerich JE (1999) Is insulin resistance the principal cause of type 2 diabetes? Diabetes Obes Metab 1:257–263

Golay A, Ybarra J (2005) Link between obesity and type 2 diabetes. Best Pract Res Clin Endocrinol Metab 19:649–663

Goodarzi MT, Navidi AA, Rezaei M, Babahmadi-Rezaei H (2010) Oxidative damage to DNA and lipids: correlation with protein glycation in patients with type 1 diabetes. J Clin Lab Anal 24:72–76

Grey F, Tirabassi R, Meyers H et al (2010) A viral microRNA down-regulates multiple cell cycle genes through mRNA 5′UTRs. PLoS Pathog 6:e1000967

Hafner M, Landthaler M, Burger L et al (2010) Transcriptome-wide identification of RNA-binding protein and microRNA target sites by PAR-CLIP. Cell 141:129–141

Hage Hassan R, Hainault I, Vilquin JT et al (2012) Endoplasmic reticulum stress does not mediate palmitate-induced insulin resistance in mouse and human muscle cells. Diabetologia 55:204–214

Halliwell B, Gutteridge JMC (2007) Free radicals in biology and medicine. Oxford University, Oxford

Hannon-Fletcher MP, O'Kane MJ, Moles KW et al (2000) Levels of peripheral blood cell DNA damage in insulin dependent diabetes mellitus human subjects. Mutat Res 460:53–60

Hara K, Horikoshi M, Kitazato H et al (2006) Hepatocyte nuclear factor-4alpha P2 promoter haplotypes are associated with type 2 diabetes in the Japanese population. Diabetes 55:1260–1264

Hata I, Kaji M, Hirano S et al (2006) Urinary oxidative stress markers in young patients with type 1 diabetes. Pediatr Int 48:58–61

Hayashi T, Saito A, Okuno S et al (2005) Damage to the endoplasmic reticulum and activation of apoptotic machinery by oxidative stress in ischemic neurons. J Cereb Blood Flow Metab 25:41–53

Helwak A, Kudla G, Dudnakova T, Tollervey D (2013) Mapping the human miRNA interactome by CLASH reveals frequent noncanonical binding. Cell 153:654–665

IDF (2013) The IDF diabetes atlas. international diabetes federation. sixth edition ed. brussels. http://www.idf.org/diabetesatlas. Accessed 14 March 2014

Inagi R (2011) Inhibitors of advanced glycation and endoplasmic reticulum stress. Methods Enzymol 491:361–380

Irvine KM, Gallego P, An X et al (2012) Peripheral blood monocyte gene expression profile clinically stratifies patients with recent-onset type 1 diabetes. Diabetes 61:1281–1290

Ismail-Beigi F (2012) Clinical practice. Glycemic management of type 2 diabetes mellitus. N Engl J Med 366:1319–1327

Iwasaki Y, Suganami T, Hachiya R et al (2014) Activating transcription factor 4 links metabolic stress to interleukin-6 expression in macrophages. Diabetes 63:152–161

Jiang F, Zhang Y, Dusting GJ (2011) NADPH oxidase-mediated redox signaling: roles in cellular stress response, stress tolerance, and tissue repair. Pharmacol Rev 63:218–242

Kalyanaraman B (2013) Teaching the basics of redox biology to medical and graduate students: oxidants, antioxidants and disease mechanisms. Redox Biol 1:244–257

Katsuoka F, Motohashi H, Ishii T et al (2005) Genetic evidence that small maf proteins are essential for the activation of antioxidant response element-dependent genes. Mol Cell Biol 25:8044–8051

Kimura K, Yamada T, Matsumoto M et al (2012) Endoplasmic reticulum stress inhibits STAT3-dependent suppression of hepatic gluconeogenesis via dephosphorylation and deacetylation. Diabetes 61:61–73

Kohn LD, Wallace B, Schwartz F, McCall K (2005) Is type 2 diabetes an autoimmune-inflammatory disorder of the innate immune system? Endocrinology 146:4189–4191

Komura T, Sakai Y, Honda M et al (2010) CD14 + monocytes are vulnerable and functionally impaired under endoplasmic reticulum stress in patients with type 2 diabetes. Diabetes 59:634–643

Krol J, Loedige I, Filipowicz W (2010) The widespread regulation of microRNA biogenesis, function and decay. Nat Rev Genet 11:597–610

Kruger DF, Martin CL, Sadler CE (2006) New insights into glucose regulation. Diabetes Educ 32:221–228

Lam TK, Carpentier A, Lewis GF et al (2003) Mechanisms of the free fatty acid-induced increase in hepatic glucose production. Am J Physiol Endocrinol Metab 284:E863–E873

Lebrun P, Van Obberghen E (2008) SOCS proteins causing trouble in insulin action. Acta Physiol (Oxf) 192:29–36

Lee HC, Wei YH (2007) Oxidative stress, mitochondrial DNA mutation, and apoptosis in aging. Exp Biol Med (Maywood) 232:592–606

Lee RC, Feinbaum RL, Ambros V (1993) The C. elegans heterochronic gene lin-4 encodes small RNAs with antisense complementarity to lin-14. Cell 75:843–854

Lenzen S (2008) Oxidative stress: the vulnerable beta-cell. Biochem Soc Trans 36:343–347

Lim KS, Jeyaseelan K, Whiteman M et al (2005a) Oxidative damage in mitochondrial DNA is not extensive. Ann N Y Acad Sci 1042:210–220

Lim LP, Lau NC, Garrett-Engele P et al (2005b) Microarray analysis shows that some microRNAs downregulate large numbers of target mRNAs. Nature 433:769–773

Lin CC, Huang HH, Hu CW et al (2014) Trace elements, oxidative stress and glycemic control in young people with type 1 diabetes mellitus. J Trace Elem Med Biol 28:18–22

Lodovici M, Giovannelli L, Pitozzi V et al (2008) Oxidative DNA damage and plasma antioxidant capacity in type 2 diabetic patients with good and poor glycaemic control. Mutat Res 638:98–102

Lowe CE, Cooper JD, Brusko T et al (2007) Large-scale genetic fine mapping and genotype-phenotype associations implicate polymorphism in the IL2RA region in type 1 diabetes. Nat Genet 39:1074–1082

Manoel-Caetano FS, Xavier DJ, Evangelista AF et al (2012) Gene expression profiles displayed by peripheral blood mononuclear cells from patients with type 2 diabetes mellitus focusing on biological processes implicated on the pathogenesis of the disease. Gene 511:151–160

Mao Y, Mohan R, Zhang S, Tang X (2013) MicroRNAs as pharmacological targets in diabetes. Pharmacol Res 75:37–47

Marselli L, Thorne J, Dahiya S et al (2010) Gene expression profiles of Beta-cell enriched tissue obtained by laser capture microdissection from subjects with type 2 diabetes. PLoS One 5:e11499

Martinon F, Glimcher LH (2011) Regulation of innate immunity by signaling pathways emerging from the endoplasmic reticulum. Curr Opin Immunol 23:35–40

Meyre D, Bouatia-Naji N, Tounian A et al (2005) Variants of ENPP1 are associated with childhood and adult obesity and increase the risk of glucose intolerance and type 2 diabetes. Nat Genet 37:863–867

Moller DE (2001) New drug targets for type 2 diabetes and the metabolic syndrome. Nature 414:821–827

Moynihan KA, Grimm AA, Plueger MM et al (2005) Increased dosage of mammalian Sir2 in pancreatic beta cells enhances glucose-stimulated insulin secretion in mice. Cell Metab 2:105–117

Nathan C, Cunningham-Bussel A (2013) Beyond oxidative stress: an immunologist's guide to reactive oxygen species. Nat Rev Immunol 13:349–361

Nishikawa T, Edelstein D, Du XL et al (2000) Normalizing mitochondrial superoxide production blocks three pathways of hyperglycaemic damage. Nature 404:787–790

Nisticò L, Buzzetti R, Pritchard LE et al (1996) The CTLA-4 gene region of chromosome 2q33 is linked to, and associated with, type 1 diabetes. Belgian Diabetes Registry. Hum Mol Genet 5:1075–1080

Oslowski CM, Hara T, O'Sullivan-Murphy B et al (2012) Thioredoxin-interacting protein mediates ER stress-induced beta cell death through initiation of the inflammasome. Cell Metab 16:265–273

Pacal L, Varvarovska J, Rusavy Z et al (2011) Parameters of oxidative stress, DNA damage and DNA repair in type 1 and type 2 diabetes mellitus. Arch Physiol Biochem 117:222–230

Palsgaard J, Brons C, Friedrichsen M et al (2009) Gene expression in skeletal muscle biopsies from people with type 2 diabetes and relatives: differential regulation of insulin signaling pathways. PLoS One 4:e6575

Piperi C, Adamopoulos C, Dalagiorgou G et al (2012) Crosstalk between advanced glycation and endoplasmic reticulum stress: emerging therapeutic targeting for metabolic diseases. J Clin Endocrinol Metab 97:2231–2242

Planas R, Carrillo J, Sanchez A et al (2010) Gene expression profiles for the human pancreas and purified islets in type 1 diabetes: new findings at clinical onset and in long-standing diabetes. Clin Exp Immunol 159:23–44

Rains JL, Jain SK (2011) Oxidative stress, insulin signaling, and diabetes. Free Radic Biol Med 50:567–575

Reczko M, Maragkakis M, Alexiou P et al (2012) Functional microRNA targets in protein coding sequences. Bioinformatics 28:771–776

Robertson RP, Harmon JS (2007) Pancreatic islet beta-cell and oxidative stress: the importance of glutathione peroxidase. FEBS Lett 581:3743–3748

Ron D, Walter P (2007) Signal integration in the endoplasmic reticulum unfolded protein response. Nat Rev Mol Cell Biol 8:519–529

Ruiz-García A, Monsalve E, Novellasdemunt L et al (2011) Cooperation of adenosine with macrophage Toll-4 receptor agonists leads to increased glycolytic flux through the enhanced expression of PFKFB3 gene. J Biol Chem 286:19247–19258

Sano R, Reed JC (2013) ER stress-induced cell death mechanisms. Biochim Biophys Acta 1833:3460–3470

Santos JH, Hunakova L, Chen Y et al (2003) Cell sorting experiments link persistent mitochondrial DNA damage with loss of mitochondrial membrane potential and apoptotic cell death. J Biol Chem 278:1728–1734

Schena M, Shalon D, Davis RW, Brown PO (1995) Quantitative monitoring of gene expression patterns with a complementary DNA microarray. Science 270:467–470

Scheuner D, Vander Mierde D, Song B et al (2005) Control of mRNA translation preserves endoplasmic reticulum function in beta cells and maintains glucose homeostasis. Nat Med 11:757–764

Schwanstecher C, Schwanstecher M (2002) Nucleotide sensitivity of pancreatic ATP-sensitive potassium channels and type 2 diabetes. Diabetes 51(Suppl 3):358–362

Seghrouchni I, Drai J, Bannier E et al (2002) Oxidative stress parameters in type I, type II and insulin-treated type 2 diabetes mellitus; insulin treatment efficiency. Clin Chim Acta 321:89–96

Sethi JK, Hotamisligil GS (1999) The role of TNF alpha in adipocyte metabolism. Semin Cell Dev Biol 10:19–29

Slatter DA, Bolton CH, Bailey AJ (2000) The importance of lipid-derived malondialdehyde in diabetes mellitus. Diabetologia 43:550–557

Song F, Jia W, Yao Y et al (2007) Oxidative stress, antioxidant status and DNA damage in patients with impaired glucose regulation and newly diagnosed Type 2 diabetes. Clin Sci (Lond) 112:599–606

Stechova K, Kolar M, Blatny R et al (2011) Healthy first degree relatives of patients with type 1 diabetes exhibit significant differences in basal gene expression pattern of immunocompetent cells compared to controls: expression pattern as predeterminant of autoimmune diabetes. Scand J Immunol 75:210–219

Stolar MW, Hoogwerf BJ, Gorshow SM et al. (2008) Managing type 2 diabetes: going beyond glycemic control. J Manag Care Pharm 14:2–19

Storr SJ, Woolston CM, Zhang Y, Martin SG (2013) Redox environment, free radical, and oxidative DNA damage. Antioxid Redox Signal 18:2399–2408

Stumvoll M, Goldstein BJ, van Haeften TW (2005) Type 2 diabetes: principles of pathogenesis and therapy. The Lancet 365:1333–1346

Takahashi P, Xavier DJ, Evangelista AF et al (2014) MicroRNA expression profiling and functional annotation analysis of their targets in patients with type 1 diabetes mellitus. Gene 539:213–223

Takamura T, Honda M, Sakai Y et al (2007) Gene expression profiles in peripheral blood mononuclear cells reflect the pathophysiology of type 2 diabetes. Biochem Biophys Res Commun 361:379–384

Taylor SI (1999) Deconstructing type 2 diabetes. Cell 97:9–12

Tonooka N, Oseid E, Zhou H et al (2007) Glutathione peroxidase protein expression and activity in human islets isolated for transplantation. Clin Transplant 21:767–772

van Oostrom O, de Kleijn DP, Fledderus JO et al. (2009) Folic acid supplementation normalizes the endothelial progenitor cell transcriptome of patients with type 1 diabetes: a case-control pilot study. Cardiovasc Diabetol 8:47

von Herrath M (2009) Can we learn from viruses how to prevent type 1 diabetes?: the role of viral infections in the pathogenesis of type 1 diabetes and the development of novel combination therapies. Diabetes 58:2–11

Wållberg M, Cooke A (2013) Immune mechanisms in type 1 diabetes. Trends Immunol 34:583–591

Wang Y, Vera L, Fischer WH, Montminy M (2009) The CREB coactivator CRTC2 links hepatic ER stress and fasting gluconeogenesis. Nature 460:534–537

Wightman B, Ha I, Ruvkun G (1993) Posttranscriptional regulation of the heterochronic gene lin-14 by lin-4 mediates temporal pattern formation in C. elegans. Cell 75:855–862

Yang J, Robert CE, Burkhardt BR et al (2005) Mechanisms of glucose-induced secretion of pancreatic-derived factor (PANDER or FAM3B) in pancreatic beta-cells. Diabetes 54:3217–3228

Zaccone P, Cooke A (2011) Infectious triggers protect from autoimmunity. Semin Immunol 23:122–129

Zipitis CS, Akobeng AK (2008) Vitamin D supplementation in early childhood and risk of type 1 diabetes: a systematic review and meta-analysis. Arch Dis Child 93:512–517

Chapter 10
MicroRNAs in Cancer

Adriane F. Evangelista and Marcia M. C. Marques

Abstract A frequent problem in the diagnosis of human tumors is the lack of biomarkers associated with biological processes that drive the initiation and progression of cancer. Evidences indicate that microRNAs (miRNAs) regulate the expression of different genes (post-transcriptional regulation) associated with carcinogenesis process. Because of this, small RNAs have become a crucial point in the molecular dissection of human cancer a few years ago. MiRNAs are small non-coding RNAs (19–24 nucleotides) responsible for fine-tuning of gene expression influencing the stability and efficiency of translation of messenger RNAs (mRNA) targets. Recently it has been proposed that the pathogenesis of cancer involves, among other macromolecules, miRNAs whose expression profiles are associated with prognosis and therapeutic results in various human cancers. The mechanisms mediating tumor progression exerted by miRNAs have been addressed only recently. MiRNAs exhibits a dual role in all types of cancer depending on their expression status i.e. up or down regulated. Thus, miRNAs may act as oncogenes (oncomirs) or tumor-suppressors. However, if the interest is to use miRNAs as biomarkers they should be ideally easily assayed with minimally invasive medical procedures but at the same time offering high sensitivity and specificity. In this context, recent findings suggest the potential of circulating miRNAs in the screening or monitoring cancer treatment.

10.1 MicroRNAs: Characterization and Biogenesis

MicroRNAs, or miRNAs, are small, non-protein-coding RNAs (19–24 nucleotides) produced from precursor hairpin RNAs of approximately 60–110 nucleotides that are involved in the post-transcriptional regulation of gene expression (Ambros

M. M. C. Marques (✉) · A. F. Evangelista
Molecular Oncology Research Center, Barretos Cancer Hospital,
14784-400 Barretos, São Paulo, Brazil
e-mail: mmcmsilveira@gmail.com

M. M. C. Marques
Barretos School of Health Sciences—FACISB, 14784-400 Barretos, São Paulo, Brazil

© Springer International Publishing Switzerland 2014
G. A. Passos (ed.), *Transcriptomics in Health and Disease,*
DOI 10.1007/978-3-319-11985-4_10

2004; Bartel 2004). Canonical miRNA biogenesis occurs through the sequential processing of primary transcripts (pri-miRNA) that is mediated by two ribonuclease III (RNase III) enzymes, Dicer and Drosha (Ketting et al. 2001; Lee et al. 2003). In addition to the canonical pathway, several alternative biogenesis pathways have been described in recent years, in which other cellular ribonucleases replace Drosha to generate the pre-miRNA hairpins (Yang and Lai 2011). The main Drosha-independent pathways include the splicing of mirtron derivatives (Okamura et al. 2007; Ruby et al. 2007) and mirtrons (Babiarz et al. 2008), miRNA biogenesis mediated by RNaseZ (Bogerd et al. 2010) and miRNA biogenesis mediated by integrators (Cazalla et al. 2011).

The mechanism of action of miRNAs involves binding of the miRNA to the 3' untranslated region (3' UTR) of the target mRNA, resulting in the regulation of mRNA stability and protein synthesis (Ambros 2004; Bartel 2004). This post-transcriptional regulation by miRNAs occurs through interaction (base pairing) in the 3' Untranslated Region of the mRNA (3' UTR) and depends on the degree of complementarity between the miRNA and the target mRNA. This interaction can lead to the inhibition of translation or mRNA degradation (Sevignani et al. 2006). Imperfect pairing leads to the inhibition of translation of the target mRNA, which is the major mechanism of action of miRNAs in mammals (He and Hannon 2004). The small size of miRNAs and their ability to function without complete base pairing mean that a single miRNA can regulate several mRNA targets and that multiple miRNAs can cooperate to control a single mRNA (Sevignani et al. 2006).

Recent studies have reported that miRNAs can regulate gene expression by binding the 5' UTR of target mRNAs, which can either allow or inhibit translation. This new mechanism of gene regulation by miRNAs has been demonstrated by several in vitro, in vivo and in silico approaches, and a variety of tools have been developed to predict new targets and the functions of miRNAs bound to the 5' UTR (Da Sacco and Masotti 2012). The first study that demonstrated a mechanism for the post-transcriptional regulation of gene expression by a 5' UTR-bound miRNA was conducted in Drosophila melanogaster. In this study, the authors reported that the interaction of miR-2 with the 5' UTR region of the human β-globin gene inhibits the translation of the corresponding protein. However, this inhibition is only partially regulated by binding in the 5'UTR and could also be a result of miR-2 binding to the 3' UTR of the target mRNA (Moretti et al. 2010).

In humans, endogenous miRNAs can regulate approximately 30 % of all protein coding genes (Lewis et al. 2005) and thus coordinate key cellular processes such as cell proliferation, DNA repair, differentiation, metabolism and apoptosis (Croce and Calin 2005). Since the discovery of miRNAs in 1993, several studies have been performed to understand how these molecules are involved in normal physiological processes and the onset of a wide variety of diseases. Over 200 different types of human miRNAs have been described, and this number is increasing rapidly (http://www.mirbase.org/).

Indeed, the deregulation of cellular miRNA expression has been consistently observed in several pathologies, including cancer (Lu et al. 2005). Each miRNA consists of a unique sequence and exhibits a cell type-dependent pattern of expression (Landgraf et al. 2007; Lu et al. 2005).

10.2   Dysregulation of miRNA Expression in Human Cancer

In recent years, miRNAs have become very important for the molecular understanding of human cancers (He and Hannon 2004). The small size of miRNAs and their ability to act without complete base pairing mean that a single miRNA can regulate several mRNA targets and that multiple miRNAs can cooperate to control a single mRNA (Koturbash et al. 2011). Recently, it has been proposed that the pathogenesis of cancer involves, among other macromolecules, the miRNAs, the expression profiles of which are associated with the prognosis and therapeutic responses of a variety of human cancers (Ma et al. 2007).

The miRNAs involved in the neoplastic process can be divided into two groups, oncomiRs and anti-oncomiRs, that negatively regulate tumor suppressor genes and oncogenes, respectively (Fabbri et al. 2007). For example, a group of miRNAs, known as the miR-17-92 cluster, is considered to be potentially oncogenic because it is frequently overexpressed in solid tumors (Garzon et al. 2009). However, the miRNA let-7 is considered to be an anti-oncomiR ecause it blocks the action of the RAS oncogene (Johnson et al. 2005).

Approximately 50 % of miRNAs are located at fragile sites or regions associated with cancer (Calin et al. 2004a, b, 2002). Studies analyzing miRNA expression profiles on a large scale have shown that these molecules can be used as molecular biomarkers because their expression signatures can differentiate, at a molecular level, normal cells from neoplastic cells and can be used to classify several cancer types (Calin and Croce 2006).

The first studies addressing miRNAs in tumors described the identification of sequences encoding miR-15 and miR-16 in the chromosomal region 13q14 that are deleted in more than half of chronic myeloid leukemia patients (Calin et al. 2002).

Since then, altered miRNA expression has been observed in several tumors including breast, colorectal, prostate, lung and liver tumors. In breast cancer, Iorio et al. (2005) performed the first study evaluating miRNA expression in tumor samples from breast cancer patients using microarray technology. In this study, the miRNAs miR-10b, miR-125b and miR-145 were upregulated in breast cancer patients compared with normal individuals, while miR-155 and miR-21 were downregulated.

Another relevant study assessed the expression profiles of 540 miRNAs in samples of six different solid tumors (colon, lung, breast, stomach, pancreas, and prostate). Initially, a comparison between tumor tissues and normal tissues was performed that identified 26 upregulated and 17 downregulated miRNAs. The results showed that miR-21 and miR-17-5p were upregulated in all cancers but that miR-155 was upregulated only in breast, lung and colon cancers. Furthermore, it was observed that miR-106a was upregulated in colon cancer, but weakly expressed in breast cancer (Volinia et al. 2006).

Several studies showed that the expression of the miR-34 family is induced in response to DNA damage in a p53-dependent manner (Bommer et al. 2007; Chang et al. 2007). The p53 transcription factor is a tumor suppressor activated after DNA damage, oxidative stress or oncogene activation that acts as a guardian of the genome (Vousden and Lane 2007).

The miR-34 family is evolutionarily conserved and, in the vertebrate genome, is composed of three members (miR-34a, miR-34b and miR-34c) that are expressed from two different loci: miR-34a is located on chromosome 1p16, while miR-34b and miR-34c are located on chromosome 11q23A (Bommer et al. 2007). Several stimuli, such as DNA damage and oncogene activation, lead to the expression of miR-34a and miR-34b/c, which are directly regulated by p53. Thus, the identification of the miR-34 family as a p53 target expands the repertoire of regulation by this gene, indicating that p53 is important in the regulation of both protein-coding genes and miRNAs.

An increasing number of studies have shown that miRNAs play an important role in the onset, progression and subsequent invasion and metastasis of different types of cancers (Almeida et al. 2010; Calin and Croce 2006). The involvement of miRNAs in tumor metastasis has been under intense investigation in recent years. The role of miRNAs in metastasis was initially discovered by Ma et al. (2007) who found that miR-10b initiates invasion and metastasis in breast cancer. Later, Tavazoie et al. (2008) revealed that miR-335 suppressing metastasis and migration by targeting the transcription factor SOX4 and extracellular matriz components such as tenascin C. At the same time, it was reported that miR-373 and miR-520c stimulate tumor cell migration and invasion, which was suggested to result from the suppression of CD44, a gene encoding a surface receptor for hyaluronic acid that seems to be a candidate marker for early cancer detection (Negrini and Calin 2008).

Together, these studies revealed a balance between non-coding RNAs as both stimulators and inhibitors of metastasis, leading to the identification of several potential targets that represent a molecular link between the loss of miRNA expression and the specific behavior of a given tumor.

These findings are important not only because they represent a new field of research, but also because the authors finely dissected the molecular pathways that are involved in the metastasis of mammary tumors. The abnormal expression of miRNAs in tumors, which is characterized by different expression levels of the mature miRNA or miRNA precursor sequences compared to normal cells, has proven to be the main abnormality of the "miRNome" (the genome-wide set of miRNAs) in cancer cells.

10.3   Circulating miRNAs: Novel Biomarkers for Cancer

Considering that dysregulated miRNA expression is tissue specific, several studies have explored the potential use of miRNA expression profiles as biomarkers for the diagnosis, prognosis and response to treatment of several types of cancers. The first tumor-specific miRNAs were discovered in the serum of patients with B-cell lymphoma, in which the increased expression of miR-21 was associated with an increase in disease-free survival (Lawrie et al. 2008). Since then, various studies have assessed the potential use of serum or plasma miRNAs as potential biomarkers for different types of human cancers.

An interesting prospective study showed that tumor-specific microRNAs such as miR-195 are differentially expressed in the circulation of women with breast cancer when compared with women at same age in a control group. Furthermore, it was observed that after resection of the tumor, serum levels of miR-195 and let-7a were reduced (Heneghan et al. 2010). Another study showed that analysis of the combined expression of miR-21, miR-210, miR-155 and miR-196a in plasma could distinguish patients with breast adenocarcinoma from control subjects (Wang et al. 2009).

The ability to use miRNAs as biomarkers in the diagnosis and prognosis of cancer is primarily due to their stability and resistance to long periods of storage and conditions that would normally cause degradation of other RNAs. Recent studies have shown that miRNAs are preserved in serum samples that have been stored for 10 years (Patnaik et al. 2010). This stability can be partially explained by the discovery of lipoprotein complexes, including small vesicles of endocytic origin called exosomes or microvesicles, which carry miRNAs (Valadi et al. 2007), messenger RNAs (El-Hefnawy et al. 2004) and proteins (Smalheiser 2007).

These microvesicles are generally characterized by size into two major classes: a larger class of approximately 200–1000 nm and a smaller class of microvesicles of approximately 30–200 nm, called exosomes. These microvesicles are formed by internalization of the endosomal membrane to form multivesicular bodies (MVBs) that can subsequently merge with the plasma membrane, releasing the exosomes to the outside of the cell (Théry et al. 2002). When in circulation, these exosomes can export their miRNAs to recipient cells through endocytosis. After entering the cell, the delivered miRNAs are processed by the same machinery used for their biogenesis and can regulate gene expression in the recipient cell, leading to a physiological change in that cell (Fig. 10.1).

Exosomes carrying miRNAs can be found not only in blood but also in other fluids, such as saliva and urine (Michael et al. 2010). Recently, exosomes have emerged as important mediators of cellular communication that are involved in normal physiological processes such as the immune response, lactation and neuronal function (Admyre et al. 2007), as well as in the development and progression of diseases such as cancer (Record 2013).

In the context of cancer, this mechanism was clearly demonstrated in glioblastoma patients in whom tumor cells exported exosomes containing mRNA, miRNA and angiogenic proteins that were captured through EGFRvIII receptors by normal cells, such as brain microvascular endothelial cells (Skog et al. 2008). In this study, it was shown that the messages delivered by tumor-derived exosomes containing miRNAs could promote tumor progression. Furthermore, the results of this study showed that patients with cancer had higher levels of exosomes in their plasma than control subjects.

The use of miRNAs as cancer biomarkers is dependent on scientific evidence and studies that aim to identify tissue-specific miRNAs detectable in fluids and to establish molecular signatures capable of characterizing the health status of pa-

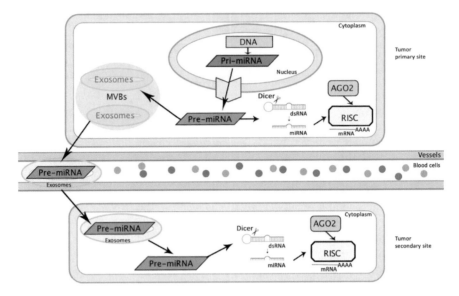

Fig. 10.1 The model proposed for the biogenesis and mechanism of action of circulating miR-NAs. The miRNAs are initially transcribed in the nucleus and subsequently processed by Dicer in the cytoplasm. There are at least two pathways in which the pre-miRNA can be packaged in microparticles: transported by exosomes and MVBs or other ways not entirely clear. After fusion with the plasma membrane, the exosomes are released into the circulation. When is located inside the recipient cell (or secondary sites), miRNAs can be processed by the same miRNA biogenesis machinery thereby regulating the gene expression of this new cell

tients. Therefore, circulating miRNAs in body fluids and in extracellular compartments can act as hormones, triggering changes in cellular gene expression through components secreted by a donor cell at the primary tumor site.

10.4 Computational Approaches for miRNA Target Prediction and Their Application in Cancer Research

As described in previous sections, miRNAs are a novel, important class of regulatory molecules (Sevignani et al. 2006). The study of miRNAs is in its infancy, and the characterization of potential targets usually provides clues to understand the roles of miRNAs. MiRNAs have the ability to regulate genes involved in diverse cellular processes such as growth, proliferation, and cellular differentiation as well as a variety of diseases, such as cancer (Bartel 2004).

Studies in this field have shown that each miRNA can bind to several transcripts, with an estimated average of approximately 200 targets per miRNA (Friedman et al. 2009). Two main strategies have been used to identify the targets of miRNAs: direct cloning and computational predictions (Bentwich et al. 2005). The computational strategy for target identification has been questioned because of the high rate of false positive results provided by the prediction algorithms. Recently, new experimental methods for large-scale target validation have emerged, such as Stable Isotope Labeling by Amino acids in Cell culture (SILAC), which is a mass spectrometry (MS)-based quantitative proteomics used for miRNA target screening (Vinther et al. 2006), and Photoactivatable-Ribonucleoside-Enhanced Crosslinking and Immunoprecipitation (PAR-CLIP) which is a biochemical method used for identifying microRNA-containing ribonucleoprotein complexes (miRNPs) (Hafner et al. 2012), among others. These new methods have provided more robust results in this field and may lead to the creation of future computational tools.

This section mainly addresses the computational tools for target prediction in general and strategies that have been applied to the study of cancer. In general, the available target prediction tools can currently be divided into two categories. The first includes those with pre-computed predictions, in which the user does not need to perform all the steps but can simply search by miRNA name or identification. The second consists of a server that allows the user to add their own sequences for the analysis, making the prediction more versatile (Lindow 2011).

The first database designed to catalog the sequences of miRNAs as soon as they are identified was miRBase. This database serves as a repository for sequences and annotations, providing access to virtually all published miRNAs (Griffiths-Jones et al. 2006). Currently, the database is in version 20 (June 2013) and consists of 24521 entries representing hairpin precursor miRNAs and 30424 entries for mature miRNAs in 206 species (www.mirbase.org). It is important to emphasize that with the advance of large-scale technologies, especially next-generation sequencing, new miRNAs have been reported at a very high rate (Cordero et al. 2012; Stäehler et al. 2012).

One of the basic principles widely used by these algorithms is complementary base pairing. In plants, complementary base pairing between miRNAs and their targets is almost perfect. In mammals, only a portion of the miRNA binds directly to the target, making prediction more difficult. Thus, several rules have been established to identify targets based on sequence complementarity; complementarity is especially important in positions 2–8 (seed region) at the 5' end of the miRNA, or a high degree of similarity in the 3' end of the miRNA can compensate for low complementarity in the seed region (Bartel 2009; Rajewsky 2006). Some additional criteria vary with the type of algorithm and may include RNA-RNA interactions based on thermodynamic principles or permission of G:U pairing, among others (Bartel 2009).

Furthermore, several algorithms use patterns of conservation for target prediction. However, because not all sites are necessarily conserved, some programs are no longer including this step (Thadani and Tammi 2006). Nevertheless, conservation is of great relevance in cancer. MicroRNAs reported to be oncogenes or tumor suppressors are frequently conserved across species (Wang et al. 2010). Furthermore, the transcription of ultra-conserved regions (UCRs) among humans, rats and mice also appears to be related to miRNAs, and there are even databases that compute this information (Goymer 2007; Taccioli et al. 2009).

Overall, the most popular computational tools that follow these rules and predict targets based mainly on sites in the 3' UTR of the target are DIANA-microT (Maragkakis et al. 2011), miRanda (Betel et al. 2010), PicTar (Krek et al. 2005), PITA (Kertesz et al. 2007), RNAhybrid (Rehmsmeier et al. 2004) and TargetScan (Grimson et al. 2007). Moreover, several tools have recently been developed to study the interactions between miRNAs and the 5' UTR (or CDS) of target genes, such as miBridge, miRTar, miRWalk and Sfold-STarMirDB (Da Sacco and Masotti 2012). Other widely used tools that allow the combination of several popular algorithms for target prediction are MAMI (Liang 2008), MirGen (Megraw et al. 2007) and miRDip (Shirdel et al. 2011).

Other types of algorithms combine target prediction with gene expression data. Among the most popular of these tools is GenMiR, which uses a Bayesian method to combine results obtained from prediction algorithms, such as TargetScan, with expression data obtained from miRNA and mRNA microarrays (Huang et al. 2007). Other methods such as correlations, probabilistic methods, regression and associations with transcription factors have been proposed to assess miRNA-mRNA networks based on gene expression data (Joung et al. 2007; Li et al. 2010).

Therefore, the use of a variety of tools is recommended to computationally search for the most representative targets and avoid false-positive data. In general, prediction methods considered efficient contain the following steps: (i) use of several algorithms with different methods for confirmation, (ii) comparisons between mRNA and microRNA expression profiles, (iii) consideration of nearby sites that may act synergistically, and (iv) experimental validation or subsequent functional assays (Witkos et al. 2011). It is important to consider that although in silico analysis is extremely useful for the prediction of microRNA targets, experimental validation is necessary to evaluate the real role of these putative interactions.

Furthermore, it is important to address the large amounts of data that are generated from large-scale transcriptome studies or large-scale target prediction. Usually, they can be summarized using functional enrichment analysis. The goal of this strategy is to provide a statistical method to estimate the enrichment, i.e., the higher-than-expected representation, of certain functional categories, excluding functional terms that could be identified by chance. Several tools use the Fisher's exact test to estimate enrichment. Databases such as DAVID (Database for Annotation, Visualization and Integrated Discovery) analyze data based on Gene Ontology functional categories and pathways from databases such as KEGG and others (Huang et al. 2008). This strategy of enrichment has been widely applied, as seen in recent work from our group (Takahashi et al. 2014).

Finally, there are some databases, which are generally manually curated from the literature, that compile information about diseases to derive biologically relevant information from lists of miRNAs. Among the most well-known are miR2Disease, which addresses 163 diseases, and the Human microRNA Disease Database (HMDD), which provides information about microRNAs for more than 100 diseases and approximately 40 tissues. In the case of cancer, there are several databases available such as oncomiRDB, which computes oncogenic and tumor suppressor miRNAs (Wang et al. 2014), miRCancer, which includes 236 miRNAs in 79 cancer types (Xie et al. 2013), and TUMIR, which includes 1163 studies addressing the relationship between miRNAs and cancer (Dong et al. 2013). In summary, these tools provide evidence of the role of microRNAs from their targets, with approaches that can be used in cancer research.

References

Admyre C, Johansson SM, Qazi KR, Filén J-J, Lahesmaa R, Norman M, Neve EPA, Scheynius A, Gabrielsson S (2007) Exosomes with immune modulatory features are present in human breast milk. J Immunol 179:1969–1978

Almeida MI, Reis RM, Calin GA (2010) MicroRNAs and metastases—the neuroblastoma link. Cancer Biol Ther 9:453–454

Ambros V (2004) The functions of animal microRNAs. Nature 431:350–355. doi:10.1038/nature02871

Babiarz JE, Ruby JG, Wang Y, Bartel DP, Blelloch R (2008) Mouse ES cells express endogenous shRNAs, siRNAs, and other microprocessor-independent, dicer-dependent small RNAs. Genes Dev 22:2773–2785. doi:10.1101/gad.1705308

Bartel DP (2004) MicroRNAs: genomics, biogenesis, mechanism, and function. Cell 116:281–297

Bartel DP (2009) MicroRNAs: target recognition and regulatory functions. Cell 136:215–233. doi:10.1016/j.cell.2009.01.002

Bentwich I, Avniel A, Karov Y, Aharonov R, Gilad S, Barad O, Barzilai A, Einat P, Einav U, Meiri E, Sharon E, Spector Y, Bentwich Z (2005) Identification of hundreds of conserved and non-conserved human microRNAs. Nat Genet 37:766–770. doi:10.1038/ng1590

Betel D, Koppal A, Agius P, Sander C, Leslie C (2010) Comprehensive modeling of microRNA targets predicts functional non-conserved and non-canonical sites. Genome Biol 11:R90. doi:10.1186/gb-2010-11-8-r90

Bogerd HP, Karnowski HW, Cai X, Shin J, Pohlers M, Cullen BR (2010) A mammalian herpesvirus uses noncanonical expression and processing mechanisms to generate viral microRNAs. Mol Cell 37:135–142. doi:10.1016/j.molcel.2009.12.016

Bommer GT, Gerin I, Feng Y, Kaczorowski AJ, Kuick R, Love RE, Zhai Y, Giordano TJ, Qin ZS, Moore BB, MacDougald OA, Cho KR, Fearon ER (2007) p53-mediated activation of miRNA34 candidate tumor-suppressor genes. Curr Biol 17:1298–1307. doi:10.1016/j.cub.2007.06.068

Calin GA, Croce CM (2006) MicroRNA signatures in human cancers. Nat Rev Cancer 6:857–866

Calin GA, Dumitru CD, Shimizu M, Bichi R, Zupo S, Noch E, Aldler H, Rattan S, Keating M, Rai K, Rassenti L, Kipps T, Negrini M, Bullrich F, Croce CM (2002) Frequent deletions and down-regulation of micro- RNA genes miR15 and miR16 at 13q14 in chronic lymphocytic leukemia. Proc Natl Acad Sci U S A 99:15524–15529. doi:10.1073/pnas.242606799

Calin GA, Liu C-G, Sevignani C, Ferracin M, Felli N, Dumitru CD, Shimizu M, Cimmino A, Zupo S, Dono M, Dell'Aquila ML, Alder H, Rassenti L, Kipps TJ, Bullrich F, Negrini M, Croce CM (2004a) MicroRNA profiling reveals distinct signatures in B cell chronic lymphocytic leukemias. Proc Natl Acad Sci U S A 101:11755–11760. doi:10.1073/pnas.0404432101

Calin GA, Sevignani C, Dumitru CD, Hyslop T, Noch E, Yendamuri S, Shimizu M, Rattan S, Bullrich F, Negrini M, Croce CM (2004b) Human microRNA genes are frequently located at fragile sites and genomic regions involved in cancers. Proc Natl Acad Sci U S A 101:2999–3004. doi:10.1073/pnas.0307323101

Cazalla D, Xie M, Steitz JA (2011) A primate herpesvirus uses the integrator complex to generate viral microRNAs. Mol Cell 43:982–992. doi:10.1016/j.molcel.2011.07.025

Chang T-C, Wentzel EA, Kent OA, Ramachandran K, Mullendore M, Lee KH, Feldmann G, Yamakuchi M, Ferlito M, Lowenstein CJ, Arking DE, Beer MA, Maitra A, Mendell JT (2007)

Transactivation of miR-34a by p53 broadly influences gene expression and promotes apoptosis. Mol Cell 26:745–752. doi:10.1016/j.molcel.2007.05.010

Cordero F, Beccuti M, Arigoni M, Donatelli S, Calogero RA (2012) Optimizing a massive parallel sequencing workflow for quantitative miRNA expression analysis. PLoS One 7:e31630. doi:10.1371/journal.pone.0031630

Croce CM, Calin GA (2005) miRNAs, cancer, and stem cell division. Cell 122:6–7. doi:10.1016/j.cell.2005.06.036

Da Sacco L, Masotti A (2012) Recent insights and novel bioinformatics tools to understand the role of microRNAs binding to 5' untranslated region. Int J Mol Sci 14:480–495. doi:10.3390/ijms14010480

Dong L, Luo M, Wang F, Zhang J, Li T, Yu J (2013) TUMIR: an experimentally supported database of microRNA deregulation in various cancers. J Clin Bioinforma 3:7. doi:10.1186/2043-9113-3-7

El-Hefnawy T, Raja S, Kelly L, Bigbee WL, Kirkwood JM, Luketich JD, Godfrey TE (2004) Characterization of amplifiable, circulating RNA in plasma and its potential as a tool for cancer diagnostics. Clin Chem 50:564–573. doi:10.1373/clinchem.2003.028506

Fabbri M, Ivan M, Cimmino A, Negrini M, Calin GA (2007) Regulatory mechanisms of microRNAs involvement in cancer. Expert Opin Biol Ther 7:1009–1019. doi:10.1517/14712598.7.7.1009

Friedman RC, Farh KK-H, Burge CB, Bartel DP (2009) Most mammalian mRNAs are conserved targets of microRNAs. Genome Res 19:92–105. doi:10.1101/gr.082701.108

Garzon R, Calin GA, Croce CM (2009) MicroRNAs in Cancer. Annu Rev Med 60:167–179. doi:10.1146/annurev.med.59.053006.104707

Goymer P (2007) Genetics: conserved by evolution, but altered in cancer. Nat Rev Cancer 7:812–813. doi:10.1038/nrc2261

Griffiths-Jones S, Grocock RJ, van Dongen S, Bateman A, Enright AJ (2006) miRBase: microRNA sequences, targets and gene nomenclature. Nucleic Acids Res 34:D140–144. doi:10.1093/nar/gkj112

Grimson A, Farh KK-H, Johnston WK, Garrett-Engele P, Lim LP, Bartel DP (2007) MicroRNA targeting specificity in mammals: determinants beyond seed pairing. Mol Cell 27:91–105. doi:10.1016/j.molcel.2007.06.017

Hafner M, Lianoglou S, Tuschl T, Betel D (2012) Genome-wide identification of miRNA targets by PAR-CLIP. Methods 58:94–105. doi:10.1016/j.ymeth.2012.08.006

He L, Hannon GJ (2004) MicroRNAs: small RNAs with a big role in gene regulation. Nat Rev Genet 5:522–531. doi:10.1038/nrg1379

Heneghan HM, Miller N, Lowery AJ, Sweeney KJ, Newell J, Kerin MJ (2010) Circulating microRNAs as novel minimally invasive biomarkers for breast cancer. Ann Surg 251:499–505. doi:10.1097/SLA.0b013e3181cc939f

Huang JC, Babak T, Corson TW, Chua G, Khan S, Gallie BL, Hughes TR, Blencowe BJ, Frey BJ, Morris QD (2007) Using expression profiling data to identify human microRNA targets. Nat Methods 4:1045–1049. doi:10.1038/nmeth1130

Huang DW, Sherman BT, Lempicki RA (2008) Systematic and integrative analysis of large gene lists using DAVID bioinformatics resources. Nat Protoc 4:44–57. doi:10.1038/nprot.2008.211

Iorio MV, Ferracin M, Liu C-G, Veronese A, Spizzo R, Sabbioni S, Magri E, Pedriali M, Fabbri M, Campiglio M, Ménard S, Palazzo JP, Rosenberg A, Musiani P, Volinia S, Nenci I, Calin GA, Querzoli P, Negrini M, Croce CM (2005) MicroRNA gene expression deregulation in human breast cancer. Cancer Res 65:7065–7070. doi:10.1158/0008-5472.CAN-05-1783

Johnson SM, Grosshans H, Shingara J, Byrom M, Jarvis R, Cheng A, Labourier E, Reinert KL, Brown D, Slack FJ (2005) RAS is regulated by the let-7 microRNA family. Cell 120:635–647. doi:10.1016/j.cell.2005.01.014

Joung J-G, Hwang K-B, Nam J-W, Kim S-J, Zhang B-T (2007) Discovery of microRNA-mRNA modules via population-based probabilistic learning. Bioinformatics 23:1141–1147. doi:10.1093/bioinformatics/btm045

Kertesz M, Iovino N, Unnerstall U, Gaul U, Segal E (2007) The role of site accessibility in microRNA target recognition. Nat Genet 39:1278–1284. doi:10.1038/ng2135

Ketting RF, Fischer SE, Bernstein E, Sijen T, Hannon GJ, Plasterk RH (2001) Dicer functions in RNA interference and in synthesis of small RNA involved in developmental timing in C. elegans. Genes Dev 15:2654–2659. doi:10.1101/gad.927801

Koturbash I, Zemp FJ, Pogribny I, Kovalchuk O (2011) Small molecules with big effects: the role of the microRNAome in cancer and carcinogenesis. Mutat Res 722:94–105. doi:10.1016/j.mrgentox.2010.05.006

Krek A, Grün D, Poy MN, Wolf R, Rosenberg L, Epstein EJ, MacMenamin P, da Piedade I, Gunsalus KC, Stoffel M, Rajewsky N (2005) Combinatorial microRNA target predictions. Nat Genet 37:495–500. doi:10.1038/ng1536

Landgraf P, Rusu M, Sheridan R, Sewer A, Iovino N et al (2007) A mammalian microRNA expression atlas based on small RNA library sequencing. Cell 129:1401–1414. doi:10.1016/j.cell.2007.04.040

Lawrie CH, Gal S, Dunlop HM, Pushkaran B, Liggins AP, Pulford K, Banham AH, Pezzella F, Boultwood J, Wainscoat JS, Hatton CSR, Harris AL (2008) Detection of elevated levels of tumour-associated microRNAs in serum of patients with diffuse large B-cell lymphoma. Br J Haematol 141:672–675. doi:10.1111/j.1365-2141.2008.07077.x

Lee Y, Ahn C, Han J, Choi H, Kim J, Yim J, Lee J, Provost P, Rådmark O, Kim S, Kim VN (2003) The nuclear RNase III Drosha initiates microRNA processing. Nature 425:415–419. doi:10.1038/nature01957

Lewis BP, Burge CB, Bartel DP (2005) Conserved seed pairing, often flanked by adenosines, indicates that thousands of human genes are microRNA targets. Cell 120:15–20. doi:10.1016/j.cell.2004.12.035

Li L, Xu J, Yang D, Tan X, Wang H (2010) Computational approaches for microRNA studies: a review. Mamm Genome 21:1–12. doi:10.1007/s00335-009-9241-2

Liang Y (2008) An expression meta-analysis of predicted microRNA targets identifies a diagnostic signature for lung cancer. BMC Med Genomics 1:61. doi:10.1186/1755-8794-1-61

Lindow M (2011) Prediction of targets for microRNAs. Methods Mol Biol 703:311–317. doi:10.1007/978-1-59745-248-9_21

Lu J, Getz G, Miska EA, Alvarez-Saavedra E, Lamb J, Peck D, Sweet-Cordero A, Ebert BL, Mak RH, Ferrando AA, Downing JR, Jacks T, Horvitz HR, Golub TR (2005) MicroRNA expression profiles classify human cancers. Nature 435:834–838. doi:10.1038/nature03702

Ma L, Teruya-Feldstein J, Weinberg RA (2007) Tumour invasion and metastasis initiated by microRNA-10b in breast cancer. Nature 449:682–688. doi:10.1038/nature06174

Maragkakis M, Vergoulis T, Alexiou P, Reczko M, Plomaritou K, Gousis M, Kourtis K, Koziris N, Dalamagas T, Hatzigeorgiou AG (2011) DIANA-microT web server upgrade supports fly and worm miRNA target prediction and bibliographic miRNA to disease association. Nucleic Acids Res 39:W145–148. doi:10.1093/nar/gkr294

Megraw M, Sethupathy P, Corda B, Hatzigeorgiou AG (2007) miRGen: a database for the study of animal microRNA genomic organization and function. Nucleic Acids Res 35:D149–155. doi:10.1093/nar/gkl904

Michael A, Bajracharya SD, Yuen PST, Zhou H, Star RA, Illei GG, Alevizos I (2010) Exosomes from human saliva as a source of microRNA biomarkers. Oral Dis 16:34–38. doi:10.1111/j.1601-0825.2009.01604.x

Moretti F, Thermann R, Hentze MW (2010) Mechanism of translational regulation by miR-2 from sites in the 5' untranslated region or the open reading frame. RNA 16:2493–2502. doi:10.1261/rna.2384610

Negrini M, Calin GA (2008) Breast cancer metastasis: a microRNA story. Breast Cancer Res 10:203. doi:10.1186/bcr1867

Okamura K, Hagen JW, Duan H, Tyler DM, Lai EC (2007) The mirtron pathway generates microRNA-class regulatory RNAs in Drosophila. Cell 130:89–100. doi:10.1016/j.cell.2007.06.028

Patnaik SK, Mallick R, Yendamuri S (2010) Detection of microRNAs in dried serum blots. Anal Biochem 407:147–149. doi:10.1016/j.ab.2010.08.004

Rajewsky N (2006) MicroRNA target predictions in animals. Nat Genet (Suppl. 38):S8–13. doi:10.1038/ng1798

Record M (2013) Exosomal lipids in cell–cell communication. In: Zhang H-G (ed) Emerging concepts of tumor exosome-mediated cell-cell communication, Springer, New York, pp 47–68

Rehmsmeier M, Steffen P, Hochsmann M, Giegerich R (2004) Fast and effective prediction of microRNA/target duplexes. RNA 10:1507–1517. doi:10.1261/rna.5248604

Ruby JG, Jan CH, Bartel DP (2007) Intronic microRNA precursors that bypass Drosha processing. Nature 448:83–86. doi:10.1038/nature05983

Sevignani C, Calin GA, Siracusa LD, Croce CM (2006) Mammalian microRNAs: a small world for fine-tuning gene expression. Mamm Genome 17:189–202. doi:10.1007/s00335-005-0066-3

Shirdel EA, Xie W, Mak TW, Jurisica I (2011) NAViGaTing the micronome—using multiple microRNA prediction databases to identify signalling pathway-associated microRNAs. PLoS One 6:e17429. doi:10.1371/journal.pone.0017429

Skog J, Würdinger T, van Rijn S, Meijer DH, Gainche L, Sena-Esteves M, Curry WT Jr, Carter BS, Krichevsky AM, Breakefield XO (2008) Glioblastoma microvesicles transport RNA and proteins that promote tumour growth and provide diagnostic biomarkers. Nat Cell Biol 10:1470–1476. doi:10.1038/ncb1800

Smalheiser NR (2007) Exosomal transfer of proteins and RNAs at synapses in the nervous system. Biol Direct 2:35. doi:10.1186/1745-6150-2-35

Stäehler CF, Keller A, Leidinger P, Backes C, Chandran A, Wischhusen J, Meder B, Meese E (2012) Whole miRNome-wide differential co-expression of microRNAs. Genomic Proteomics Bioinform 10:285–294. doi:10.1016/j.gpb.2012.08.003

Taccioli C, Fabbri E, Visone R, Volinia S, Calin GA, Fong LY, Gambari R, Bottoni A, Acunzo M, Hagan J, Iorio MV, Piovan C, Romano G, Croce CM (2009) UCbase & miRfunc: a database of ultraconserved sequences and microRNA function. Nucleic Acids Res 37:D41–D48. doi:10.1093/nar/gkn702

Takahashi P, Xavier DJ, Evangelista AF, Manoel-Caetano FS, Macedo C, Collares CVA, Foss-Freitas MC, Foss MC, Rassi DM, Donadi EA, Passos GA, Sakamoto-Hojo ET (2014) MicroRNA expression profiling and functional annotation analysis of their targets in patients with type 1 diabetes mellitus. Gene 539:213–223. doi:10.1016/j.gene.2014.01.075

Tavazoie SF, Alarcón C, Oskarsson T, Padua D, Wang Q, Bos PD, Gerald WL, Massagué J (2008) Endogenous human microRNAs that suppress breast cancer metastasis. Nature 451:147–152. doi:10.1038/nature06487

Thadani R, Tammi MT (2006) MicroTar: predicting microRNA targets from RNA duplexes. BMC Bioinformatics 7(Suppl. 5):S20. doi:10.1186/1471-2105-7-S5-S20

Théry C, Zitvogel L, Amigorena S (2002) Exosomes: composition, biogenesis and function. Nat Rev Immunol 2:569–579. doi:10.1038/nri855

Valadi H, Ekström K, Bossios A, Sjöstrand M, Lee JJ, Lötvall JO (2007) Exosome-mediated transfer of mRNAs and microRNAs is a novel mechanism of genetic exchange between cells. Nat Cell Biol 9:654–659. doi:10.1038/ncb1596

Vinther J, Hedegaard MM, Gardner PP, Andersen JS, Arctander P (2006) Identification of miRNA targets with stable isotope labeling by amino acids in cell culture. Nucleic Acids Res 34:e107. doi:10.1093/nar/gkl590

Volinia S, Calin GA, Liu C-G, Ambs S, Cimmino A, Petrocca F, Visone R, Iorio M, Roldo C, Ferracin M, Prueitt RL, Yanaihara N, Lanza G, Scarpa A, Vecchione A, Negrini M, Harris CC, Croce CM (2006) A microRNA expression signature of human solid tumors defines cancer gene targets. Proc Natl Acad Sci U S A 103:2257–2261. doi:10.1073/pnas.0510565103

Vousden KH, Lane DP (2007) p53 in health and disease. Nat Rev Mol Cell Biol 8:275–283. doi:10.1038/nrm2147

Wang J, Chen J, Chang P, LeBlanc A, Li D, Abbruzzesse JL, Frazier ML, Killary AM, Sen S (2009) MicroRNAs in plasma of pancreatic ductal adenocarcinoma patients as novel blood-based biomarkers of disease. Cancer Prev Res (Phila) 2:807–813. doi:10.1158/1940-6207.CAPR-09-0094

Wang D, Qiu C, Zhang H, Wang J, Cui Q, Yin Y (2010) Human microRNA oncogenes and tumor suppressors show significantly different biological patterns: from functions to targets. PLoS One 5:e13067. doi:10.1371/journal.pone.0013067

Wang D, Gu J, Wang T, Ding Z (2014) OncomiRDB: a database for the experimentally verified oncogenic and tumor-suppressive microRNAs. Bioinformatics Btu155. doi:10.1093/bioinformatics/btu155

Witkos TM, Koscianska E, Krzyzosiak WJ (2011) Practical aspects of microRNA target prediction. Curr Mol Med 11:93–109

Xie B, Ding Q, Han H, Wu D (2013) miRCancer: a microRNA-cancer association database constructed by text mining on literature. Bioinformatics 29:638–644. doi:10.1093/bioinformatics/btt014

Yang J-S, Lai EC (2011) Alternative miRNA biogenesis pathways and the interpretation of core miRNA pathway mutants. Mol Cell 43:892–903. doi:10.1016/j.molcel.2011.07.024

Chapter 11
Transcriptome Profiling in Chronic Inflammatory Diseases of the Musculoskeletal System

Renê Donizeti Ribeiro de Oliveira and Paulo Louzada-Júnior

Abstract The musculoskeletal system may be affected by a myriad of conditions. Many of them are inflammatory and chronicle diseases, with functional impairment and limiting pain as major symptoms. In common, these diseases have the addressing of immune cells to musculoskeletal system structures, leading to a persistent inflammation and, in some situations, autoimmune responses. This chapter will verse about rheumatoid arthritis and the spondyloarthritis. We discuss the knowledge according to some key points on their pathophysiology, attempts of phenotypic sub-classification, therapy response prediction and elucidations on drugs mechanisms of action, with emphasis on the advantages and gains from studies using microarray technology, its limitations and future possibilities. Moreover, we emphasize the distinct patterns obtained by gene expression studies according to the RNA source, like peripheral blood and synovia, the subjects studied, whether patients are compared with healthy controls or patients of mechanic-degenerative or of other autoimmune diseases, and the stage of the diseases.

Why has it been so difficult to find the place for microarray in daily medical decisions? Is it only regarded to the heterogeneity of autoimmune diseases, mainly because these are conditions highly influenced by an enormous amount of external and internal stimuli? The answer for this question is neither affirmative nor negative. Apart of uncertainties, we can rely on some data, like good candidates for a peripheral blood rheumatoid arthritis gene chip, which are summarized in the final section along with the authors' opinions and the perspectives for advances in this intriguing and exciting field.

11.1 Introduction

The musculoskeletal system may be affected by a myriad of conditions, compromising muscles and tendons, ligaments, synovia, bursas and bones (Felson 2005). Many of them are inflammatory and chronicle diseases, with functional impairment

R. D. R. de Oliveira (✉) · P. Louzada-Júnior
Department of Clinical Medicine, Ribeirão Preto Medical School,
University of São Paulo, 14049-900, Ribeirão Preto, São paulo, Brazil
e-mail: reneimuno@yahoo.com.br

© Springer International Publishing Switzerland 2014
G. A. Passos (ed.), *Transcriptomics in Health and Disease,*
DOI 10.1007/978-3-319-11985-4_11

195

and limiting pain as major symptoms (McCarty 2005). In common, these diseases have the addressing of immune cells to those structures, leading to a persistent inflammation and, in some situations, autoimmune responses. The dysregulation of immune system is a landmark and all the attempts to achieve clinical remission are based on the control of immune cells activity.

Unfortunately, a not negligible number of patients faces uncontrollable disease for months to years, resulting in persistent disabilities or fatal evolution. Much of the inability to control the chronic inflammatory diseases comes from some hiatus in their knowledge.

Recent years have brought several changes in our understanding on pathophysiology and treatment possibilities, guided mainly by advances in molecular biology techniques. Clinicians certainly need more accurate diagnosis and classification criteria beside the best decision making scenarios, where patients are the more appropriated candidates for previous and new therapies. Molecular approaches can become an outstanding tool for both, pathophysiology investigations and therapies best choices.

Gene expression studies offer a plenty of information on several molecular processes occurring at the same time, being able, indeed, of giving us clues about striking features of a disease and, as crucial, how these processes change during pathophysiological development (Lequerré et al. 2003).

Rheumatoid arthritis (RA) and the spondyloarthritis (SpA) will be the scope of this chapter, with emphasis to the knowledge from cDNA microarray studies.

11.2   Rheumatoid arthritis

RA is a chronic systemic autoimmune disease affecting mainly synovial tissue, whose etiology still remains uncertain, although some risk factors are known, like genetic (HLD-DRB1, PTPN22, PADI and STAT4) and environmental (trauma, tobacco smoke exposure and infections) (Klareskog et al. 2006).

A pre-clinic stage is recognized (Klareskog et al. 2009), when physiopathologic mechanisms are present but not an inflammation in such a magnitude to be assessed by physical examination or complimentary exams. The first clinic is an early stage (early rheumatoid arthritis, ERA) with distinct clinical presentation and molecular (cytokines, chemokines and other inflammatory mediators) pattern, but a cutting edge between this and the second stage, long-standing disease (LSRA), is to be achieved. The identification of at least these two stages is pivotal because the therapy success varies widely for the drugs available with the disease progression. An important concept is the "window of opportunity" (O'Dell 2002), namely the first 6 months of onset, a period with the higher chance of disease control and none or minimal structural damage.

One could argue that different inflammatory and immunopathological processes govern each stage of RA and some data already confirm this, but we are still waiting for the moment when this information will be enough to clearly sub classify patients and, maybe the most important, guide the choice for the best individualized therapy.

The last assertion goes by the order of the day in RA, because for up to 50 % of the patients, functionality has been lost in a painful manner before they receive an efficient therapy (Möttönen et al. 2002). Moreover, from a public healthy funding point of view, hundreds of million dollars (Schoels et al. 2010) are expended each year in unsuccessful treatments, especially with the advent of biological drugs—TNF-blockers, Abatacept, Rituximab and Tocilizumab.

Based on all above exposed, it is peremptory to deepen the knowledge in each one of these disease aspects. Some cDNA microarray studies started contributing this task, as summarized in the following subsections.

11.2.1 Pathophysiology

The majority of the studies on RA pathophysiology compared the patients with three main groups: healthy controls, patients of other autoimmune diseases or patients of osteoarthritis (OA). Few studies bring only an RA group.

11.2.1.1 RA vs healthy control

Peripheral Blood Sampling Studies Using peripheral blood, some groups studied RA signature compared with healthy controls. Teixeira et al. (2009) selected RA patients in use of Disease Modifying Anti-rheumatic Drugs (DMARD) and healthy controls and profiled gene expression of peripheral blood mononuclear cells (PBMC). The dendrogram completely separated patients from controls, finding some processes differentially regulated between the groups, like inflammation, anti-microbial activity, immunomodulatory function and cellular stress. Batliwalla et al. (2005a) studied PBMC from RA patients (all with active disease, 7 with no previous therapy and 22 non-responders to DMARD and before TNF-blocker therapy) and healthy controls and found 52 upregulated genes in RA, being 21 monocyte enriched transcripts. Although there was no perfect distinction into 2 groups by the tree view, RA patients expressed high levels of S100A12 and GAB2, both involved in disease-related pathways. Two groups studied B cells, Szodoray et al. (2006) and Haas et al. (2006). In common, they found highly expressed genes in RA related to functional classes like cell cycle, metabolism, cytokines and apoptosis, with remarkable overexpression of anti-apoptotic and underexpression of pro-apoptotic genes.

Synovial Sampling Studies RA synovial fibroblasts (RASF) were also compared with healthy synovial fibroblasts under hypoxic conditions (Del Rey et al. 2010). This approach intended to evaluate whether RASF respond abnormally to hypoxia, a common environmental characteristic of RA synovia. Several genes, mainly related to anaerobic energy production, cytokines and chemokines were found differentially expressed.

11.2.1.2   RA vs other autoimmune diseas

Peripheral Blood Sampling Studies Silva et al. (2007) showed similarities in PBMC gene expression profile between RA and systemic lupus erythematosus (SLE), with some genes probably involved in the pathogenesis of both diseases. In other study, Maas et al. (2002), with the intention of investigate whether autoimmune diseases might share a gene expression signature, evaluated PBMC from patients of four diseases (SLE, RA, multiple sclerosis and type I diabetes mellitus) comparing to healthy controls before and after influenza vaccination. The comparison between the autoimmune response and a normal response to immunization yielded a set of genes (underexpressed in autoimmunity: pro-apoptotic, enzyme and cycle inhibitors; overexpressed in autoimmunity: intracellular signaling, autoantigens and inflammatory mediators) suggesting the existence of genes primarily involved in autoimmunity.

A shared type I interferon (IFN) signature was described for many autoimmune diseases, including RA (van der Pouw Kraan et al. 2007) and SLE (Baechler et al. 2003). Reynier et al. (2011) studied RA and SLE patients and healthy controls, finding 2 subtypes of RA, one with high expression of type I INF-induced genes, comparable to SLE levels and other with low expression levels, comparable to healthy subjects.

11.2.1.3   RA vs OA

Synovial Sampling Studies Devauchelle et al. (2004) obtained by cDNA microarray of synovial tissues a set of 63 genes able to separate every RA patient from OA patient. Galligan et al. (2007) were able to identify in RASF a signature comparing them to synovial OA fibroblasts in an approach having healthy synovial fibroblasts as control. Several clinic and laboratory aspects (HAQ—Health Assessment Questionnaire score, C-reactive protein, erythrocyte sedimentation rate, rheumatoid factor positivity and methotrexate [MTX]/prednisone treatment) separately studied brought a set of genes able to separate RA from OA patients, albeit these signatures were shared by 30–50 % of the RA or OA patients, frustrating the possibility of a single gene signature for all patients and reinforcing the heterogeneity of such multifactorial conditions like the rheumatic diseases.

In an elegant study, Yoshida et al. (2012) studied the gene expression of synovial specimens from RA and OA patients and correlate them with histologic techniques. The samples were scored as high or low histological synovitis, which was related only to C-reactive protein level. The microarray showed 197 differentially expressed genes, being 47 related to inflammatory response (chemokines and INF-related genes). Pathway analysis demonstrated direct or indirect relationship among TNF, the chemokines and type I IFN. Separation of the samples on dendrogram fit exactly the laboratory and histological sub-classification.

Another possible approach for synovial tissue is to isolate RASF, culture and stimulate them with molecules with recognized influence on RA pathophysiology, e.g. TNF and IL-17. Gallagher et al. (2003) compared the gene expression of RASF and OA synovial fibroblasts stimulated with TNF for 4 or 24 h and these cells with unstimulated ones. The TNF stimulated some processes in RA comparing to OA cells, like cell cycle, apoptotic mediators, cytokines and inflammatory mediators. The authors highlighted the upregulation of IFN-induced genes. Having the primary aim to study apoptosis-related genes, Qingchun et al. (2008) obtained synovial tissue from RA and OA patients and found 8 genes, those anti-apoptotic upregulated and those apoptotic downregulated.

11.2.1.4 RA Patients Only

Synovial Sampling Studies Taberner et al. (2005) isolated RASF and stimulated them with IL-1 or TNF for 4 h. Both stimulations resulted in many commonly regulated processes, like apoptosis, chemokine and cytokines regulation, cell cycle and cell signaling, meaning that these two important cytokines may share distal signaling pathways in RA.

Another group, Zrioual et al. (2009), using similar study design, stimulated synovial fibroblasts for 48 h with IL-17A or IL-17F alone or in combination with TNF, finding that IL-17A and IL-17F share most of regulated genes, as demonstrated by unsupervised hierarchical clustering analysis, and that the typical expression patterns were not significantly changed by the costimulation with TNF. These results may reassure the synergistic roles of the TNF and IL-17 driving the tissue damage.

11.2.2 Stage of disease comparison

As stated previously, RA persists a subclinical disease for a variable period, when many patients manifest arthralgia without joint inflammatory signals. If the joint pain is associated with Rheumatoid factors (RF) and Anti-citrullinated peptide antibodies (ACPA), 40 % of these patients will develop clinical RA (Goekoop-Ruiterman et al. 2007). Van Baarsen et al. (2010) studied gene expression of PBMC from a group of subjects facing arthralgia and positive RF and/or ACPA test comparing it to healthy controls and established RA. Those subjects clustering with RA patients developed RA in a 1 year follow-up and markedly expressed genes involved in IFN-mediated immunity, cytokine-mediated immunity and hematopoiesis, and interestingly, those subjects expressing high levels of genes involved in B cell activation did not develop RA.

Lequerré et al. (2009) compared synovial specimens from ERA, LSRA (treated with MTX) and healthy control, resulting in a clear distinction between ERA and LSRA, being the upregulated processes in ERA more specifically involved in stress responses, defense mechanisms and apoptosis, while in LSRA they were related to cell surface receptor-mediated signal transduction, cell cycle control, apoptosis inhibition, and granulocyte-mediated immunity, suggesting the involvement of a proliferative process. By these data, authors suggest that each stage holds different processes and pathways with some overlapping but with opposite regulations.

11.2.3 Phenotype sub-classification

The RA heterogeneity is confirmed molecularly by some studies in this section.

Peripheral Blood Sampling Studies Junta et al. (2008) studied the gene expression of PBMC from LSRA patients by grouping them taking into account molecular (HLA-DR shared epitope carriage), laboratory (ACPA positivity) and clinical aspects (disease activity and drug therapy), what resulted in different sets of genes specifically regulated in each evaluation and some genes and processes shared by these different ways of dividing patients.

With similar idea, two groups studied RA patients dividing them according to RF positivity (van der Pouw Kraan et al. 2007) or treatment with MTX (Bovin et al. 2004), comparing the gene expression levels between the RA groups and with healthy controls. In both cases, the patient subdivisions did not find gene expression correspondence. Although frustrating, it is interesting to know that in some cases what clinicians see is dissociated from what molecular biologists see.

Synovial Sampling Studies Studying LSRA synovial specimens, van der Pouw Kraan et al. (2003a) found 3 subgroups of patients, namely RA-Ia, with genes involved in antigen processing and presentation, profiling a high inflammatory gene expression signature, and RA-Ib and RA-II, with genes involved in tissue remodeling, particularly through the Wnt pathway.

Ungethuem et al. (2010) proposed an RA sub-classification based on the expression of one gene, PGR4, the most distinctive gene in a set of 7 co-regulated (PRG4, CLU, TIMP3, TIMP4, GPX3, TXNIP and BMP4). RA samples were compared with OA and healthy control samples, suggesting that an RA phenotype of aggressive disease correlates with low expression of PRG4.

A good panorama of the relationship among clinical presentation, histopathologic features and gene expression was found by Yoshida et al. (2012), evaluating synovial tissue from RA compared with those from OA patients. All RA samples were scored according to a synovitis grade in high and low, which kept correlation with clinical score of high and low disease activity. The microarray efficiently separated all RA samples into 2 groups corresponding to the histological and clinical classifications. The samples with high synovitis presented 48 inflammatory response related genes; the most highlighted were CCL5, CXCL9, CXCL10, STAT1 and IRF1. Pathway analysis showed direct or indirect relationship among these chemokines, TNF and type I IFN.

Like with peripheral blood samples, not all studies are concordant. Despite of the finding of three subsets of RA patients based on the expression of INF-induced genes, van der Pouw Kraan et al. (2003b) could not associate the molecular sub-classification to those from clinical data.

11.2.4 Therapy Outcome Prediction

Some of the most pursued aims in the rheumatology field are the disease biomark-ers. A main issue about treatment in RA is the non-response to every therapy avail-able at some extension. This lack of efficacy happens when treating patients either with MTX or the other DMARD or with biological drugs. We are hardly in need of biomarkers, mainly those for therapy prediction. Some gene expression studies tried to help this task.

About Infliximab, a TNF-blocker, the different sources of RNA brought conflict-ing results. Two groups found a different set of 8 genes able to predict the response to the drug using blood cells as source of RNA. For Lequerré et al. (2006), the genes were MTCBP-1, AKAP9, RASGRP3, PTPN12, RSP28, HLA-DPB1, MRPL22 and EPS15, with 90 % sensitivity and 70 % specificity of prediction. The genes for Julià et al. (2009b) were HLA-DRB3, SH2D1, GNLY, CAMP, SLC2A3, IL2RB, MXD4 and TLR5, with estimated 94.4 % sensitivity and 85.7 % specificity. In a highly controlled situation, Oliveira et al. (2012) used PBMC gene expression profiling to evaluate whether a molecular discrimination is possible between responders and non-responders to MTX and TNF-blockers combined therapy, what resulted in a group of 3 differentially expressed genes (BCL2A1, CCL4 and CD83) sharply sep-arating patients by the response to therapy, although the set of genes was not further independently validated.

Data from LSRA synovial specimens showed no difference comparing gene ex-pression profiles of responders and non-responders to Infliximab (Lindberg et al. 2010). In the study previously cited (Reynier et al. 2011), the authors tried to corre-late the phenotypes high and low expression of IFN-related genes with the response to TNF-blockers therapy, but were not able to do it.

One study tested the possibility of discriminating responders and non-responders to Etanercept (Koczan et al. 2008), another TNF-blocker. Twenty out of 42 dif-ferentially expressed genes, the majority down-regulated in responders in the first week of therapy, were chosen for qPCR, being 8 validated (NFKBIA, CCL4, IL8, IL1B, PDE4B, TNFAIP3, PPP1R15A and ADM). These 8 genes were combined in pairs and triplets, in a model for testing the accuracy of the response prediction. It was not possible to predict the response by analyzing the gene expression prior to therapy, but only after 3 days. Even though the model may be useful, it is difficult to imagine the clinicians interrupting the treatment after only 3 days supported by a gene expression study.

The response to Rituximab, an anti-B cell therapy, was evaluated by two studies, both using whole-blood. Raterman et al. (2012), seeking a profile of non-responders to the drug, described a set of 8 type I INF-related genes (LY6E, HERC5, IFI44L, ISG15, MxA, MxB, EPSTI1 and RSAD2) whose high expression was associated with poor clinical response. The set accuracy evaluation was performed by area under ROC curves, resulting in 100 % specificity and 44 % sensitivity. A high specificity is suitable when looking for the correct prediction of non-response. Julià et al. (2009a) found a pair of genes (ARG1 and TRAF1) with a strong inverse expression correlation capable of responder and non-responders separation, i.e. when ARG1 exhibited expression levels higher than TRAF1 the patient presented poor clinical response to Rituximab, and when the contrary occurred, the patient achieved good clinical response. Although the number of patients needs to be augmented, the validation of the pair by quantitative polymerase chain reaction (q-PCR) resulted the same data, corroborating the potential for a biomarker.

11.2.5 Elucidations on drugs mechanisms of action

Blits et al. (2013) investigated 17 folate pathway genes in RA under MTX treatment, RA treatment-naïve and healthy controls. Compared to controls, RA treatment-naïve patients exhibited up-regulated folate pathway genes while patients treated with MTX exhibited no difference, suggesting that the drug normalized the pattern of folate pathway up-regulation in RA. These data confirm and bring new insights to the MTX mechanisms of action, but a limitation should be addressed—both groups of RA patients had the same mean of disease activity score, i.e., the disease activity was comparable in individuals using and not using the drug. From this result we could speculate that changes in gene expression caused by MTX in folate pathway solely are not sufficient to explain its effects when therapy is successful.

Cutolo et al. (2011) used the gene expression profiling of PBMC to evaluate the effect of the treatment with Leflunomide and Prednisone in treatment-naïve ERA patients, comparing them to healthy controls. The most differentially expressed genes between patients and controls were related to inflammatory process (MAPK9 and HIF1A), genetic susceptibility to RA (STAT4) and resistance or inhibition of apoptosis (MIF, STAT6, NFKB1 and TNFRSF1B), all up-regulated. The treatment caused a significant reduction in the expression level of all these genes, showing that the reversion of an anti-apoptotic state is important for the Leflunomide mechanism of action.

In an *in vitro* condition, Häupl et al. (2007) studied the effect of culturing RASF and normal synovial fibroblasts with MTX in their gene expression. After 36 h of incubation, MTX caused differential expression of 29 genes in RASF. Ten out of 29 genes had lowering of expression level compared to those found in untreated normal synovial fibroblasts, mostly genes involved in cell growth and apoptosis. Concomitantly, authors assessed the viability of treated RASF, finding it decreased. It may corroborate what was cited previously in this chapter about the apoptosis as an important process influenced by MTX and tightly related to its efficacy.

11.3 Spondyloarthritis

The term spondyloarthritis encompasses a group of diseases sharing pathophysiology, clinical and radiographic aspects, upon a common genetic inheritance (HLA-B27) (van der Linden et al. 2008)—ankylosing spondylitis (AS), psoriatic arthritis (PsA), enteropathic arthropathy, reactive arthritis, undifferentiated spondyloarthritis (uSpA) and juvenile-onset ankylosing spondylitis. Typical clinical features are the inflammatory involvement, with bone erosions and ossifications of the axial skeleton and enthesis, but many patients suffer with peripheral arthritis and eye involviment (Gladman 1998). Axial disease is treated primarily with non-steroidal anti-inflammatory drugs and DMARD and TNF-blockers are useful for many clinical presentations (Sieper et al. 2009).

The initial microarray studies on SpA had small number of subjects and comparisons with RA and healthy controls. Gu et al. (2002a) used a microarray specific for processes like cell signaling and adhesion, cytokines, chemokines and their receptors, aiming to compare gene expression of four groups—SpA (AS, uSpA and reactive arthritis), PsA, RA and healthy controls. PBMC were the source of RNA. Although this microarray did not allow hierarchic clustering, there were some discriminative genes comparing pairs of groups, e.g. MNDA, IL-12, IFN-γ and IL-7Rα for SpA vs healthy; ETR103, PDGF, RANTES and CCR1 for PsA vs healthy and M1P2α, IL-2Rα and MAP for SpA vs PsA. Some of them were validated by qPCR. The cytokines, chemokines and related receptors highlighted in this study are known to take part in the pathophysiology of the SpA.

Gu et al. (2002b) studied the gene expression of synovial fluid mononuclear cells of SpA (all diseases) and RA patients comparing them with PBMC from healthy controls. From 300 differentially expressed genes, authors arbitrarily took 23, mainly cytokines (IL-1β, TNF, TGF-β, IL-6) chemokines (IL-8, MCP-1), receptors (CCR1, IL-2Rα, CXCR4, TNFR2) and cell signaling genes, all up-regulated in the pathological samples. The qPCR validation proved the microarray results, with values quite similar for SpA and RA.

T cells and macrophage are the most common infiltrating cells in AS sacroiliitis and their abundance correlates with clinic and laboratory parameters (Bollow et al. 2000). This was the rationale for Smith et al. (2008) to study the gene expression of monocytes-derived macrophages from SA compared with healthy controls in two situations, without stimulus and after stimulation with INF-γ followed by LPS stimulation. A set of 141 differentially expressed genes was found, being 78 IFN-γ responsive genes. The authors discuss the result of a "reverse" IFN signature, meaning that genes normally up-regulated by IFN were underexpressed and those normally down-regulated were overexpressed. This differential gene expression was abrogated when the cells were treated with IFN-γ. After evaluation of IFN-γ mRNA, found in low levels, the authors suggested that the low level of IFN-γ may be responsible for the altered gene expression of INF-γ responsive genes.

In one study, a group evaluated gene expression of PBMC (Duan et al. 2010) and whole-blood (Pimentel-Santos et al. 2011) from AS patients and healthy controls, using the same methods. For both cell populations, hundreds of genes were differentially expressed and the most relevant were chosen, taking into account p value, fold-change and biological relevance. In both cases, hierarchical clustering was not able to segregate patients and controls and the set of genes were different, though many of them related to SpA pathophysiology. Still on peripheral blood studies, Assassi et al. (2011) compared whole-blood gene expression of AS and healthy subjects, finding a dysregulation of Toll-like receptors (TLRs) pathways related genes, remarkably TLR4 and TLR5. The data, like that on macrophage study, corroborates the importance of innate immunity in SpA.

One study sought for gene expression data using hip joint ligament specimens of AS patients undergoing total hip replacement compared with specimens from the same surgery in individuals with femoral neck fracture (Xu et al. 2012). The microarray generated a highly distinctive set of 43 genes involved in processes like cell proliferation and differentiation, cell adhesion, angiogenesis and synthesis of lysosomes, but their function in SpA pathogenesis is uncertain.

Similarly to RA, the search for biomarkers is an important issue in SpA. In fact, the situation tends to be worse for SpA than for RA, because we do not have any laboratory test for SpA diagnosis (although the HLA-B typing has been used now for diagnosis purposes). The differential diagnosis between inflammatory and non-inflammatory low back pain is a crucial distinction in daily practice and sometimes difficult to achieve (Jois et al. 2008). Gu et al. (2009) used gene expression profiling of PBMC from AS, uSpA and individuals with non-inflammatory low back pain, proposing that RGS1 can become a biomarker since its expression was the most discriminative in the microarray and, validated by qPCR, separated uSpA and AS from mechanical back pain with an area under ROC curve of, respectively, 0.99 and 0.84.

PsA had its pathophysiology investigated by Batliwalla et al. (2005b). Patients were compared with healthy controls by the evaluation of whole-blood gene expression, generating a set of 56 overexpressed and 257 underexpressed genes in PsA, mainly involved with inflammation, apoptosis, cell cycle, cell signaling and regulation of transcription. Logistic regression ranked the most discriminatory gene, NUP62, whose expression correctly classified all controls and 94.7% of the PsA patients.

11.4 Conclusions

In conclusion, microarray opened up a new era in our possibilities on understanding autoimmune diseases. Differences in gene expression profiles may provide a unique perspective allowing us to distinguish different pathogenic mechanisms and biomarkers of diagnosis, prognosis and drug responsiveness.

If expression-based profiling is to be of practical importance and widely used, the ease of sample accessibility is crucial. In early stages of the autoimmune diseases an appropriated tissue sample is not always available, making the study of peripheral blood cells a good and easy option to address the disease status at any time. Synovial specimens used in the majority of the studies have the disadvantage of representing a very long-standing disease, because frequently come from joint replacement. The heterogeneity of the disease is reflected in peripheral blood cells, making them good sentinels.

By now, we can say that the studies have brought many discordances, but some important conclusions, in part because each disease aspect studied is able to yield a different gene signature, and each patient exhibits slightly different gene expression and, consequently, different phenotype.

Why has it been so difficult to find the place for microarray in daily medical decisions? Is it only regarded to the heterogeneity of autoimmune diseases, mainly because these are conditions highly influenced by an enormous amount of external and internal stimuli? The answer for this question is neither affirmative nor negative. Must be considered the fact that the study designs are so distinct—different probe sets, platforms, data processing, statistical programs and data processing, differences in cell types examined, frequently small sample sizes and, in some studies, lack of appropriate controls. We would like to reinforce that is necessary to know as much as possible about clinical features in order to create the best relationship between clinical and molecular knowledge, because it is essential to know exactly for whom it will be applied. Moreover, the study group should be the most homogenous, with phenotypes narrowly defined.

Apart of these uncertainties, we can rely on some data.

Based on microarray studies, some gene set and equations were proposed to discriminate an autoimmune background.

It is possible to separate RA patients in all disease stage from healthy subjects. Moreover, patients with pre-clinic stage (e.g. arthralgia with positive autoantibodies) that share the gene expression profile with patients with clinic stages should be considered for RA treatment.

The type I INF-related genes were found to be important in sub classifying RA patients, elucidating some pathophysiological aspects that may be predictive of the disease onset, response to therapy and prognosis. In some studies, STAT1 pathway was strictly related to type I INF, TNF and chemokines.

A tissue distinction of disease stage for RA is possible, based on gene expression profiling.

In a well-controlled situation it is possible to find close relationship among the synovitis grade, clinical scores of disease activity and synovial gene expression. This point should be intensely explored in the future years as it raises many possibilities for therapeutic interventions.

We have really good candidates for an AR gene chip, a therapy outcome prediction on the way for the next years.

From microarray evaluation of different cells, apoptosis is a pivotal, and probably shared, process for the efficacy of many drugs. It put the intervention on apoptosis as a big deal for targeted therapy.

In SpA, the studies confirmed some known pathophysiological aspects, highlighting some cytokines/chemokines pathways. The tissue samples and whole-blood gave clues on the crucial importance of innate immunity for these diseases, what should also be tested for therapeutic purposes. Interestingly, a possible opposite effect of IFN-related genes comparing with RA. The idea of biomarkers is always desirable, but the candidates need more accurate studies (Fig. 11.1)

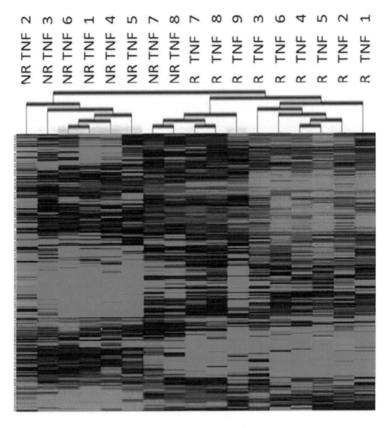

Fig. 11.1 Hierarchical clustering of differential gene expression in patients with RA stratified according to response to MTX plus infliximab, a TNF-blocker, combined therapy. (Figure from Oliveira et al. (2012) J Rheumatol 39:1524 with permission from the Publisher)

References

Assassi S, Reveille JD, Arnet FC et al (2011) Whole-blood gene expression profiling in ankylosing spondylitis shows upregulation of toll-like receptor 4 and 5. J Rheumatol 38:87–98
Baechler EC, Batliwalla FM, Karypis G et al (2003) Interferon-inducible gene expression signature in peripheral blood cells of patients with severe lupus. Proc Natl Acad Sci U S A 100:2610–2615
Batliwalla FM, Baechler EC, Xiao X et al (2005a) Peripheral blood gene expression profiling in rheumatoid arthritis. Genes Immun 6:388–397

Batliwalla FM, Li W, Ritchlin CT et al (2005b) Microarray analyses of peripheral blood cells identifies unique gene expression signature in psoriatic arthritis. Mol Med 11:21–29

Blits M, Jansen G, Assaraf TG et al (2013) Methotrexate normalizes up-regulated folate pathway
genes in rheumatoid arthritis. Arthritis Rheum 65:2791–2802

Bollow M, Fischer T, Reisshauer H et al (2000) Quantitative analyses of sacroiliac biopsies in
spondyloarthropathies: T cells and macrophages predominate in early and active sacroiliitis—
cellularity correlates with the degree of enhancement detected by magnetic resonance imaging.
Ann Rheum Dis 59:135–140

Bovin LF, Rieneck K, Workman C et al (2004) Blood cell gene expression profiling in rheumatoid
arthritis. Discriminative genes and effect of rheumatoid factor. Immunol Lett 93:217–226

Cutolo M, Villaggio B, Brizzolara R et al (2011) Identification and quantification of selected inflammatory genes modulated by leflunomide and prednisone treatment in early rheumatoid
arthritis patients. Clin Exp Rheumatol 29:72–79

Del Rey MJ, Izquierdo E, Usategui A et al (2010) The transcriptional response of normal and rheumatoid arthritis synovial fibroblasts to hypoxia. Arthritis Rheum 62:3584–3594

Devauchelle V, Marion S, Cagnard N et al (2004) DNA microarray allows molecular profiling of
rheumatoid arthritis and identification of pathophysiological targets. Genes Immun 5:597–608

Duan R, Leo P, Bradbury L et al (2010) Gene expression profiling reveals a downregulation in
immune-associated genes in patients with AS. Ann Rheum Dis 69:1724–1729

Felson DT (2005) Epidemiology of the rheumatic diseases. In: Koopman WJ, Moreland LW (eds)
Arthritis and allied conditions. Lippincott Williams & Wilkins, Philadelphia, pp 1–36

Gallagher J, Howlin J, McCarthy C et al (2003) Identification of Naf1/ABIN-1 among TNF-alpha-
induced expressed genes in human synoviocytes using oligonucleotide microarrays. FEBS Lett
551:8–12

Galligan CL, Baig E, Bykerk V et al (2007) Distinctive gene expression signatures in rheumatoid arthritis synovial tissue fibroblast cells: correlates with disease activity. Genes Immun
8:480–491

Gladman DD (1998) Clinical aspects of the spondyloarthropathies. Am J Med Sci 316:234–328

Goekoop-Ruiterman YP, de Vries-Bouwstra JK, Allaart CF et al (2007) Comparison of treatment
strategies in early rheumatoid arthritis: a randomized trial. Ann Intern Med 146:406–415

Gu J, Märker-Hermann E, Baeten D et al (2002a) A 588-gene microarray analysis of the peripheral
blood mononuclear cells of spondyloarthropathy patients. Rheumatology (Oxford) 41:759–766

Gu J, Märker-Hermann E, Baeten D et al (2002b) Clues to pathogenesis of spondyloarthropathy
derived from synovial fluid mononuclear cell gene expression profiles. J Rheumatol 29:2159–
2164

Gu J, Wei YL, Wei JC et al (2009) Identification of RGS1 as a candidate biomarker for undifferentiated spondylarthritis by genome-wide expression profiling and real-time polymerase chain
reaction. Arthritis Rheum 60:3269–3279

Haas CS, Creighton CJ, Pi X et al (2006) Identification of genes modulated in rheumatoid arthritis
using complementary DNA microarray analysis of lymphoblastoid B cell lines from disease-
discordant monozygotic twins. Arthritis Rheum 54:2047–2060

Häupl T, Yahyawi M, Lübke C et al (2007) Gene expression profiling of rheumatoid arthritis synovial cells treated with antirheumatic drugs. J Biomol Screen 12:328–340

Jois RN, Macgregor AJ, Gaffney K (2008) Recognition of inflammatory back pain and ankylosing
spondylitis in primary care. Rheumatology (Oxford) 47:1364–1366

Julià A, Barceló M, Erra A et al (2009a) Identification of candidate genes for rituximab response
in rheumatoid arthritis patients by microarray expression profiling in blood cells. Pharmacogenomics 10:1697–1708

Julià A, Erra A, Palacio C et al (2009b) An eight-gene blood expression profile predicts the response to infliximab in rheumatoid arthritis. PLoS One 4:e7556

Junta CM, Sandrin-Garcia P, Fachin-Saltoratto AL et al (2008) Differential gene expression of
peripheral blood mononuclear cells from rheumatoid arthritis patients may discriminate im-
munogenetic, pathogenic and treatment features. Immunology 127:365–372

Klareskog L, Padyukov L, Rönnelid J et al (2006) Genes, environment and immunity in the development of rheumatoid arthritis. Curr Opin Immunol 18:650–655

Klareskog L, Catrina AI, Paget S (2009) Rheumatoid arthritis. Lancet 373:659–672

Koczan D, Drynda S, Hecker M et al (2008) Molecular discrimination of responders and nonresponders to anti-TNF alpha therapy in rheumatoid arthritis by etanercept. Arthritis Res Ther 10:R50

Lequerré T, Coulouarn C, Derambure C et al (2003) A new tool for rheumatology: large-scale analysis of gene expression. Joint Bone Spine 70:248–256

Lequerré T, Gauthier-Jauneau AC, Bansard C et al (2006) Gene profiling in white blood cells predicts infliximab responsiveness in rheumatoid arthritis. Arthritis Res Ther 8:R105

Lequerré T, Bansard C, Vittecoq O et al (2009) Early and long-standing rheumatoid arthritis: distinct molecular signatures identified by gene-expression profiling in synovia. Arthritis Res Ther 11:R99

Lindberg J, Wijbrandts CA, van Baarsen LG et al (2010) The gene expression profile in the synovium as a predictor of the clinical response to infliximab treatment in rheumatoid arthritis. PLoS One 5:e11310

Maas K, Chan S, Parker J et al (2002) Cutting edge: molecular portrait of human autoimmune disease. J Immunol 169:5–9

McCarty DJ (2005) Differential diagnosis of arthritis: analysis of signs and symptoms. In: Koopman WJ, Moreland LW (eds) Arthritis and allied conditions. Lippincott Williams & Wilkins, Philadelphia, pp 37–50

Möttönen T, Hannonen P, Korpela M et al (2002) Delay to institution of therapy and induction of remission using single-drug or combination-disease-modifying antirheumatic drug therapy in early rheumatoid arthritis. Arthritis Rheum 46:894–898

O'Dell JR (2002) Treating rheumatoid arthritis early: a window of opportunity?. Arthritis Rheum 46:283–285

Oliveira RD, Fontana V, Junta CM et al (2012) Differential gene expression profiles may differentiate responder and nonresponder patients with rheumatoid arthritis for methotrexate (MTX) monotherapy and MTX plus tumor necrosis factor inhibitor combined therapy. J Rheumatol 39:1524–1532

Pimentel-Santos FM, Ligeiro D, Matos M et al (2011) Whole blood transcriptional profiling in ankylosing spondylitis identifies novel candidate genes that might contribute to the inflammatory and tissue-destructive disease aspects. Arthritis Res Ther 13:R57

Qingchun H, Runyue H, LiGang J et al (2008) Comparison of the expression profile of apoptosis-associated genes in rheumatoid arthritis and osteoarthritis. Rheumatol Int 28:697–701

Raterman HG, Vosslamber S, de Ridder S et al (2012) The interferon type I signature towards prediction of non-response to rituximab in rheumatoid arthritis patients. Arthritis Res Ther 14:R95

Reynier F, Petit F, Paye M et al (2011) Importance of correlation between gene expression levels: application to the type I interferon signature in rheumatoid arthritis. PLoS One 6:e24828

Schoels M, Wong J, Scott DL et al (2010) Economic aspects of treatment options in rheumatoid arthritis: a systematic literature review informing the EULAR recommendations for the management of rheumatoid arthritis. Ann Rheum Dis 69:995–1003

Sieper J, Rudwaleit M, Baraliakos X et al (2009) The Assessment of SpondyloArthritis international Society (ASAS) handbook: a guide to assess spondyloarthritis. Ann Rheum Dis 68(Suppl. 2):ii1–44

Silva GL, Junta CM, Mello SS et al (2007) Profiling meta-analysis reveals primarily gene coexpression concordance between systemic lupus erythematosus and rheumatoid arthritis. Ann N Y Acad Sci 1110:33–46

Smith JA, Barnes MD, Hong D et al (2008) Gene expression analysis of macrophages derived from ankylosing spondylitis patients reveals interferon-gamma dysregulation. Arthritis Rheum 58:1640–1649

Szodoray P, Alex P, Frank MB et al (2006) A genome-scale assessment of peripheral blood B-cell molecular homeostasis in patients with rheumatoid arthritis. Rheumatology 45:1466–1476

Taberner M, Scott KF, Weininger L et al (2005) Overlapping gene expression profiles in rheumatoid fibroblast-like synoviocytes induced by the proinflammatory cytokines interleukin-1 beta and tumor necrosis factor. Inflamm Res 54:10–16

Teixeira VH, Olaso R, Martin-Magniette ML et al (2009) Transcriptome analysis describing new immunity and defense genes in peripheral blood mononuclear cells of rheumatoid arthritis patients. PLoS One 4:e6803

Ungethuem U, Haeupl T, Witt H et al (2010) Molecular signatures and new candidates to target the pathogenesis of rheumatoid arthritis. Physiol Genomics 42A:267–282

van Baarsen LGM, Bos WH, Rustenburg F et al (2010) Gene expression profiling in autoantibody-positive patients with arthralgia predicts development of arthritis. Arthritis Rheum 62:694–704

van der Linden S, van der Heijde D, Landewé R (2008) Classification and epidemiology of spondyloarthritis. In: Hochberg MC, Silman AJ, Smolen JS, Weinblatt ME, Weisman MH (eds) Rheumatology. Elsevier, Philadelphia, pp 1103–1107

van der Pouw Kraan TC, van Gaalen FA, Kasperkovitz PV et al (2003a) Rheumatoid arthritis is a heterogeneous disease: evidence for differences in the activation of the STAT-1 pathway between rheumatoid tissues. Arthritis Rheum 48:2132–2145

van der Pouw Kraan TCTM, van Gaalen FA, Huizinga TWJ et al (2003b) Discovery of distinctive gene expression profiles in rheumatoid synovium using cDNA microarray technology: evidence for the existence of multiple pathways of tissue destruction and repair. Genes Immun 4:187–196

van der Pouw Krann TC, Wijbrandts CA, van Baarsen LG et al (2007) Rheumatoid arthritis subtypes identified by genomic profiling of peripheral blood cells: assignment of a type I interferon signature in a subpopulation of patients. Ann Rheum Dis 66:1008–1014

Xu L, Sun Q, Jiang S et al (2012) Changes in gene expression profiles of the hip joint ligament of patients with ankylosing spondylitis revealed by DNA chip. Clin Rheumatol 31:1479–1491

Yoshida S, Arakawa F, Higuchi F et al (2012) Gene expression analysis of rheumatoid arthritis synovial lining regions by cDNA microarray combined with laser microdissection: up-regulation of inflammation-associated STAT1, IRF1, CXCL9, CXCL10, and CCL5 Scand. J Rheumatol 41:170–179

Zrioual S, Ecochard R, Tournadre A et al (2009) Genome-wide comparison between IL-17A- and IL-17F-induced effects in human rheumatoid arthritis synoviocytes. J Immunol 182:3112–3120

Chapter 12
Transcriptome Profiling in Experimental Inflammatory Arthritis

Olga Martinez Ibañez, José Ricardo Jensen and Marcelo De Franco

Abstract Rheumatoid arthritis (RA) is a chronic inflammatory autoimmune disease that affects 0.5 to 1% of the human population. Gene expression profiling studies of tissues from RA patients showed marked variation in gene expression profiles that allowed identifying distinct molecular disease mechanisms involved in RA pathology. The relative contribution of the different mechanisms may vary among patients and in different stages of disease. Thus, the broad goals of expression profiling in RA are the improvement of understanding of the pathogenic mechanisms underlying RA, the identification of disease subsets and new drug targets and the assessment of disease activity, such as: responsiveness to therapy, overall disease severity and organ specific risk and development of new diagnostic tests. Genetic and environmental factors contribute to the development of this disease and numerous studies have indicated the participation of the major histocompatibility complex (MHC) class II alleles and non-MHC genes. Therefore, identification of the major roles of the participating cells and of candidate genes has been an important subject of study to the understanding of RA pathogenesis.

12.1 Introduction

Rheumatoid arthritis (RA) is a chronic inflammatory autoimmune disease that affects 0.5 to 1% of the human population. RA is a complex pathology characterized by systemic chronic inflammation with the accumulation into synovium and periarticular spaces of activated T and B lymphocytes, innate immune cells such as neutrophils, mast cells, dendritic cells, natural killer cells and macrophages, and endothelial cells. Rheumatoid fibroblast-like synoviocytes, which exhibit invasive characteristics and synovial macrophages with pro-inflammatory properties are crucial for the progression of arthritis causing proliferation of synovial membranes

M. De Franco (✉) · O. Martinez Ibañez · J. Ricardo Jensen
Laboratory of Immunogenetics, Butantan Institute, Avenida Vital Brasil 1500,
05503-900 São Paulo, São Paulo, Brazil
e-mail: marcelo.franco@butantan.gov.br

© Springer International Publishing Switzerland 2014
G. A. Passos (ed.), *Transcriptomics in Health and Disease,*
DOI 10.1007/978-3-319-11985-4_12

and the formation of the invasive pannus that erodes cartilage and bone. In human patients the clinical signs of RA are largely heterogeneous but the disease is considered to be autoimmune (You et al. 2014). RA heterogeneity is demonstrated by the presence of distinct autoantibody specificities, such as antibodies against immunoglobulins, the rheumatoid factor (RF), and anti-cyclic citrullinated peptide antibodies (ACPA) in the serum, the differential responsiveness to treatment, and by the variability in clinical signs (Silman and Pearson 2002). The precise etiology of RA remains poorly understood, but the main symptoms are chronic synovitis, joint erosion, and several immune abnormalities in both the innate and adaptive compartments.

Given the complexity of RA, systems biology approaches designed to give a general view of different aspects of the disease are required to better understand the basis of arthritis. Oligonucleotide-based microarray technology for global gene expression profiling has arisen as a powerful tool to investigate the molecular complexity and pathogenesis of arthritis and other complex pathologies. This genomic or transcriptomic method combined with post-genomic techniques provides an opportunity to monitor the complex interactions between genes and environment, the regulation of genes and of RNA transcripts and proteins that constitute the basis for the etiology or progression of the diseases (Jarvis and Frank 2010).

Gene expression profiling studies of tissues from RA patients showed marked variation in gene expression profiles that allowed to identify distinct molecular disease mechanisms involved in RA pathology (Baechler et al. 2006). The relative contribution of the different mechanisms may vary among patients and in different stages of disease. Thus, the broad goals of expression profiling in RA are the improvement of understanding of the pathogenic mechanisms underlying RA, the identification of disease subsets and new drug targets and the assessment of disease activity, such as: responsiveness to therapy, overall disease severity and organ specific risk and development of new diagnostic tests (Teixeira et al. 2009).

Genetic and environmental factors contribute to the development of this disease. Numerous studies have indicated the participation of the major histocompatibility complex (MHC) class II alleles and non-MHC genes, such as the *solute carrier family 11a member 1—SLC11A1* (formerly named *NRAMP1-* Natural resistance associated macrophage protein 1) related to macrophage activation (Runstadler et al. 2005). Identification of the major roles of the participating cells and of candidate genes has been an important subject of study to the understanding of RA pathogenesis (Kurko et al. 2013).

12.2   Experimental Models of Rheumatoid Arthritis

The initial or preclinical stages of RA are difficult to be studied in humans but numerous arthritis experimental models have been developed which are valuable tools for in-depth investigation of pathogenic pathways that are involved in the several phases of the disease (Kobezda et al. 2014). Regarding ethical procedures, in these models the animals can be submitted to immunizations with arthritogenic substances or antigens, to cell transfer or depletion, to phenotypic selective crosses, to genetic manipulations for the production of transgenic or knockout individuals, etc. Most importantly, these models have been useful for the candidacy of targets for preventive or therapeutic strategies (Asquith et al. 2009).

Several studies have used different animal models for arthritis, generally induced by the injection of adjuvants (AIA), proteoglycan (PGIA), type II collagen (CIA) or pristane (PIA) (Kannan et al. 2005).

Collagen-induced arthritis (CIA). Type II collagen (CII) is expressed exclusively in the articular joint. Although the relationship between anti-CII immunity and human rheumatoid arthritis (RA) has been studied for a long time, definitive conclusions have not been established. CII, as an autoantigen, has been studied extensively in small animal models, such as mice and rats, and the collagen-induced arthritis (CIA) model has increased our understanding of the pathogenesis of human RA (Cho et al. 2007). The disease is class II MHC-restricted but mouse strains with permissive haplotypes vary in their susceptibility to CIA. Arthritis development is associated with B and T lymphocyte responses and the generation of anti-collagen antibodies and T-cells.

Collagen antibody-induced arthritis (CAIA) in mice has demonstrated the role of humoral immunity in arthritis development. It has been useful for the identification of collagen epitopes for the generation of arthritogenic antibody cocktails that represent humoral auto-immunity in RA. The disease is characterized by macrophage and polymorphonuclear cell infiltration and no T- and B-cell involvement and is non-MHC class II restricted (Hirose and Tanaka 2011).

Proteoglycan-induced arthritis (PGIA) is based in the immunization of mice with human cartilage-derived proteoglycans which induces the development of severe polyarthritis and spondylitis (Glant et al. 2003).

Pristane-induced arthritis (PIA) has proven to be a valuable experimental model for inflammatory RA. The natural saturated terpenoid alkane 2,4,6,10-tetramethyl pentadecane induces an acute inflammation followed by a chronic relapsing phase. The reaction is T-cell dependent with edema and articular infiltration of mononuclear and polymorphonuclear cells (Potter and Wax 1981).

There are also genetically manipulated models that develop RA spontaneously. For example, transgenic mice over-expressing human TNF-α develop chronic inflammatory erosive polyarthritis (Li and Schwarz 2003). This model highlights the importance of TNF-α in cytokine network in RA. Another example is the IL-1 receptor antagonist deficient mouse that develops inflammatory arthritis mediated by a polarized TH17 response (van den Berg 2009; Lubberts et al. 2005).

In experimental models, microarray analysis should optimally be carried out in isolated populations of cells. However, in complex diseases such as RA there is extensive tissue damage with the contribution of several cell types. Hence the analysis of rodent whole ankle joints or of footpads which comprise heterogeneous cell types, gives a better global view of differential gene expression during the several phases of arthritis onset and development. Differential expression of genes encoding tissue repair factors, signal transduction molecules, transcription factors, and DNA repair enzymes as well as cell cycle regulators, have been observed in multiple microarray experiments. An interesting observation in these experiments is the transcriptome map of the differentially expressed genes; in different models of arthritis there is a functional grouping of dysregulated genes forming clusters in the chromosomes. Examples are the MHC class I and class II gene clusters, known to affect susceptibility to a variety of autoimmune diseases and the chemoattractant gene clusters such as CC or CXC chemokine ligands and receptors which mediate infiltration of leukocytes into synovial tissue, a hallmark of RA (Fujikado et al. 2006). Some studies attempt to link differentially expressed genes into interactive regulatory networks (Silva et al. 2009). This approach is quite powerful to identify new targets for therapy by looking at the network structures, the places (genes) with highest connectivity in which disruption would have a larger impact.

12.3   Loci Regulating Inflammatory Arthritis

The identification of the loci influencing inflammatory arthritis in animal models is important for parallel genetic studies in humans. Individual genetic constitution of experimental animals involving major histocompatibility complex (MHC) or non-MHC genes has been associated with variations in rheumatoid arthritis susceptibility. In mice or rats, genome wide linkage studies with DNA polymorphism markers such as microsatellites or single nucleotide polymorphisms (SNPs), have been carried out using intercross progenies of resistant and susceptible strains. These studies, in which environmental effects and genetic backgrounds are controlled, have been useful for the study of the genetic basis of RA (Ibrahim and Yu 2006).

Several QTL (Quantitative Trait Loci) were identified in different models of experimental arthritis. The first locus controlling pristane induced arthritis (PIA) detected in mice was *Prtia1* on chromosome 3, in an intercross population from mice selected for high and low antibody production (Jensen et al. 2006). QTL were also mapped in other arthritis models such as those induced by *Borrelia burgdorferi* (Roper et al. 2001), PGIA (Glant et al. 2004) and by collagen (Adarichev et al. 2003). Non overlapping sets of QTLs were identified, generating a heterogeneous picture of risk alleles (Besenyei et al. 2012; Kurko et al. 2013). The results evidence the genetic heterogeneity in the control of the different stages and phenotypes of the disease. Table 12.1 presents some relevant coincident susceptibility QTLs in rheumatoid arthritis, according to GWAS studies carried out in mice and humans.

Numerous RA QTLs have been mapped but few of the associated polymorphisms were identified in protein-coding regions of genes causing changes in protein structure or function. This suggests that polymorphisms in non-coding regions which might affect gene expression largely contribute to variations in RA susceptibility. In this way, transcriptome technology can also be used to detect genetic polymorphisms that regulate gene expression levels.

Table 12.1 Common arthrtitis associated QTL (Non-MHC regions) mapped by GWAS in mice and humans

Mouse			Human	
Chr	Locus name	Candidate gene	Chr position	Locus name
1	*Cia14*	Aff3: expressed in lymphoid cells, encodes a nuclear factor that contains transcriptional activation domains	2q11	AFF3
1	*Cia9, Pgia1*	Fcgr2b: a variant allele alters dendritic cell behavior, suggesting a role for dendritic cells in RA pathology	1q23	FCGR2A
2	*Cia2, Cia4, Pgia2*	Traf1/Hc: Genetic variants associated to risk of anti-CCP antibody-positive RA	9q33	TRAF1/C5
3	*Cia21, Cia22, Pgia26 Prtia1*	Cd2: encodes a co-stimulatory molecule found on natural killer and T cells	1p23	CD2
		Ptpn22: the gene is a negative regulator of T cells. Allele variant affects binding to an intracellular signaling molecule (Csk) resulting in a failure to switch off T cells or to delete auto-reactive T cells during thymic development	1p13	PTPN22
5	*Pgia16, Cia13*	Rbjp: The gene encodes a transcription factor involved in the notch signaling pathway and in regulation of T-cell development	4p15	RBJP
6	*Pgia19*	Irf5: transcription factor involved in antiviral and anti-inflammatory responses and in differentiation of B cells regulation	7q32	IRF5
10	*Pgia6,*	Tnfaip3: knock-out mice develop severe inflammation	6q23	TNFAIP3

Table 12.1 (continued)

Mouse			Human	
Chr	Locus name	Candidate gene	Chr position	Locus name
	Pgia6b	Prdm1: The gene product is a transcription factor involved in B cell regulation	6q21	PRDM1
10	*Cia8*	Kif5a: gene encodes a kinesin-heavy chain	12q13	KIF5A
		Pip4k2c: phosphatidyl inositol kinase		PIP4K2C
13	*Pgia15,*	IL6st/Ankrd55: IL-6	5q11	ANKRD55
	Cia19	Signal transduction gene region		IL6ST
15	*Pgia9, Cia35, Cia37*	IL2rb	22q12	IL2RB
		Bik: apoptosis-inducing,BCL2-interacting killer	8p3	BIK
18	*Pgia11*	Ptpn2: KO mice have increased susceptibility to inflammatory diseases	18p11.3-p11.2	PTPN2

Gene names: *Affr* AF4/FMR2 family, member 3; *Fcgr2b* Fc receptor IgG, low affinity IIb; *Traf1* TNF receptor-associated factor 1; *Ptpn22* protein tyrosine phosphatase, non-receptor type 22 (lymphoid); *Rbjp* recombination signal binding protein for immunoglobulin kappa J region; *Irf5* interferon regulatory factor 5; *Tnfaip3* tumor necrosis factor, alpha-induced protein 3; *Prdm* PR domain containing 1, with ZNF domain; *Kif5a* kinesin family member 5A; *Pip4k2c* phosphatidylinositol-5-phosphate 4-kinase, type II, gamma; *Ankrd55* ankyrin repeat domain 55; *Ptpn2* protein tyrosine phosphatase, non-receptor type 2

12.4 Combining Transcriptome and Genome Screening to Identify Genes That Control Arthritis

The two genomic approaches, that is, transcriptome and genome screening (GWAS) have been combined in studies where the locations of differently expressed genes during RA are compared with those mapping at QTLs for arthritis, for immune or inflammatory responses or for other autoimmune diseases (Yu et al. 2007). The approach has been useful to candidate genes inside the QTLs. The coincidence of chromosomal locations of genes in QTLs in different model systems with the locations of the corresponding human orthologue is a good indicator of their implication in RA control.

Furthermore, the modulation of common genes during RA, irrespective of etiology and of species indicates the importance of these mediators in the pathogenesis of arthritis. For example, the augmented expression of chemokines and receptors, which recruit neutrophils or naïve and memory T cells to inflammatory sites. Che-mokines and ligands are found in the synovial tissue of patients with RA; proin-flammatory cytokines and their cognate receptors, such as IL-1β, IL-1RI, TNF-α R, IL-6Rα, IL-2Rγ, and IL-17R, are up-regulated in several RA models as well as in arthritis patients; IL-1β induces serum amyloid A3 (Saa3) and the matrix metal-loproteinases Mmp-3 and Mmp-9 that are also upregulated in several models. High upregulation in runt-related transcription factor 1 (RUNX1) and a group of trans-porter genes such as solute carrier 11 family A1 (*Slc11a1*, formerly *Nramp1*), is also a common feature in RA models. In synthesis, a remarkable feature that originated from numerous transcriptome or genomic studies of arthritis has been the demon-stration of gene expression signatures associated with inflammation. The results evidence that besides being an antigen-driven event there is an important interplay between innate and adaptive immunity systems in the etiology of RA (Jarvis and Frank 2010).

12.5   A Model to Study Inflammatory Rheumatoid Arthritis: Airmax and Airmin Phenotypically Selected Mouse Lines

Heterogeneous mice selected for maximal (AIRmax) or minimal (AIRmin) acute inflammatory reaction appeared to be useful models for studying the mechanisms involved in rheumatoid arthritis susceptibility (Vigar et al. 2000).

AIRmax and AIRmin mice were produced by bidirectional selection, starting from a highly polymorphic population (F0) derived from the intercrossing of eight inbred mouse strains (Fig. 12.1). The selection phenotypes chosen were localized leukocyte influx and exudated plasma proteins 24 h after the subcutaneous injec-tion of polyacrylamide beads (Biogel), a non-antigenic, insoluble, and chemically inert substance (Ibanez et al. 1992). The progressive divergence of the AIRmax and AIRmin lines during successive generations of selective breeding reached 20- and 2.5-fold differences in leukocyte infiltration and exudated protein concentrations respectively. These differences resulted from the accumulation of alleles in quanti-tative trait loci endowed with opposite and additive effects on the inflammatory re-sponse. Inbreeding was avoided for the selective breeding, and as such AIRmax and AIRmin mice are outbred mice that maintain a heterogeneous genetic background but are homozygous in acute inflammation modifier loci. Analysis of the selective processes indicated that the AIR phenotype is regulated by at least 11 QTL (Biozzi et al. 1998).

Pristane-induced arthritis (PIA) has proven to be a valuable experimental mod-el for inflammatory RA for its delayed onset, chronicity and independence from xenoantigen administration. Thus, arthritis ensues from a sensitization over time and pristane has been described to improve autoimmunity by the activation of the immune response against cross-reactive microbiota antigens (Patten et al. 2004). AIRmax mice are extremely susceptible whereas AIRmin mice are resistant to PIA

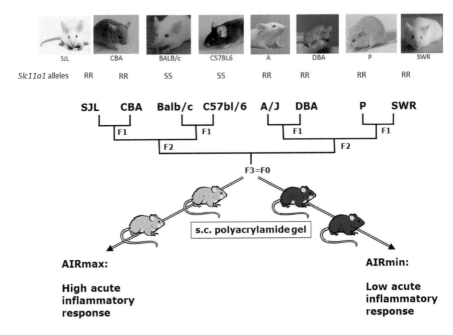

Fig. 12.1 Scheme used for the production of the foundation population (F0) by the intercrossing of eight inbred strains of mice for the production of AIRmax and AIRmin mice by bidirectional phenotypic selection

(Fig. 12.2a). The incidence of PIA in AIRmax mice was similar to that of inbred DBA/1 and BALB/c mice although with higher severity. The incidence and severity were more intense than in the CBA/Igb model since 15 to 25 % of these mice develop inflammation of the ankle and wrist joints approximately 200 days after pristane injection. PIA is accompanied by markedly elevated humoral agalactosyl IgG levels mediated by IL6 production (Thompson et al. 1992) and CD4+ T cell (Th) dependent (Stasiuk et al. 1997) immune responses to mycobacterial 65-kDa heat shock protein (hsp65). Moreover, the protection against PIA is mediated by Th2-associated cytokines produced after hsp65 pre-immunization (Thompson et al. 1998; Thompson et al. 1990). In contrast to the immune response profile observed in inbred mice, high IgG1 anti-hsp65 levels were observed in susceptible AIRmax mice, whereas IgG2a was the predominant isotype in the resistant AIRmin mice. Additionally, it was shown that IL-4, IL-6 and TNF secreting splenic cells were significantly more abundant in AIRmax than in AIRmin animals. IFNg-producing cells, on the other hand, increased in AIRmin mice only. Specific pathogen-free susceptible mice do not develop this disease, but when transferred to a conventional environment, they reacquire arthritis susceptibility, indicating the involvement of environmental factors in PIA (Thompson and Elson 1993).

The results in the AIRmax and AIRmin PIA model, when compared to those obtained in inbred mice, evidence the interference of genetic background in the

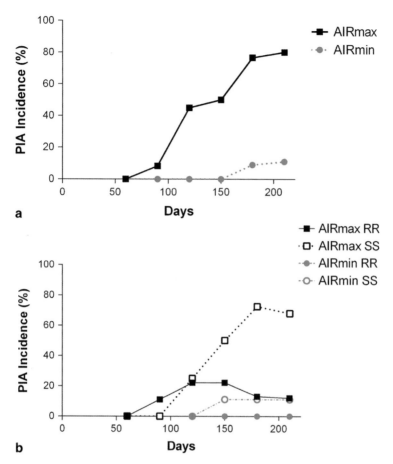

Fig. 12.2 PIA incidence in AIRmax and AIRmin mice and their sublines homozygous for the Slc11a1 gene R and S alleles. Mice receitved two ip injections of pristane with 60 days interval

mechanisms underlying arthritis susceptibility and severity. Interaction of arthritis controlling genes with heterogeneous genetic backgrounds and variability in gut microbiota might contribute to the variable signs of arthritis occurring in humans.

The transporter gene *Solute carrier 11 family a1* (*Slc11a1*) has been described in mice as a major modulator of susceptibility to infectious diseases, and is expressed in macrophages and neutrophils. *Slc11a1* is pleiotropic, interfering with macrophage activation, oxidative and nitrosamine bursts, TNF, IFNg, and IL-1 production, and the expression of MHC class II molecules. In mice, the mutation corresponding to the *Slc11a1* S allele associated with susceptibility determines a gly169asp substitution resulting in a nonfunctional protein that promotes an accumulation of ions inside the phagosome of macrophages that favors pathogen replication (Vidal et al. 1992). In the experiment for the production of AIRmax and AIRmin mouse lines, the frequency of the *Slc11a1* S allele was 25% in the founder population (F0), but shifted to 60% in AIRmin and to 9% in AIRmax after 30 generations of selective breeding. The results suggest that these changes in allele frequencies were the result of the selection process for acute inflammatory response (Araujo et al. 1998).

The effect of the *Slc11a1 R* and *S* alleles during PIA development was studied in AIRmax and AIRmin mice that were rendered homozygous for the *Slc11a1* alleles by genotype-assisted breeding (Fig. 12.2b). AIRmax mice homozygous for the *S* allele (AIRmax[SS]) were significantly more susceptible (80 % incidence) to RA than AIRmax[RR] mice (30 % incidence) evidencing the influence of this polymorphism in RA (Peters et al. 2007). The involvement of this gene in this study as well as in other murine arthritis models constituted the basis for the study of *Slc11a1* involvement in human RA. In fact, several authors reported linkage of *SLC11A1* alleles to human RA probably associated to a polymorphic repeat in the RUNX1-containing promoter region of the gene (Ates et al. 2009).

12.6 Mapping of QTL Controlling PIA in Airmax and Airmin Mice

A genome wide linkage study was carried out in a large F2 population of inter-crossed AIRmax and AIRmin F2(AIRmax x AIRmin) mice through linkage analysis of PIA severity phenotype with a panel of SNPs. Two new PIA QTL (*Prtia* 2 and *Prtia*3) were mapped on chromosomes 5 and 8, respectively, and three suggestive QTL were detected on chromosomes 7, 17 and 19 (De Franco et al. 2014). In this same F2(AIRmax x AIRmin) population, loci that regulate the intensity of the acute inflammatory response were mapped on chromosomes 5, 7, 8 and 17, which overlap the QTL that control PIA severity, suggesting common regulations (Vorraro et al. 2010; Galvan et al. 2011). Co-located chromosome 5 QTL controlling arthritis severity and humoral responses during B. burgdorferi infection were identified in the F2 intercross of C3H/HeNCr and C57BL/6NCr mice (Weis et al. 1999), suggesting the involvement of the chemokine Cxcl9 gene, which maps to the QTL peak in this model (Ma et al. 2009).

In order to candidate genes within the QTL detected in the AIRmax and AIRmin model, transcriptome studies were performed using tissues or cells from normal or arthritic individuals. In this model, the total number of up- and down-regulated genes in each line was distinct, as can be seen in Fig. 12.3. More genes were modulated in AIRmax than in AIRmin mice, although a gene ontology analysis revealed an over-representation of genes related to inflammatory reaction and chemotaxis biological themes in both lines (Fig. 12.4). Global gene expression analysis indicated 419 differentially expressed genes between AIRmax and AIRmin mice. Figs. 12.5 and 12.6 show genes differentially expressed on chromosomes 5 and 8 respectively. Several genes related to inflammation, cell adhesion, and chemotaxis could be observed on chromosome 5, while tissue antigens, cell differentiation, hemeoxigenase and scavenger receptor genes were observed on chromosome 8 (De Franco et al. 2014).

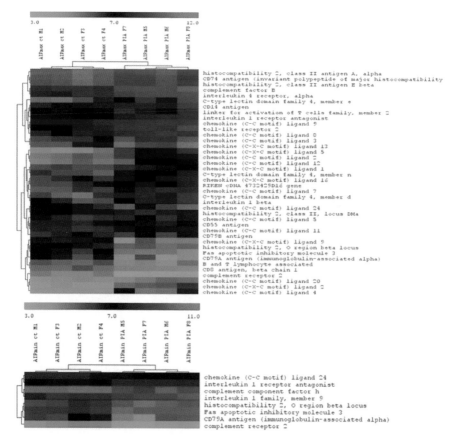

Fig. 12.3 Up- and down-modulated inflammatory and chemokine genes in AIRmax and AIRmin mice. Total RNA was extracted from arthritic paws at 160 days after pristane injection

Ibrahim and collaborators investigated the gene expression profiles of inflamed paws in DBA/1 inbred mice using a similar approach for collagen-induced arthritis (Ibrahim et al. 2002). In their work, inflammation resulted in increased gene expression of matrix metalloproteinases, and immune-related extra-cellular matrix and cell-adhesion molecules, as well as molecules involved in cell division and transcription, in a manner very similar to the AIRmax/AIRmin model. However, the total number of differentially-expressed genes involved in the inbred mice model (223) was lower than in our model (419), suggesting that the heterogeneous background of AIRmax and AIRmin mice permitted a larger genome involvement in this phenotype. Among the differentially-expressed genes, inflammatory and chemokine genes on chromosome 5 and *macrophage scavenger receptor 1* (*Msr1*) and *hemeoxigenase 1* (*Hmox1*) genes on chromosome 8 appear to be the major candidates.

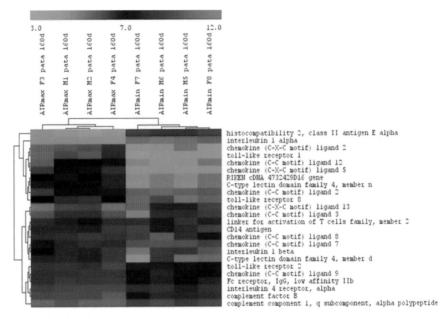

Fig. 12.4 Differentially expressed inflammatory and chemokine genes between AIRmax and AIRmin mice

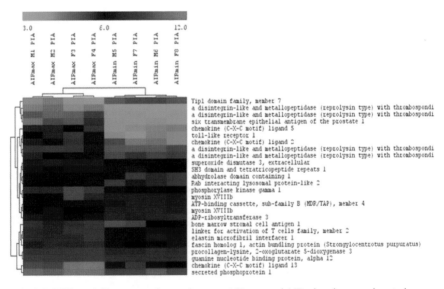

Fig. 12.5 Differentially expressed genes between AIRmax and AIRmin mice mapping at chromosome 5

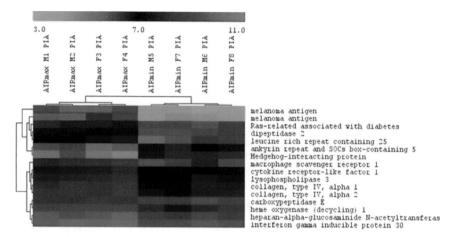

Fig. 12.6 Differentially expressed genes between AIRmax and AIRmin mice mapping at chromosome 8

Chemokines are involved in leukocyte recruitment to inflammatory sites, such as to synovial tissue in rheumatoid arthritis (RA). However, they may also be homeostatic as these functions often overlap (Ibrahim et al. 2001). Chemokines have essential roles in the recruitment and activation of leucocyte subsets within tissue microenvironments, and stromal cells actively contribute to these networks. Macrophages play a central role in the pathogenesis of rheumatoid arthritis (RA), which is marked by an imbalance of inflammatory and anti-inflammatory macrophages in RA synovium. Although the polarization and heterogeneity of macrophages in RA have not been fully elucidated, the identities of macrophages in RA can potentially be defined by their products, including co-stimulatory molecules, scavenger receptors, cytokines/chemokines and their receptors, and transcription factors (Li et al. 2012). It has been demonstrated that inappropriate constitutive chemokine expression contributes to the persistence of inflammation by actively blocking its resolution (Filer et al. 2008). This was also observed in urethane induced lung carcinogenesis, where transcriptome analysis revealed that the genes involved in transendotelial migration and chemokine-cell adhesion were differently-expressed in normal lungs of susceptible AIRmin and resistant AIRmax mice (De Franco et al. 2010), suggesting important roles for these phenotypes in chronic diseases.

12.7 Concluding Remarks

Recent advances in the field of genetics have dramatically changed our understanding of autoimmune disease. Candidate gene and, more recently, genome-wide association (GWA) and linkage studies have led to an explosion in the number of loci and pathways known to contribute to autoimmune phenotypes, confirming a major role for the MHC region and, more recently, identifying risk loci involving both the innate and adaptive immune responses. However, most regions found through GWA scans have yet to isolate the association to the causal allele(s) responsible for conferring disease risk. A role for rare variants (allele frequencies of < 1 %) has begun to emerge. The study of the abundant long intergenic non-coding RNAs and of small interfering RNA, (microRNAs) has also become a powerful tool to understand the mechanisms that modulate the gene expression profiles in RA and other autoimmune diseases (Jarvis and Frank 2010; Donate et al. 2013). Future research will also use next generation sequencing (NGS) technology to comprehensively evaluate the human genome for risk variants. Whole transcriptome sequencing (e.g. RNA-Seq), which combines gene expression, sequence and splice variant analysis, will provide much more detailed gene expression data. Regardless of the current or future technology, the versatility of murine models will continue to be required to advance our understanding of human diseases.

References

Adarichev VA, Valdez JC, Bardos T et al (2003) Combined autoimmune models of arthritis reveal shared and independent qualitative (binary) and quantitative trait loci. J Immunol 170(5):2283–2292

Araujo LM, Ribeiro OG, Siqueira M et al (1998) Innate resistance to infection by intracellular bacterial pathogens differs in mice selected for maximal or minimal acute inflammatory response. Eur J Immunol 28(9):2913–2920

Asquith DL, Miller AM, Mcinnes IB et al (2009) Animal models of rheumatoid arthritis. Eur J Immunol 39(8):2040–2044

Ates O, Dalyan L, Musellim B et al (2009) NRAMP1 (SLC11A1) gene polymorphisms that correlate with autoimmune versus infectious disease susceptibility in tuberculosis and rheumatoid arthritis. Int J Immunogenet 36(1):15–19

Baechler EC, Batliwalla FM, Reed AM et al (2006) Gene expression profiling in human autoimmunity. Immunol Rev 210:120–137

Besenyei T, Kadar A, Tryniszewska B et al (2012) Non-MHC risk alleles in rheumatoid arthritis and in the syntenic chromosome regions of corresponding animal models. Clin Dev Immunol. doi:10.1155/2012/284751

Biozzi G, Ribeiro OG, Saran A et al (1998) Effect of genetic modification of acute inflammatory responsiveness on tumorigenesis in the mouse. Carcinogenesis 19(2):337–346

Cho YG, Cho ML, Min SY et al (2007) Type II collagen autoimmunity in a mouse model of human rheumatoid arthritis. Autoimmun Rev 7(1):65–70

De Franco M, Colombo F, Galvan A et al (2010) Transcriptome of normal lung distinguishes mouse lines with different susceptibility to inflammation and to lung tumorigenesis. Cancer Lett 294(2):187–194

De Franco M, Peters LC, Correa MA et al (2014) Pristane-induced arthritis loci interact with the *Slc11a1* gene to determine susceptibility in mice selected for high inflammation. PLoS One 9(2):e88302

Donate PB, Fornari TA, Macedo C et al (2013) T cell post-transcriptional miRNA-mRNA interaction networks identify targets associated with susceptibility/resistance to collagen-induced arthritis. PLoS One 8(1):e54803

Filer A, Raza K, Salmon M et al (2008) The role of chemokines in leucocyte-stromal interactions in rheumatoid arthritis. Front Biosci 13:2674–2685

Fujikado N, Saijo S, Iwakura Y (2006) Identification of arthritis-related gene clusters by microarray analysis of two independent mouse models for rheumatoid arthritis. Arthritis Res Ther 8(4):100–125

Galvan A, Vorraro F, Cabrera W et al (2011) Association study by genetic clustering detects multiple inflammatory response loci in non-inbred mice. Genes Immun 12(5):390–394

Glant TT, Finnegan A, Mikecz K (2003) Proteoglycan-induced arthritis: immune regulation, cellular mechanisms, and genetics. Crit Rev Immunol 23(3):199–250

Glant TT, Adarichev VA, Nesterovitch AB et al (2004) Disease-associated qualitative and quantitative trait loci in proteoglycan-induced arthritis and collagen-induced arthritis. Am J Med Sci 327(4):188–195

Hirose J, Tanaka S (2011) Animal models for bone and joint disease. CIA, CAIA model. Clin Calcium 21(2):253–259

Ibanez OM, Stiffel C, Ribeiro OG et al (1992) Genetics of nonspecific immunity: I. Bidirectional selective breeding of lines of mice endowed with maximal or minimal inflammatory responsiveness. Eur J Immunol 22(10):2555–2563

Ibrahim SM, Yu X (2006) Dissecting the genetic basis of rheumatoid arthritis in mouse models. Curr Pharm Des 12(29):3753–3759

Ibrahim SM, Mix E, Bottcher T et al. (2001) Gene expression profiling of the nervous system in murine experimental autoimmune encephalomyelitis. Brain 124:1927–1938

Ibrahim SM, Koczan D, Thiesen HJ (2002) Gene-expression profile of collagen-induced arthritis. J Autoimmun 18(2):159–167

Jarvis JN, Frank MB (2010) Functional genomics and rheumatoid arthritis: where have we been and where should we go? Genome Med 2(7):44–59

Jensen JR, Peters LC, Borrego A et al (2006) Involvement of antibody production quantitative trait loci in the susceptibility to pristane-induced arthritis in the mouse. Genes Immun 7(1):44–50

Kannan K, Ortmann RA, Kimpel D (2005) Animal models of rheumatoid arthritis and their relevance to human disease. Pathophysiology 12(3):167–181

Kobezda T, Ghassemi-Nejad S, Mikecz K et al (2014) Of mice and men: how animal models advance our understanding of T-cell function in RA. Nat Rev Rheumatol 10(3):160–170

Kurko J, Besenyei T, Laki J et al (2013) Genetics of rheumatoid arthritis—a comprehensive review. Clin Rev Allergy Immunol 45(2):170–179

Li P, Schwarz EM (2003) The TNF-alpha transgenic mouse model of inflammatory arthritis. Springer Semin Immunopathol 25(1):19–33

Li J, Hsu HC, Mountz JD (2012) Managing macrophages in rheumatoid arthritis by reform or removal. Curr Rheumatol Rep 14(5):445–454

Lubberts E, Koenders MI, Van Den Berg WB (2005) The role of T-cell interleukin-17 in conducting destructive arthritis: lessons from animal models. Arthritis Res Ther 7(1):29–37

Ma Y, Miller JC, Crandall H et al (2009) Interval-specific congenic lines reveal quantitative trait Loci with penetrant lyme arthritis phenotypes on chromosomes 5, 11, and 12. Infect Immun 77(8):3302–3311

Patten C, Bush K, Rioja I et al (2004) Characterization of pristane-induced arthritis, a murine model of chronic disease: response to antirheumatic agents, expression of joint cytokines, and immunopathology. Arthritis Rheum 50(10):3334–3345

Peters LC, Jensen JR, Borrego A et al (2007) Slc11a1 (formerly NRAMP1) gene modulates both acute inflammatory reactions and pristane-induced arthritis in mice. Genes Immun 8(1):51–56

Potter M, Wax JS (1981) Genetics of susceptibility to pristane-induced plasmacytomas in BALB/cAn: reduced susceptibility in BALB/cJ with a brief description of pristane-induced arthritis. J Immunol 127(4):1591–1595

Roper RJ, Weis JJ, Mccracken BA et al (2001) Genetic control of susceptibility to experimental Lyme arthritis is polygenic and exhibits consistent linkage to multiple loci on chromosome 5 in four independent mouse crosses. Genes Immun 2(7):388–397

Runstadler JA, Saila H, Savolainen A et al (2005) Association of SLC11A1 (NRAMP1) with persistent oligoarticular and polyarticular rheumatoid factor-negative juvenile idiopathic arthritis in Finnish patients: haplotype analysis in Finnish families. Arthritis Rheum 52(1):247–256

Silman AJ, Pearson JE (2002) Epidemiology and genetics of rheumatoid arthritis. Arthritis Res 4 (Suppl 3):S265–S272

Silva GL, Junta CM, Sakamoto-Hojo ET et al (2009) Genetic susceptibility loci in rheumatoid arthritis establish transcriptional regulatory networks with other genes. Ann N Y Acad Sci 1173:521–537

Stasiuk LM, Ghoraishian M, Elson CJ et al (1997) Pristane-induced arthritis is CD4+ T-cell dependent. Immunology 90(1):81–86

Teixeira VH, Olaso R, Martin-Magniette ML et al (2009) Transcriptome analysis describing new immunity and defense genes in peripheral blood mononuclear cells of rheumatoid arthritis patients. PLoS One 4(8):e6803

Thompson SJ, Elson CJ (1993) Susceptibility to pristane-induced arthritis is altered with changes in bowel flora. Immunol Lett 36(2):227–231

Thompson SJ, Rook GA, Brealey RJ et al (1990) Autoimmune reactions to heat-shock proteins in pristane-induced arthritis. Eur J Immunol 20(11):2479–2484

Thompson SJ, Hitsumoto Y, Zhang YW et al (1992) Agalactosyl IgG in pristane-induced arthritis. Pregnancy affects the incidence and severity of arthritis and the glycosylation status of IgG. Clin Exp Immunol 89(3):434–438

Thompson SJ, Francis JN, Siew LK et al (1998) An immunodominant epitope from mycobacterial 65-kDa heat shock protein protects against pristane-induced arthritis. J Immunol 160(9):4628–4634

Van Den Berg WB (2009) Lessons from animal models of arthritis over the past decade. Arthritis Res Ther 11(5):250–259

Vidal SM, Epstein DJ, Malo D et al (1992) Identification and mapping of six microdissected genomic DNA probes to the proximal region of mouse chromosome 1. Genomics 14(1):32–37

Vigar ND, Cabrera WH, Araujo LM et al (2000) Pristane-induced arthritis in mice selected for maximal or minimal acute inflammatory reaction. Eur J Immunol 30(2):431–437

Vorraro F, Galvan A, Cabrera WH et al (2010) Genetic control of IL-1 beta production and inflammatory response by the mouse Irm1 locus. J Immunol 185(3):1616–1621

Weis JJ, Mccracken BA, Ma Y et al (1999) Identification of quantitative trait loci governing arthritis severity and humoral responses in the murine model of Lyme disease. J Immunol 162(2):948–956

You S, Yoo SA, Choi S et al (2014) Identification of key regulators for the migration and invasion of rheumatoid synoviocytes through a systems approach. Proc Natl Acad Sci U S A 111(1):550–555

Yu X, Bauer K, Koczan D et al (2007) Combining global genome and transcriptome approaches to identify the candidate genes of small-effect quantitative trait loci in collagen-induced arthritis. Arthritis Res Ther 9(1):3–17

Chapter 13
Transcriptome in Human Mycoses

Nalu T. A. Peres, Gabriela F. Persinoti, Elza A. S. Lang, Antonio Rossi and Nilce M. Martinez-Rossi

Abstract Mycoses are infectious diseases caused by fungi, which incidence has increased in recent decades due to the increasing number of immunocompromised patients and improved diagnostic tests. As eukaryotes, fungi share many similarities with human cells, making it difficult to design drugs without side effects. Commercially available drugs act on a limited number of targets, and has been reported fungal resistance to commonly used antifungal drugs. Therefore, elucidating the pathogenesis of fungal infections, the fungal strategies to overcome the hostile environment of the host, and the action of antifungal drugs is essential for developing new therapeutic approaches and diagnostic tests. Large-scale transcriptional analyses using microarrays and RNA sequencing (RNA-seq), combined with improvements in molecular biology techniques, have improved the study of fungal pathogenicity. Such techniques have provided insights into the infective process by identifying molecular strategies used by the host and pathogen during the course of human mycoses. In this chapter, the latest discoveries about the transcriptome of major human fungal pathogens will be discussed. Genes that are essential for host-pathogen interactions, immune response, invasion, infection, antifungal drug response and resistance will be highlighted. Finally, their importance to the discovery of new molecular targets for antifungal drugs will be discussed.

13.1 Introduction

Fungi are eukaryotic microorganisms widely distributed in nature, existing as yeasts, molds, and mushrooms. Fungi are important decomposers of biomass and are useful in baking and wine fermentation. However, fungi can also cause severe, life-threatening infections in humans, animals, and vegetables, which can result in enormous economic losses. Humans are constantly in contact with fungi by inhaling spores in the air and ingesting them as nutritional sources. Worldwide, human

N. M. Martinez-Rossi (✉) · N. T. A. Peres · G. F. Persinoti · E. A. S. Lang · A. Rossi
Department of Genetics, Ribeirão Preto Medical School, University of São Paulo,
14049-900 Ribeirão Preto, São Paulo, Brazil
e-mail: nmmrossi@usp.br

© Springer International Publishing Switzerland 2014
G. A. Passos (ed.), *Transcriptomics in Health and Disease,*
DOI 10.1007/978-3-319-11985-4_13

mycoses have increased in incidence due to the high prevalence of immunocompromised patients, becoming a major public-health concern. Although fungal infections are widespread, they are often overlooked, and in general, public-health agencies perform little surveillance of fungal infections (Brown et al. 2012). Fungi cause a wide spectrum of disease, ranging from asymptomatic infection to disseminated and fatal diseases. Nevertheless, fungal infections are not frequently diagnosed, which impairs the proper epidemiological surveillance of these diseases. Therefore, more data about the life cycle of pathogenic fungi and the pathogenesis of their infections will aid the development of therapeutic approaches and diagnostic tests. Although research funding for human mycoses remains lower than that for other areas of medical microbiology, the number of publications in the field of medical mycology has increased over the past several decades.

Fungi can infect several anatomical sites, resulting in different clinical symptoms. The most prevalent are cutaneous, mucosal, and pulmonary diseases. These infections can be acquired from trauma to the skin and mucosa, direct or indirect contact with infected humans and animals, contact with contaminated fomites, or inhalation. Airborne fungal infections typically result in pulmonary diseases. Skin and nail infections affect both healthy and immunocompromised individuals, decreasing quality of life by causing discomfort and pruritus. Cutaneous infections are most commonly caused by dermatophytes, a closely related group of keratinophilic molds that infect humans and animals. They are directly or indirectly transmitted between infected organisms and contaminated objects, such as towels and manicure appliances (Peres et al. 2010a). *Candida* species can also cause skin and nail infections, but they more commonly cause oropharyngeal (thrush) and vulvovaginal candidiasis. Many *Candida* species are harmless and are commensal microorganisms. However, immune-system impairment favors their pathogenicity, and they can cause opportunistic infections. *C. albicans* is part of the normal microbiota of mucous membranes of the respiratory, gastrointestinal, and female genital tracts. Changes in the host's immunological status enable its invasive behavior, leading to tissue damage and dissemination through the bloodstream to other organs (Mayer et al. 2013).

Deep fungal infections are mainly caused by *Aspergillus fumigatus*, *Cryptococcus neoformans*, *Coccidioides immitis*, *Paracoccidioides brasiliensis*, *Histoplasma capsulatum*, and *Blastomyces dermatitidis*. Some fungal diseases are endemic, such as blastomycosis (*B. dermatitidis*) and histoplasmosis (*H. capsulatum*), which are mainly found in the United States, and paracoccidioidomycosis (*P. brasiliensis*), which is primarily found in Latin America. Others are cosmopolitan and are encountered worldwide (Brown et al. 2012). Fungal spores are present in the environment and can be inhaled; upon reaching the lungs, they adhere to the parenchyma and initiate the infectious process. From the lungs, they can enter the bloodstream and disseminate to other organs, mainly the liver and spleen. Table 13.1 summarizes the major human fungal pathogens and their associated diseases. Overall, treatment of clinical mycoses can be a very long and expensive process that is often associated with uncomfortable side effects that lead to treatment interruption (Martinez-Rossi et al. 2008).

Table 13.1 Main fungal pathogens and their associated diseases in humans

Fungi	Main species	Disease
Aspergillus	*A. fumigauts*	Pulmonary infections (invasive aspergilosis and aspergilloma)
		Allergy
Blastomyces	*B. dermatitis*	Skin lesions
		Pulmonary infections
Candida	*C. albicans*	Cutaneous infections—skin and nails
	C. glabrata	Oropharyngeal candidiasis (thrush)
	C. parapsilosis	Vulvovaginal candidiasis
Coccidioides	*C. immitis*	Pulmonary infections
Cryptococcus	*C. neoformans*	Meningitis
		Pulmonary infections (pneumonia)
Dermatophytes	*Trichophyton rubrum*	Cutaneous infections—skin, nail, and hair (tinea or ringworms)
	Trichophyton mentagrophytes	
	Microsporum canis	
Histoplasma	*H. capsulatum*	Pulmonary infections
Malassezia	*M. furfur*	Cutaneous infections—skin (pityriasis versicolor)
Paracoccidioides	*P. brasiliensis*	Pulmonary and systemic infections
Penicillium	*P. marneffeii*	Pulmonary infections
Pneumocystis	*P. carinii*	Pneumonia

Some fungal species are found as filamentous or yeast forms, while others are dimorphic (i.e., found in both forms). In dimorphic fungi, the yeast form represents the parasitic phase, and the hyphae form represents the saprophytic phase. The filament (or hyphae) is a tubular multi-cellular structure, and cells are divided into compartments by the formation of a septum. Yeasts are round, single cells that reproduce by budding and some species can form pseudohyphae, a chain of interconnected yeast cells. Fungi can undergo sexual or asexual reproduction, producing spores that can be inhaled or enter the body at sites of tissue damage. Once the spores reach their appropriate niche, they develop into hyphae that invade the tissue in search of nutrients.

The genomes of several fungal pathogens have been sequenced, enabling the design and analysis of microarrays, representing a large amount of transcriptomic data to be explored. Fungal transcriptomics have been used to analyze gene expression and regulation in response to antifungal exposure, environmental changes, and interaction with the host during infection. The transcriptional profile may help elucidate several aspects of fungal biology, including signaling pathways that enable fungal survival, and help predict molecular targets for the development of novel antifungal drugs (Peres et al. 2010b; Cairns et al. 2010). Transcriptional and proteomic analyses have been used to identify connections among signaling and metabolic pathways that govern fungal development, morphogenesis, and pathogenicity as well as the host's immune response. Recent advances in molecular biology methods, especially RNA-seq by next-generation sequencing, have enabled the study of the whole transcriptome. These analyses provide insights into the functionality of the genome, revealing molecular components of cells and tissues involved in physiological and pathological processes. Transcriptome allows the analysis of all transcript species, including mRNAs, which encode proteins, as well as non-coding RNAs (ncRNAs) and small RNAs (sRNAs), which regulate gene expression and maintain cellular homeostasis. Furthermore, transcriptional profiling by RNA-seq is useful to determine gene structures at transcription initiation sites, 5'- and 3'-ends, and introns as well as splicing patterns (Wang et al. 2009).

This chapter will discuss recent advances in fungal transcriptomics arising from microarray and RNA-seq analyses. The contribution of these findings to the understanding of fungal biology and fungal diseases will be highlighted. The intrinsic relationship of the outcome of fungal infections and the immunological status of the host stresses the need to evaluate the host immune response to fungi. Furthermore, knowledge of the genes expressed in response to stressful environmental conditions and the gene networks that regulate the transcriptome during a fungal infection will help elucidate the pathogenesis of fungal infections and identify possible molecular targets for the development of novel therapeutic agents. Such information will aid both the treatment and prevention of fungal infections.

13.2   Host Immune Response to Fungal Infections

The host's immunological status is the primary determinant of the severity of fungal infections, which can range from asymptomatic to severe and disseminated. Immunocompromised patients often suffer from severe, disseminated, and fatal fungal infections. Host-pathogen interactions are complex and involve several molecules on the surface of both host and fungal cells. These surface molecules participate in pathogen recognition, fungal adhesion to host cells, downstream intracellular signaling, and metabolic pathways that mediate pathogen survival and host immune responses. An understanding of the infective process requires molecular knowledge of the pathogen strategies for infecting the tissue as well as the host responses aimed at eliminating the pathogen and maintaining cellular integrity. The development of experimental models has improved the study of infectious diseases, and most of these models utilize immunosuppressed mice, because most fungal species cause opportunistic infections. However, for some pathogens, such as anthropophilic dermatophytes, these models are not suitable and *ex vivo* and *in vitro* assays have being performed, providing insights into the pathogenic process and immune response triggered by the fungus.

Fungal diseases can result from poor immune response or from exacerbated activation of the immune system, such as the inflammatory response. Therefore, the interplay of the innate and adaptive immune systems and their appropriate activation are crucial for successful pathogen clearance and cellular homeostasis. Recent studies have characterized different mechanisms underlying the host's immune response to fungi. The innate immune response is comprised of the epithelial barrier, mucosa, and phagocytes (i.e., neutrophils, macrophages, and dendritic cells [DCs]), which play essential roles in preventing the entry of pathogenic microorganisms and rapidly killing these pathogens, as well as activating the adaptive immune response. Complement and other molecules, such as antimicrobial peptides and mannose-binding lectin, are also important host defense mechanisms. Pattern-recognition receptors (PRRs) on the surface of host cells interact with pathogen-associated molecular patterns (PAMPs) on the surface of pathogens, such as mannoproteins and β-glucan in fungi. The molecular interaction between PRRs and PAMPs triggers intracellular signaling pathways that initiate early inflammatory and non-specific responses in the host and upregulates virulence factors in the pathogen that enhance survival. PRRs include the toll-like receptors (TLRs) TLR2, TLR4, and TLR9, complement receptor 3, mannose receptor, Fcγ receptor, and Dectin-1 (Romani 2011).

In general, a Th1 response is correlated with protective immunity against fungi. The Th1 response is characterized by the production of interferon gamma (IFN-γ) and leads to cell-mediated immunity. Antigen-presenting cells (APCs), such as macrophages and DCs, initiate the Th1 response once their PRRs engage with fungal PAMPS, which leads to cellular activation and elicits effector properties. Th1 cells are essential for optimal activation of phagocytes at the site of infection through producing signature cytokines, such as like IFN-γ. Moreover, Th17 cells support Th1 cellular responses in experimental mucosal candidiasis, indicating an important role for Th17 cells in promoting neutrophil recruitment and Th1 immune responses (Conti et al. 2009; Romani 2011; Bedoya et al. 2013). Whole-genome transcriptional analyses have identified specific transcriptional profiles of host cells in response to various fungal species. Different cell types respond to fungal stimuli by activating distinct intracellular signaling pathways downstream of different PRRs. This mechanism confers plasticity to immune cells, such as DCs and macrophages, which shapes T-cell responses during fungal infections. The distinct signaling pathways in phagocytes influence the balance between innate and adaptive immune responses and the balance between CD4$^+$ T cells and regulatory T cells. Together, the balance between these processes establishes the outcome of the infection (Romani 2011).

H. capsulatum is a dimorphic fungus that causes respiratory infections and disseminated disease in immunocompromised hosts. The *H. capsulatum* hyphae produce spores (conidia) in the environment, which can be inhaled by humans. Inside the host, the conidia undergo morphological changes to form yeast cells. Once inhaled, the conidia are captured by phagocytic cells, such as macrophages, and trigger the host immune response. On the other hand, yeast cells use alveolar macrophages as vehicles to spread to different organs, such as the liver, spleen, lymph nodes, and bone marrow (Deepe 2000). Transcriptomic analysis by microarrays has revealed that conidia and yeast cells induce different transcriptional responses in macrophages. In response to infection with *H. capsulatum* conidia, macrophages specifically upregulated type I IFN-induced genes, including IFN-β and a classic type 1 IFN secondary response signature, in addition to general inflammatory genes. This effect was dependent on interferon regulatory factor 3 (IRF3) and independent of the TLR signaling pathway. This expression profile suggested that type 1 IFN contributes to the outcome of infection with *H. capsulatum*. Indeed, although ELISA failed to detect IFN-β protein, macrophages lacking the type 1 IFN receptor IFNAR1 did not express IFN-β in response to conidia exposure. Furthermore, IFNAR1 knockout mice showed a decreased fungal burden in the lungs and spleen after intranasal infection with conidia and yeast cells as compared to wild type mice. Therefore, IFNAR1 signaling might contribute to the virulence and disease burden of *H. capsulatum* infection rather than conferring protection. The authors suggested that type 1 IFNs might modulate cytokine production, apoptosis of infected macrophages, or specific aspects of the adaptive immune response to *H. capsulatum* (Inglis et al. 2010).

However, in the pathogenic yeast *C. neoformans*, which causes severe meningoencephalitis in immunocompromised patients, type 1 IFN signaling directs cytokine responses toward a protective type 1 pattern during murine cryptococcosis. IFNAR1 and IFN-β knockout infected mice displayed higher fungal burdens in the lungs and brain as well as decreased survival as compared to wild type mice (Biondo et al. 2008). Likewise, *C. albicans* induced the expression of type 1 IFN genes and proteins in DCs but not in macrophages. This pathway relied on Myd88 signaling, which is a TLR adaptor molecule. IFNAR1 and IFN-β knockout mice also displayed a lower survival rate and increased fungal burden in the kidneys. Therefore, the type 1 IFN response was protective against *C. albicans* by stimulating the host's innate immune defense mechanisms (Biondo et al. 2011). These studies highlight the complexity of the immune response to fungal pathogens and demonstrate the various roles that type 1 IFN plays.

The major virulence factor of *C. neoformans* is its polysaccharide capsule, which interferes with recognition by immune cells. Microarray analysis showed profound differences in gene expression profiles of DCs during phagocytosis of encapsulated versus non-encapsulated isogenic strains of *C. neoformans* opsonized with mouse serum. In general, the non-encapsulated strain induced the expression of genes involved in DC maturation, chemokines, and cytokines, characterizing an immunostimulatory response. In contrast, the encapsulated strain caused a downregulation or no change in the expression of these genes, indicating that the capsule prevented the activation of immune response-related genes. Among the proteins encoded by the genes upregulated in response to the non-encapsulated *C. neoformans* strain were CD86, CD83, the transcription factor Relb, ICAM1, major histocompatibility complex class II (MHC-II)-related genes (H2-D1, H2-Q7, and H2-Q8), and the pro-inflammatory cytokines IL-12, TNF-α, and IL-1. These proteins are involved in DC activation, maturation, and migration through the endothelium as well as induction of an inflammatory response. Several chemokines were also upregulated in DCs stimulated with the non-encapsulated strain, including CCL3, CCL4, CCL7, CCL12, CXCL10, CCL22, and the chemokine receptor CCR7, which contributes to the accumulation of inflammatory cells at the site of infection. Among the proteins encoded by the genes downregulated by the encapsulated strain were E74-like factor 1 (Elf1) and sequestosome 1 (Sqstm1), which regulate the expression of cytokines genes and the induction of NF-κB signaling, respectively (Lupo et al. 2008).

P. brasiliensis causes pulmonary and systemic infections. Once in the lungs, the yeast cells interact with resident phagocytic cells, such as macrophages and DCs. Microarray analyses of murine macrophages and DCs after phagocytosis of P. brasiliensis identified differential expression of genes encoding inflammatory cytokines, chemokines, signal-transduction proteins, and apoptosis-related proteins (Silva et al. 2008). Among the genes upregulated in macrophages were the pro-inflammatory chemokines CCL21, CCL22, and CXCL1. CXCL1 and CCL22 recruit neutrophils and monocytes, respectively, while CCL21 mediates the homing of lymphocytes to secondary lymphoid organs. Upregulation of the gene encoding NF-κB might account for the upregulation of pro-inflammatory chemokines and cytokines (e.g., TNF-α) that increase the cytotoxic activity of macrophages (Silva et al. 2008). Expression of the TNF-α gene and protein increased upon macrophage infection with P. brasiliensis. This was consistent with previous findings that p55KO mice, which are deficient in TNF-α, were unable to control P. brasiliensis infection, given the increased fungal burden and the absence of a well-formed granuloma (Silva et al. 2008; Souto et al. 2000). After exposure to P. brasiliensis, macrophages highly expressed apoptotic genes, including caspases 2, 3, and 8, which may represent a mechanism of eliminating the fungus without damaging host tissues. On the other hand, the fungus induced the expression of matrix metalloproteases genes, which may have facilitated fungal invasion, given their role in tissue remodeling (Silva et al. 2008).

Expression of the gene encoding IL−12 was downregulated in macrophages interacting with P. brasiliensis, which might represent a strategy of the fungi to evade the immune system. However, this gene was upregulated in DCs interacting with P. brasiliensis (Tavares et al. 2012) and C. neoformans (Lupo et al. 2008). IL-12 is associated with resistance to paracoccidioidomycosis and cryptococcosis by inducing IFN-γ production and Th1 protective responses. IL-12p140 knockout infected mice displayed decreased survival, higher fungal burden, and decreased production of IFN-γ (Decken et al. 1998; Livonesi et al. 2008). In addition to IL-12, DCs exposed to P. brasiliensis expressed genes encoding other pro-inflammatory cytokine and chemokine genes, such as TNF-α, CCL22, CCL27, CXCL10, and NF-κB, concomitantly with the downregulation of the NF-κB inhibitor Nκ-RF encoding gene. Both macrophages and DCs expressed CCL22 in response to P. brasiliensis, which might have increased the microbicidal activity of macrophages by stimulating a respiratory burst and the release of lysosomal enzymes. The chemokines expressed by macrophages and DCs in response to P. brasiliensis mediate the accumulation of leukocytes at the site of infection in order to control fungal invasion (Silva et al. 2008; Tavares et al. 2012).

Innate immune cells represent the first line of defense against pathogenic microorganisms (Mullick et al. 2004; Kim et al. 2005; Fradin et al. 2007) and include polymorphonuclear cells (PMNs, neutrophils), eosinophils, basophils, and monocytes. Monocytes release various cytokines in response to infection in order to amplify and coordinate the immune response. The dynamics of the molecular response triggered by *C. albicans* was measured in human monocytes using a time-course assay. This analysis identified a pattern of gene expression that might account for the recruitment, activation, and viability of phagocytes as well as the enhancement of chemotaxis and inflammation (Kim et al. 2005). Increased expression of genes encoding the pro-inflammatory cytokines TNF-α, IL-6, and IL-1α was correlated with neutrophil infiltration at the site of infection. In addition, there was an upregulation of genes encoding the chemokines CCL3, CCL4, CCL20, CCL18, CXCL1, CXCL-3, and IL-8, which are involved in the activation and recruitment of phagocytes and lymphocytes, as well as genes encoding the chemokine receptors CCR1, CCR5, CCR7, and CXCR5. This expression pattern likely favored the recruitment of inflammatory cells during the yeast phase of *C. albicans* life cycle. Moreover, TNF-α is essential for mounting a Th1 immune response and successfully controlling the infection. On the other hand, the expression of T cell-related genes was unchanged during the time-course analysis, suggesting that T cell regulatory molecules were not important in the early response of monocytes to *C. albicans*. It was proposed that in the early stages of infection, monocytes overexpress genes encoding various pro-inflammatory cytokines, chemokines, and chemokine receptors as well as COX2, IL-23, which are important for inflammation, and heat-shock proteins, which are implicated in the induction of inflammatory cytokines and chemokines. Thus, these changes in gene expression allow cellular recruitment and activation. Along with the pro-inflammatory response, increased expression of genes encoding anti-apoptotic molecules (XIAP and BCL2A1) may have protected the monocytes from cellular damage and death. The gene encoding the transferrin receptor (CD71) was upregulated, suggesting that iron deprivation might be a defense mechanism against infection. Indeed, iron is essential to the virulence of several pathogens (Kim et al. 2005).

Neutrophils display a potent set of hydrolytic enzymes, antimicrobial peptides, and oxidative species within their intracellular granules. These cells have an immediate and pronounced effect on *C. albicans* (Fradin et al. 2005). Granulocyte-like cells phagocytose and kill *C. albicans* and prevent hyphal growth, and undergo apoptosis after pathogen exposure. During this process, granulocytes upregulate inflammatory genes and downregulate anti-*Candida*-response genes, depending on the size of the inoculum. Among the upregulated genes were inflammatory mediators, including IL-1β, TNF-α, COX2, and the chemokine CCL3. On the other hand, genes encoding myeloperoxidase, which causes hyphal damage, and defensins, such as human neutrophil protein 1 (HPN1), were downregulated. These changes may represent mechanisms by which *C. albicans* survives the early stages of infection (Mullick et al. 2004).

In another study, a microarray of immune-related genes was used to evaluate the early response of PMN cells to *C. albicans* hyphal cells, UV-killed and live yeasts.

In PMNs, the transcriptional profiles induced by live yeasts and hyphae were more similar to one another than to that induced by dead yeasts. This suggested that fungal viability had a more significant effect on PMN gene expression than cellular morphotype. The presence of *C. albicans* did not affect the expression of genes encoding granule proteins. Nevertheless, *C. albicans* induced the upregulation of pro-inflammatory genes and cell-to-cell signaling (leukemia inhibitory factor [LIF]), signal transduction proteins, cell stimulatory factors, vascular endothelial growth factor, and PMN-recruitment chemokines (CCL3 and CXCL2). Importantly, these gene expression changes were irrespective of fungal cell type or viability. Furthermore, the few genes that were downregulated in response to *C. albicans* were involved in the regulation of cell signaling and growth (Fradin et al. 2007). In addition, exposure to viable *Candida* cells upregulated genes encoding stress-response proteins, including heat shock proteins (HSPA8, HSPCA, HSPCB, and HSPH1). This demonstrated a direct effect of live cells on PMN cells and monocytes (Kim et al. 2005). Interestingly, these genes also regulate CXC-type chemokines, indicating that this antimicrobial response amplifies the overall immune response by recruiting additional cells to the infection site. Overall, this transcriptional profile suggested that PMNs contribute to the immunological response to *C. albicans* by expressing genes involved in cellular communication, which may recruit more PMNs or other immune cells.

Systemic candidiasis is characterized by *C. albicans* entering the bloodstream, disseminating throughout the body, and causing microabscesses. In the blood vessels, the fungus must adhere to and invade endothelial cells (ECs); thus, the ECs have the potential to influence the host response to vascular invasion. Microarray transcriptional analysis of ECs in response to *C. albicans* identified the upregulation of genes with several functions, especially chemotaxis, angiogenesis, cell death, proliferation, intra- and intercellular signaling, immune response, and inflammation (Muller et al. 2007; Barker et al. 2008; Lim et al. 2011). These results support an important role for ECs in innate immunity. *C. albicans* induces several genes that are targets of the pro-inflammatory transcription factor NF-κB, which plays a central role in EC transcriptional regulation during *C. albicans* infection. In addition, the expression of chemokines, including IL-8, CXCL1, CXCL2, CXCL3, CXCL5, and CXCL6, indicates that ECs help recruit neutrophils and monocytes to the infection site (Muller et al. 2007). The overexpression of genes involved in stress and wound healing, coding for pro-inflammatory mediators such as IL-1, calgranulin C, E-selectin, and prostaglandin-endoperoxide synthase 2, correlated with the endothelial damage caused by *C. albicans*. The ECs also upregulated anti-apoptotic genes, suggesting that ECs respond to *C. albicans* by undergoing cellular proliferation (Barker et al. 2008). However, another transcriptional profile revealed that apoptotic genes were upregulated in ECs infected with a high density of *C. albicans*; the difference in the number of fungal cells may have been responsible for this difference in expression profile. In general, human umbilical vein ECs infected with high densities of *C. albicans* displayed a stronger and broader transcriptional responses than cells infected with low densities. Again, this effect may have been related to the number of cells or even to secreted molecules involved in quorum sensing. The authors hypothesized that in microenvironments with a high density of yeast cells, such as microabscesses, the fungus triggers apoptosis, which disrupts the endothelial barrier and permits fungal dissemination to different organs and tissues (Lim et al. 2011).

Given the complex and dynamic nature of host-pathogen interactions, techniques that measure both the host and pathogen responses are crucial for characterizing their interaction. Recent advances in molecular techniques have identified the molecular responses of the host and the pathogen during the infection. In dual transcriptomics, microarrays and RNA-seq have been used to identify molecular patterns of the pathogen and the host. These studies have provided insight into the dynamics of the infectious process (Westermann et al. 2012). Dual transcriptomics by RNA-seq was performed during DC phagocytosis of *C. albicans*, and gene interactions were predicted using a systems biology approach. Specifically, RNA-seq identified 545 *C. albicans* and 240 DCs genes that were differentially expressed. The genes were clustered by their expression kinetics over the duration of the interaction, and selected genes were used to infer gene interactions. After experimentally validating one of these gene interactions, the authors proposed a model in which PTX3, an opsonin secreted by DCs that facilitates phagocytosis through dectin-1, binds to the *C. albicans* cell wall, leading to its remodeling, which is mediated by the transcription factor Hap3 during invasion of innate immune cells. Remodeling of the fungal cell wall compromises the ability of immune cells to recognize fungi, thus attenuating the immune response (Tierney et al. 2012).

Microarray-based dual transcriptomics was performed on *A. fumigatus* interacting with bronchial epithelial cells. The resulting expression patterns indicated activation of the host's innate immune response (Oosthuizen et al. 2011). *A. fumigatus* is a major cause of pulmonary fungal infections, including invasive aspergillosis, aspergilloma, and allergy. During infection, environmental conidia enter the airways through inhalation. There, they germinate into hyphae and penetrate the lung parenchyma. Upon invasion, the fungus can disseminate to other organs and tissues. In response to *A. fumigatus* conidia, bronchial cells upregulated genes involved in innate immunity, chemokine activity, and inflammation. Among the overexpressed genes were those encoding the chemokines CCL3 and CCL5, which recruit leukocytes to the site of infection, matrix metalloproteinases (MMP1 and MMP3), and glutathione transferase (MGST1), which protects against oxidative damage. By comparing the expression profiles of two different cell lines, the authors identified only 17 genes in common. This demonstrates the variability in gene expression between a cell line and primary cells resulting from exposure to the same fungus (Gomez et al. 2011; Oosthuizen et al. 2011). The commonly expressed genes mainly encoded chemokines and regulators of the innate immune response. In particular, IL−6, a potent pro-inflammatory cytokine, was highly expressed in response to *A. fumigatus* conidia. This was consistent with earlier findings that IL−6-deficient mice were susceptible to invasive pulmonary aspergillosis and had impaired protective Th1 responses (Oosthuizen et al. 2011; Cenci et al. 2001). In addition, genes involved in nucleosome organization and chromatin assembly were overexpressed. Genes involved in mitosis and cell cycle progression were downregulated, suggesting decreased proliferation and cell cycle arrest during infection with *A. fumigatus* (Gomez et al. 2011).

Moreover, in response to *A. fumigatus* human monocytes presented a coordinated expression of genes involved in fungal death and invasion (Cortez et al. 2006). Among the highly expressed genes were pro-inflammatory genes, such as IL-1β, CCL3, CCL4, IL-8, PTX3, and SOD2, and regulators of inflammation, such as IL-10, COX2, and HSP40. Moreover, several anti-inflammatory genes were downregulated, such as CD14, which is involved in phagocytosis, and CCL5, which is a Th1 chemokine. This differential regulation of pro- and anti-inflammatory genes likely balanced the innate immune response. Furthermore, the coordinated expression of genes involved in oxidative response may have both eliminated the fungus and protected the cell. This was supported by the high expression of superoxide dismutase (SOD2) and dual phosphatase (DUSP1) and downregulation of catalase (CAT), glutathione peroxidase 3 (GPX3), and peroxiredoxin 5 (PRDX5) (Cortez et al. 2006).

Dermatophytes are highly specialized fungi that use keratin as a nutrient source, and thus infect keratinized structures, such as skin, hair, and nails. Upon infecting the skin, dermatophytes first encounter keratinocytes, which represents an important barrier against pathogens and help mediate the immune response (Peres et al. 2010a). The transcriptional profile of keratinocytes was measured in response to *Arthroderma benhamiae*, a zoophilic dermatophyte, and *Trichophyton tonsurans*, an anthropophilic species. During interaction, these fungi induced different cytokine expression profiles in keratinocytes, which were correlated with the inflammatory response. *Arthroderma benhamiae* induced the upregulation of 48 cytokine-related genes. In contrast, *T. tonsurans* induced the differential expression of only 12 cytokine-related genes; 4 were upregulated, and 8 were downregulated. In infected keratinocytes, the zoophilic species induced the upregulation of pro-inflammatory genes and the concomitant secretion of cytokines IL-1β, IL-6, IL-6R, and IL-17 and chemokines IL-8 and CCL2. This may promotes the infiltration of inflammatory cells in the skin during infection. Moreover, the upregulation of IL-6, IL-6R, and granulocyte-colony stimulating factor (G-CSF) may have promoted tissue remodeling and wound healing. On the other hand, the anthropophilic species induced limited cytokine expression and release, including exotoxin 2, IL-8, and IL-16. This was likely responsible for the poor inflammatory response observed in *T. tonsurans* skin infection (Shiraki et al. 2006). Moreover, mice infected with *Arthroderma benhamiae* displayed an infiltration of PMNs, macrophages, and DCs in the skin as well as increased levels of TGF-β, IL-1β, IL-6, and IL-22 mRNA in skin biopsies. Because these cytokines are involved in the establishment of the Th17 response, this pathway might regulate immunity against dermatophytes (Cambier et al. 2014).

In summary, fungal pathogens induce several changes in the host's target cells and innate immune cells. Studying the transcriptome of fungal-host interactions has elucidated the molecular patterns associated with protection from or progression of fungal infections. In general, fungi induce the upregulation of genes encoding cytokines, chemokines, and other pro-inflammatory molecules in host cells, which recruit inflammatory cells to the site of infection. Host cells exhibit different expression profiles in response to different fungal pathogens, which may account for the differences in outcomes of these infections. Moreover, some studies have identified molecular strategies by which fungi evade the host's immune system as well as host defense mechanisms that favor fungal survival. Transcriptomic analyses have generated hypotheses that can be further validated by reverse genetic approaches in order to better characterize the immune components that contribute to the outcome of fungal infections.

13.3 Metabolic Adaptation of Fungi During Infection

Fungal pathogens adapt to the host's microenvironment during infection, a process that requires dynamic responses to constantly changing conditions. In particular, nutrient availability can be limited in host niches, especially inside phagocytes. Thus, fungi undergo metabolic adaptations in order to control, for example, glycolysis, gluconeogenesis, the glyoxylate cycle, and proteolysis. This allows them to utilize diverse substrates as nutrient sources, evade the toxic conditions triggered by the immune response, and maintain their virulence despite changes in the physiological ambient (Brock 2009). Phagocytes produce reactive oxygen and nitrogen species (ROS and RNS), which induce oxidative and nitrosative stress and kill pathogens (Brown et al. 2009). Reactive species can alter or inactivate proteins, lipid membranes, and DNA. Pathogens can survive this toxic environment by producing protective enzymes, such as flavohemoglobin and S-nitrosoglutathione (GSNO) reductase, which confer resistance to nitrosative stress (de Jesús-Berríos et al. 2003), and superoxide dismutases, catalases, and peroxidases, which counteract oxidative stress. Non-enzymatic defenses include metabolites, such as melanin, mannitol, and trehalose (Missall et al. 2004). The ability of pathogens to sense and appropriately respond to environmental pH is essential for their survival in different host niches. In pathogenic fungi, the PACC/RIM signaling pathway has been implicated in survival, growth, virulence, and dissemination in different host niches (Cornet and Gaillardin 2014). The pH affects enzymatic activities, and the alkaline pH of human tissues influences nutrient uptake, because the solubility of essential elements, such as iron, is pH-dependent (Davis 2009). Iron is an essential micronutrient for both the host and pathogen, as it is required for several metabolic processes, including respiration and DNA replication. Iron, in the form of heme and iron-sulfur compounds, is an essential cofactor in various cellular enzymes, oxygen carriers, and electron-transfer systems. Iron homeostasis plays a key role in host-pathogen interactions. For instance, host tissues can restrict the availability of free iron in order to prevent infection. Accordingly, fungal pathogens have adapted strategies for iron uptake, including the production of metalloreductases, ferroxidases, and siderophores (Silva et al. 2011), in order to survive in iron-deficient niches.

Several pathways are crucial for fungal pathogens to survive in various host microenvironments during infection. *In vivo*, *ex vivo*, and *in vitro* infection models have identified the transcriptional profiles of fungal pathogens during infection and interaction with host cells. These studies have helped elucidate the pathogenesis of superficial, deep, and bloodstream fungal infections. An *in vitro* study used microarray to assess the transcriptional profile of *C. albicans* during interaction with human blood. There was an upregulation of genes involved in stress response, such as *SSA4* (a member of the *HSP70* gene family), and anti-oxidative response, such as those encoding Cu/Zn superoxide dismutase (*SOD1*), catalase (*CAT1*), and thioredoxin reductase (*TRR1*). There was a simultaneous upregulation of genes encoding the glycolytic enzymes phosphofructokinase (*PFK2*), phosphoglycerate kinase (*PGK1*), and enolase (*ENO1*) as well as those encoding the glyoxylate cycle enzymes isocitrate lyase (*ICL1*), malate synthase (*MLS1*), and acetyl-coenzyme-A-synthetase (*ACS1*). This suggested that both pathways were important for fungal dissemination. Genes involved in fermentation, such as those encoding alcohol dehydrogenases (*ADH1* and *ADH2*), were also upregulated. Importantly, *C. albicans* isolated from infected mice exhibited similar transcription profiles, thus validating some of the *in vitro* results. Moreover, these data suggested that *C. albicans* uses alternative carbon sources during blood infection and dissemination (Fradin et al. 2003).

A subsequent study investigated the utilization of the glyoxylate cycle and glycolysis by *C. albicans* interacting with different blood fractions, including erythrocytes, PMNs (mainly neutrophils), PMN-depleted blood (consisting of lymphocytes and monocytes), and plasma. *C. albicans* cells were physiologically active and displayed rapid hyphal growth while interacting with plasma, erythrocytes, and PMN-depleted blood. On the other hand, growth of *C. albicans* cells was arrested when interacting with PMNs, and only 40% of the cells interacting with whole blood produced hyphae. *C. albicans* upregulated glyoxylate cycle genes when interacting with PMNs, but not when interacted with plasma. During interaction with plasma and PMN-depleted blood, *C. albicans* upregulated genes related to glycolysis. Global cluster analysis was used to compare the transcriptional profile of *C. albicans* interacting with whole blood and blood fractions. During interaction with whole blood, the upregulation of genes related to glycolysis and the glyoxylate cycle resulted from mixed populations of fungal cells that were internalized by phagocytes, which triggers a nutrient limitation response, and not internalized (Fradin et al. 2005). Indeed, starvation inside the phagosome activated the glyoxylate cycle, which allowed nutrient uptake and survival. Accordingly, *C. albicans* deficient in the gene encoding isocitrate lyase was less virulent than the wild type strain in murine infection (Lorenz and Fink 2001).

While interacting with neutrophils, *C. albicans* also activated nitrogen- and carbohydrate-starvation responses, as indicated by the upregulation of genes encoding ammonium permeases (*MEP2* and *MEP3*), vacuolar proteases (*PRB1*, *PRB2*, and *APR1*), carboxypeptidases (*PRC1* and *PRC2*), glyoxylate cycle enzymes (*MLS1*, *ICL1*, and *ACS1*), aminoacid transporters, and proteins involved in aminoacid metabolism. Similarly, *C. albicans* activated the oxidative stress response, as indicated by the upregulation of genes encoding peroxidases and reductases, including superoxide dismutases (*SOD1* and *SOD5*) and catalase (*CAT1*) (Fradin et al. 2005). The upregulation of these genes may have contributed to fungal survival. Furthermore,

C. albicans internalized by murine macrophages *in vitro* displayed growth arrest and downregulation of genes associated with translation machinery and glycolysis. On the other hand, there was an upregulation of genes encoding enzymes involved in the gluconeogenesis pathway (phosphoenolpyruvate carboxykinase and fructose-1,6-bisphosphatase), glyoxylate cycle (isocitrate lyase and malate synthase), tricarboxylic acid cycle (aconitase, citrate synthase, and malate dehydrogenase), and β-oxidation of fatty acids, as well as several transporters. This suggested a metabolic adaptation toward the use of alternative carbon sources. Other upregulated genes included those encoding proteins important for detoxification of reactive species (flavohemoglobin, cytochrome *c* peroxidase, peroxidases, and reductases), stress response (heat shock protein HSP78), metal homeostasis, and DNA repair (Lorenz et al. 2004). Moreover, enzymes such as flavohemoglobins and superoxide dismutases, important to counteract ROS and RNS, were also implicated in *C. albicans* virulence (Missall et al. 2004)

Transcriptional profiling of *C. albicans* was performed using biopsies of infected oral mucosa from 11 HIV-positive patients. The results were compared to those obtained after *in vitro* interaction with reconstituted human epithelia (RHE). Genes related to gluconeogenesis (*PCK1*), the glyoxylate cycle (*MLS1* and *ICL1*), β-oxidation of fatty acids, and aminoacid and phosphate transport (*PHO84*) were upregulated. In the *in vitro* model, genes involved in nitric-oxide detoxification, such as those encoding flavohemoglobin (*YHB5*) and a sulfite transporter (*SSU1*), alkaline pH-responsive genes (*PHR1* and *PRA1*), and alkaline pH-induced genes of the Rim101 pathway were also upregulated. *SSU1*, *YHB1*, *YHB5*, *PHR1*, and *ICL1* were upregulated during oral mucosa infection. These changes reflected fungal responses to nitrosative stress, innate defense of epithelial cells against microbes, adaptation to the neutral-alkaline pH of the oral mucosa, and the use of alternative carbon sources at the site of infection. Despite the heterogeneity of the biopsy samples, there was a group of genes with similar expression profiles from both the patient samples and the *in vitro* RHE model. Genes related to iron acquisition (*CFL2* and *FRE4*) were upregulated in oral candidiasis but not in the *in vitro* model. In another study, Δ*icl1* was impaired to damage RHE, suggesting the importance of the glyoxylate cycle in oral candidiasis (Wachtler et al. 2011). Moreover, epithelial escape and dissemination (*EED1*), a unique species-specific *C. albicans* gene, was involved in hyphal elongation during infection (Zakikhany et al. 2007). A time-course microarray analysis of wild type and Δ*eed1* strains interacting with RHE showed that transcriptional differences between the strains increased over time. The mutant did not upregulate any genes across the entire time course, but did downregulate seven genes throughout the course of infection, including the hyphae-associated genes *ECE1* and *HYR1*. Other downregulated genes in the Δ*eed1* strain included those encoding proteins involved in polarized growth, such as CDC42, RDI1, MYO2, CDC11, CYB2, MOB1, and MLC1. A comparison of the transcriptional profile of several mutant strains suggested that EED1 was the primary component of a regulatory network controlling hyphal extension. *EED1* is positively regulated at the transcriptional level by EFG1, a member of the APSES family of transcription factors, and is repressed by NRG1, a sodium regulator. UME6, another transcriptional regulator involved in hyphal extension, is transcriptionally regulated by EED1. Both regulators participate in a pathway that controls the extension of *C. albicans* germ tubes into hyphae and the hyphae-to-yeast transition, which allows fungal dissemination within epithelial tissues (Martin et al. 2011).

In order to investigate expression changes in *C. albicans* during liver infection, transcriptional profiling was performed *in vivo* on mice infected by intraperitoneal injection as well as pig livers inoculated *ex vivo*. The upregulated genes encoded enzymes involved in glycolysis, such as phosphofructokinase (*PFK2*) and pyruvate dehydrogenase subunits (*PDA1* and *PDX1*), as well as those involved in acetyl-CoA biosynthesis and the tricarboxylic acid cycle (*KGD1* and *KGD2*). This gene expression modulation reflected the availability of carbohydrates and the utilization of glycolysis and respiration for energy production. However, the upregulation of *PCK1*, which encodes phosphoenolpyruvate carboxykinase, a key enzyme in gluconeogenesis, suggested that alternative carbon sources were also used. Other upregulated genes included *SAP2, SAP4, SAP5,* and *SAP6*, which encode the hyphae-associated aspartic proteases. Indeed, Sap2 is the major protease that enables the utilization of proteins as nitrogen sources. The upregulation of alkaline pH responsive gene (*PHR1*) suggested adaptation to an alkaline environment. Similarly, upregulation of genes encoding stress-response proteins, including heat shock proteins and molecular chaperones (*HSP78, HSP90, DDR48, HSP104, HSP12,* and *SSA4*), suggested that the heat shock response was triggered during the course of infection. However, genes related to oxidative, osmotic, and nitrosative stresses were not upregulated. On the other hand, genes related to iron, copper, zinc, and phosphate transport (*FTR1, CTR1, ZRT1, PHO84, PHO89*) were upregulated during liver infection, suggesting limited iron and phosphate in this environment. A comparison of the transcriptional profile of an invasive *C. albicans* strain with that of a noninvasive strain identified genes associated with liver invasion. One of these genes was *DFG16*, which encodes a membrane sensor in the RIM101 pathway that is crucial for pH-dependent hyphal formation, pH sensing, invasion at physiological pH, and systemic infection (Thewes et al. 2007; Martinez-Rossi et al. 2012; Rossi et al. 2013). In another study of systemic infection, rabbits were infected through the marginal ear vein, and infected kidneys were subsequently collected. *C. albicans* exhibited an upregulation of genes related to alternative pathways of carbon assimilation, such as β-oxidation of fatty acids, the glyoxylate cycle (*MLS1* and *ACS1*), and the tricarboxylic acid cycle (*CIT1, ACO1,* and *SDH12*), suggesting limited carbohydrate supply in the kidneys (Walker et al. 2009). Although genes involved in β-oxidation of fatty acids are upregulated in several models of infection, fatty-acid degradation is not essential for virulence of *C. albicans*. Nevertheless, disruption of genes involved in the glyoxylate cycle or gluconeogenesis significantly attenuated its virulence in mice (Ramirez and Lorenz 2007; Barelle et al. 2006).

C. albicans colonizes medical devices, such as intravascular catheters, by forming biofilms. Biofilms are comprised of heterogeneous microbial communities and form on biotic or abiotic surfaces embedded in an extracellular polymeric matrix. Such biofilms are associated with persistent infections and resistance to antifungal drugs and mechanical treatments. *C. albicans* forms a biofilm in four steps. First, yeast cells attach to and colonize a surface; second, yeast cells form a basal layer that anchors the biofilm; third, hyphae grow and produce pseudohyphae and extracellular matrix; finally, the yeast cells disperse. In order to characterize biofilm formation in *C. albicans*, the transcriptional regulatory network was analyzed in mutants that are unable to form biofilms. A combination of whole-genome chromatin immunoprecipitation microarray (ChIP-chip) and genome-wide transcriptional profiling identified six master regulators that control biofilm formation in *C. albicans*: BCR1, TEC1, EFG1, NDT80, ROB1, and BRG1. Each regulator controlled the other five, and most of the target genes were controlled by more than one master regulator. Moreover, the biofilm network targeted approximately 15% of the entire genome (Nobile et al. 2012).

Microarray analyses were used to profile *C. neoformans* transcription in response to murine macrophages. *C. neoformans* exhibited a downregulation of genes encoding translational machinery and an upregulation of genes associated with lipid degradation and fatty-acid catabolism (lipases and acetyl coenzyme A acetyltransferase), β-oxidation, transport of glucose and other carbohydrates, response to nitrogen starvation, the glyoxylate cycle (*ICL1*), and autophagy (*ATG3* and *ATG9*). This suggested nutritional starvation and metabolic adaptation toward the use of alternative carbon sources and nitrogen uptake. Moreover, the upregulation of several genes encoding oxidoreductases, peroxidases, and flavohemoglobin denitrosylase (FHB1), which are important for nitrosative response and virulence, indicated the presence of oxidative and nitrosative stress (de Jesús-Berríos et al. 2003). Also, there was an upregulation of genes related to endocytosis, exocytosis, and synthesis of extracellular polysaccharides and cell wall components. Genes located in the mating-type (MAT) locus and several genes associated with virulence were also upregulated. These included those encoding inositol-phosphorylceramide synthase (*IPC1*), laccases (*LAC1* and *LAC2*), genes involved in capsule formation (*CAP10, CAS31, CAS32, CAS1,* and *CAS2*), and *PKA*, a gene in the Gpa1-cAMP pathway, that is essential for virulence. In particular, the Gpa1-cAMP pathway regulates capsule formation and melanin production. Moreover, calcineurin gene (*CNA1*), which is critical for virulence, was upregulated (Fan et al. 2005).

Transcriptional analyses of *C. neoformans* isolated from cryptococcal pulmonary infection in mice revealed the upregulation of genes encoding malate synthase, phosphoenolpyruvate carboxykinase, aconitase and succinate dehydrogenase as well as those involved in β-oxidation of fatty acids. These findings suggested glucose depletion and the use of alternative carbon sources in the lungs. Genes encoding glyoxylate cycle enzymes were strongly upregulated as well as genes involved in glycolysis (e.g., fructose 1,6-biphosphate, aldolase, hexokinase, and phosphofructokinase). In addition, there was an upregulation of genes encoding transporters for monosaccharides, iron, copper, acetate, trehalose, and phosphate, enzymes involved in the production of acetyl-CoA (e.g., acetylCoA synthetase [*ACS1*]), pyruvate decarboxylase, and aldehyde dehydrogenase. The upregulation of several stress-response genes, including flavohemoglobin denitrosylase, superoxide dismutase, *HSP12*, *HSP90*, and other virulence factors, might have protected against the stressful conditions in the lungs. This profile of upregulated genes suggested the importance of generating and utilizing acetyl-CoA by *C. neoformans* during infection (Hu et al. 2008). Indeed, deletion of the *acs1* gene resulted in attenuated virulence and impaired growth on media containing acetate as a carbon source. Moreover, *ACS1* is regulated by serine/threonine protein kinase 1 (SNF1), which mediates glucose sensing, utilization of alternative carbon sources, and stress response. Deletion of the *SNF1* gene also reduced growth on acetate medium, decreased melanin production and caused loss of virulence in murine model (Hu et al. 2008). Although *C. neoformans* upregulated glyoxylate cycle genes during infection, *ICL1* and *MLS1* were not essential for establishing infection (Rude et al. 2002; Idnurm et al. 2007). On the other hand, deficits in β-oxidation pathways compromised the virulence of *C. neoformans* (Kretschmer et al. 2012).

C. neoformans var. *grubii* molecular type VNI and VNII were isolated from the cerebrospinal fluid (CSF) of two AIDS patients with cryptococcal meningitis (Chen et al. 2014). RNA was isolated from *C. neoformans* and used for transcriptome analyses by RNA-Seq. The *in vivo* transcriptomes were compared with the transcriptomes of each strain after incubation in pooled human CSF *ex vivo* or growth in YPD broth *in vitro*. As compared with growth in YPD, growth *in vivo* and *ex vivo* in CSF resulted in the upregulation of genes associated with pathogenicity, such as *CFO1*, *ENA1*, and *RIM101*. *CFO1* encodes a ferroxidase required for the utilization of transferrin, an important source of iron during brain infection and subsequent dissemination in infected mice (Jung et al. 2009). ENA1 is a sodium transporter that is important for CSF infection, intracellular survival in macrophages, and survival during meningitis in rabbits (Idnurm et al. 2009). RIM101 is a conserved pH-responsive transcription factor that regulates iron and metal homeostasis, capsule production, and cell wall formation in *C. neoformans* (O'Meara et al. 2010). Other genes, such as *SIT1* and *SRX1*, were upregulated only *in vivo*. SIT1 is a siderophore transporter that is important for growth under iron-limiting conditions, suggesting a role for iron in CSF infection. SRX1 is a sulfiredoxin that confers resistance to oxidative stress, which might improve survival in the presence of phagocytic cells in the CSF. Indeed, Rim101 controls the expression of several genes, including SIT1 and ENA1 (O'Meara et al. 2010; O'Meara et al. 2013). *ICL1* was another gene upregulated in *C. neoformans* infecting human CSF. The *ICL1* gene was also upregulated in a rabbit model of cryptococcal meningitis, although isocitrate lyase was not essential for infection (Rude et al. 2002). However, there was an upregulation of trehalose-6-phosphate synthase (TPS1) encoding gene (Steen et al. 2003), which was essential for infection (Petzold et al. 2006). Thus, in human cryptococcal meningitis, a set of genes was similarly modulated between *C. neoformans* strains isolated from infected patients. Among the differentially modulated genes, novel genes with unknown functions were also identified (Chen et al. 2014).

Recently, another transcriptional profiling study using nanoString demonstrated high concordance between the downstream targets of PKA and RIM101 in *C. neoformans*. A pairwise transcriptional analysis of the Δ*pka* and Δ*rim101* mutant strains revealed a strong correlation between the majority of RIM101-dependent and PKA1-dependent genes. This suggested that PKA1 and RIM101 interact or that they are in the same signaling pathway. In a murine model of lung infection, transcriptional and ChIPseq analyses of wild type and Δ*rim101* strains revealed genes regulated by RIM101, including *cda1* and *kre6*. CDA1 regulates levels of chitosan in the cell, and KRE6 participates in β-glucan synthesis, consistent with a role for RIM101 in regulating cell wall remodeling (O'Meara et al. 2013).

In a murine model of pulmonary aspergillosis, *A. fumigatus* exhibited a down-regulation of genes related to ribosomal biogenesis and protein biosynthesis and an upregulation of genes related to siderophore biosynthesis and transport, including ferric-chelate reductases, aminoacid permeases, GABA and proline permeases, maltose permeases and transporters, and extracellular proteases. Elastinolytic metalloprotease, an aorsin-like serine protease, and dipeptidylpeptidases are antigenic virulence factors that are important for nitrogen uptake. Several genes encoding antioxidant enzymes were also upregulated, including a Mn-superoxide dismutase and the bifunctional catalase-peroxidase CAT2. The initiation of infection was likely associated with aminoacid catabolism, as indicated by the induction of the enzyme methylcitrate synthase, which detoxifies propionyl-CoA intermediates (McDonagh et al. 2008). During invasive aspergillosis, propionyl-CoA is a toxic product generated from the degradation of valine, methionine, and isoleucine derived from host proteins. The gene encoding methylcitrate synthase was upregulated, suggesting that aminoacids are used as nutrients during invasive aspergillosis. Moreover, an *A. fumigatus* strain deficient in methylcitrate synthase displayed attenuated virulence (Ibrahim-Granet et al. 2008). During invasive aspergillosis, many of the preferentially expressed genes were located in the sub-telomeric regions of chromosomes. The coordinated expression of clustered genes included those responsible for the biosynthesis of siderophores and the secondary metabolites pseurotin and gliotoxin, which are toxins (McDonagh et al. 2008).

While interacting with human neutrophils, *A. fumigatus* conidia upregulated genes encoding proteins involved in peroxisome biogenesis, β-oxidation of fatty acids (acyl-CoA dehydrogenase and enoyl-CoA hydratase), acetate metabolism (acetyl-coenzyme A synthetase), the tricarboxylic acid cycle (aconitate, succinate dehydrogenase, and malate dehydrogenase), and the glyoxylate cycle (isocitrate lyase), suggesting a state of carbohydrate starvation (Sugui et al. 2008). There was a strong upregulation of the gene encoding formate dehydrogenase, which detoxifies formate, an indirect product of the glyoxylate cycle (Prigneau et al. 2003). Neutrophils normally produce cytotoxic ROS; however, neutrophils from patients with chronic granulomatous disease (CGD) are unable to damage *A. fumigatus* hyphae. Indeed, hyphae exposed to normal neutrophils expressed higher levels of the genes encoding glutathione peroxidase and thioredoxin reductase than those exposed to CGD neutrophils (Sugui et al. 2008). Despite the involvement of ROS released by phagocytes in killing *A. fumigatus*, a triple *SOD1/SOD2/SOD3* mutant and the parental strain were similarly virulent in experimental murine aspergillosis in immunocompromised animals (Lambou et al. 2010).

Upon contact with bronchial epithelial cells, *A. fumigatus* upregulated approximately 150 genes, most of which were related to vacuolar acidification, siderophore biosynthesis, metallopeptidase activity, and formate dehydrogenase activity (Oosthuizen et al. 2011; Kane 2007). Formate dehydrogenase (*fdh*) was also significantly upregulated upon incubation with neutrophils (Sugui et al. 2008) and biofilm formation (Bruns et al. 2010). Moreover, incubation with *A. fumigatus* induced cell death in monocyte-derived immature dendritic cells (iDC) (Morton et al. 2011). This effect was coincident with growth of the fungal germ tube and fungal upregulation of genes encoding enzymes related to oxidative stress response, ROS detoxification (e.g., superoxide dismutases), and pyomelanin biosynthesis. Additionally, *A. fumigatus* upregulated *ASPF1*, which encodes a ribotoxin that inhibits protein synthesis and induces apoptosis in iDCs *in vitro* (Ok et al. 2009).

Transcriptional profiling was performed on the dermatophyte *Arthroderma benhamiae* during an *in vivo* skin infection in guinea pigs. During acute infection, *Arthroderma benhamiae* upregulated genes encoding key enzymes of the glyoxylate cycle (MLS and ICL), formate dehydrogenase, monosaccharide transporter, oxidoreductase, opsin-related protein, and several proteases (Staib et al. 2010). The most highly upregulated gene was *SUB6* that encodes subtilisin 6, a protease previously characterized as the major allergen in another dermatophyte, *T. rubrum*. Sub6 has been shown to bind human IgE antibodies (Woodfolk and Platts-Mills 1998). The second most highly upregulated gene was that encoding an opsin-related protein with an unknown function. The same gene was previously shown to be upregulated during the parasitic *phase of* Coccidioides immitis, but its function is still unknown (Viriyakosol et al. 2013). Genes encoding proteases, such as subtilisins SUB1, SUB2, SUB6, and SUB7, the neutral protease NpII-1, and serine carboxypeptidase ScpC were also upregulated during infection (Staib et al. 2010). Proteases are the most commonly studied virulence factors of dermatophytes, and some possess keratinolytic activity, which allow them to infect the skin and nails (Monod 2008). Genes encoding SUB3, SUB5, and metalloprotease 4 (MEP4) were also upregulated in *T. rubrum* grown in keratin as the sole carbon source (Maranhão et al. 2007). Moreover, a *PACC/RIM101*-mutant strain of *T. rubrum* displayed decreased keratinolytic activity and impaired growth on the human nail *in vitro*, suggesting a role for RIM101 in the pathogenicity of *T. rubrum* (Ferreira-Nozawa et al. 2006; Silveira et al. 2010; Martinez-Rossi et al. 2012).

While interacting with human keratinocytes, *Arthroderma benhamiae* upregulated the *HYPA* gene, which encodes a hydrophobin (Burmester et al. 2011) that influences the organism's recognition by the immune system (Heddergott et al. 2012). Deletion of HYPA gene increased the susceptibility of *Arthroderma benhamiae* to human neutrophils and DCs. Compared to wild type, the Δ*hypA* mutant strain activated cellular immune defenses and increased the release of IL-6, IL-8, IL-10, and TNF-α to a higher degree. Moreover, conidia of the mutant strain were more easily killed by neutrophils (Heddergott et al. 2012). Indeed, surface expression of hydrophobin was shown to prevent immune recognition of *A. fumigatus* (Aimanianda et al. 2009).

Transcriptome data from various human fungal pathogens have identified global responses and survival strategies during interaction with host cells and infection of host niches. Accordingly, some pathways have been implicated in mycotic diseases, and fungi are able to proliferate and survive within the host by employing sophisticated mechanisms to quickly modulate gene expression and adapt to changes in the environment. Genes that are upregulated during the infective process or during interaction with host cells are potentially important for virulence (Table 13.2), which has been evaluated by the functional characterization of mutant strains. Thus, genome-wide transcriptional analyses combined with genetic approaches have provided important insight into fungal responses, adaptive processes, virulence, and pathogenesis.

13.4 Transcriptome of Drug Response and Resistance

Microorganisms respond to sub-lethal doses of chemical and physical agents by synthesizing a variety of specific proteins and low molecular weight compounds that act to promote defenses or tolerance (Fachin et al. 2001). Fungi use numerous signal transduction pathways to sense environmental stress and respond appropriately by differentially expressing cell-stress genes (Martinez-Rossi et al. 2008). Thus, analyses of transcriptional changes in response to cytotoxic drugs have elucidated the mechanisms by which fungi adapt to physiological stress as well as the mechanisms of drug action.

Although there are several commercially available antifungal drugs, the number of cellular targets is limited. Some antifungal drugs target ergosterol, a sterol analogous to cholesterol that is the main component of the fungal cell membrane and has diverse functions, including maintaining membrane stability, integrity, and permeability. Other antifungal drugs target proteins involved in the biosynthesis of ergosterol. The ergosterol biosynthetic pathway has been well described in *Saccharomyces cerevisiae* and involves approximately 20 genes, including those that convert mevalonate to squalene (Fig. 13.1), which are the primary targets of antifungal drugs. The polyenes are a class of antifungal drugs that include amphotericin B (AMB) and nystatin. They bind to ergosterol and form pores in the membrane, which causes the leakage of intracellular contents and fungal cell death. AMB also induces oxidative damage to cellular membranes through the generation of ROS.

Azoles are the most commonly used class of antifungal drugs in clinical treatment and include ketoconazole, itraconazole, fluconazole, and voriconazole. They inhibit the activity of the enzyme cytochrome P450 lanosterol 14-α demethylase (ERG11), which is responsible for the oxidative removal of the 14α-methyl group of lanosterol, an essential step in ergosterol biosynthesis. Azoles are first-line agents for the treatment of candidiasis, but their frequent use can result in resistance due to their fungistatic mechanism of action. Terbinafine (TRB) is another antifungal drug that belongs to the allylamine class and is most effective against dermatophytes. It inhibits ergosterol biosynthesis by inhibiting the enzyme squalene epoxidase (ERG1), which is responsible for converting squalene to lanosterol. Inhibition of ERG1 decreases the production of ergosterol and increases the accumulation of squalene to toxic levels.

Other antifungal drugs target DNA/RNA metabolism. Flucytosine is a cytosine analogue that was first used as an antitumor agent. It showed poor efficacy in the

Table 13.2 Putative fungal proteins associated with host interaction and pathogenesis

Protein description	Gene expression modulation and functional analysis	References
Isocitrate lyase (glyoxylate cycle enzyme)	Upregulated in *C. albicans*, *C. neoformans*, *A. fumigatus* and *A. benhamiae*. Gene inactivation attenuates virulence in *C. albicans* but not in *C. neoformans* and *A. fumigatus*	Fradin et al. (2003, 2005); Lorenz et al. (2004); Zakikhany et al. (2007); Fan et al. (2005); Chen et al. (2014); Sugui et al. (2008); Staib et al. (2010); Rude et al. (2002); Schöbel et al. (2007); Lorenz and Fink (2001); Wachtler et al. (2011)
Malate synthase (glyoxylate cycle enzyme)	Upregulated in *C. albicans*, *C. neoformans*, *A. fumigatus* and *A. benhamiae*. Gene inactivation does not attenuate virulence in *C. neoformans*	Fradin et al. (2003, 2005); Lorenz et al. (2004); Zakikhany et al. (2007); Walker et al. (2009); Hu et al. (2008); McDonagh et al. (2008); Staib et al. (2010); Idnurm et al. 2007; Cairns et al. 2010
Acetyl-coenzyme-A-synthetase (glyoxylate cycle enzyme)	Upregulated in *C. albicans*, *C. neoformans* and *A. fumigatus*. Gene inactivation attenuates virulence in *C. neoformans*	Fradin et al. (2003, 2005); Walker et al. (2009); Sugui et al. (2008); McDonagh et al. (2008); Cairns et al. (2010); Hu et al. (2008); Thewes et al. 2007; Lorenz et al. 2004
Aconitase (tricarboxylic acid cycle enzyme)	Upregulated in *C. albicans*, *C. neoformans* and *A. fumigatus*	Lorenz et al. (2004); Walker et al. (2009); Hu et al. (2008); Sugui et al. (2008)
Malate dehydrogenase (tricarboxylic acid cycle enzyme)	Upregulated in *C. albicans*, *C. neoformans* and *A. fumigatus*	Lorenz et al. (2004); Hu et al. (2008); Cairns et al. (2010); Sugui et al. (2008)
Phosphofructokinase (glycolysis enzyme)	Upregulated in *C. albicans* and *C. neoformans*	(Fradin et al. 2003; Thewes et al. 2007; Hu et al. 2008)
Enolase (glycolysis enzyme)	Upregulated in *C. albicans* and *C. neoformans*	Fradin et al. (2003); Thewes et al. (2007); Hu et al. (2008)
Phosphoenolpyruvate carboxykinase (gluconeogenesis enzyme)	Upregulated in *C. albicans* and *C. neoformans*. Gene inactivation attenuates virulence in *C. albicans*	Zakikhany et al. (2007); Lorenz et al. (2004); Thewes et al. (2007); Hu et al. (2008); Barelle et al. (2006)

Table 13.2 (continued)

Protein description	Gene expression modulation and functional analysis	References
Flavohemoglobin denitrosylases (RNS detoxification)	Upregulated in *C. albicans* and *C. neoformans*. Gene inactivation attenuates virulence in *C. albicans* and *C. neoformans*	Hu et al. (2008); Lorenz et al. (2004); Zakikhany et al. (2007); Fan et al. (2005); de Jesus-Berrios et al. (2003); Missall et al. (2004); Brown et al. (2009)
Superoxide dismutases (ROS detoxification)	Upregulated in *C. albicans*, *C. neoformans* and *A. fumigatus*. Gene inactivation attenuates virulence in *C. albicans* and *C. neoformans* but not in *A. fumigatus*	Hu et al. (2008); McDonagh et al. (2008); Morton et al. (2011); Fradin et al. (2003); Lorenz et al. (2004); Fradin et al. (2005); Lambou et al. (2010); Missall et al. (2004); Brown et al. (2009)
Hydrophobin (cell surface protein)	Upregulated in *A. fumigatus* and *A. benhamiae*. Gene inactivation in *A. fumigatus* and *A. benhamiae* increases the susceptibility to the host immune response	Cairns et al. (2010); Burmester et al. (2011); Heddergott et al. (2012); Aimanianda et al. (2009)

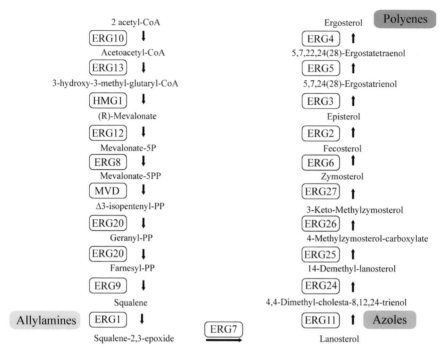

Fig. 13.1 Schematic representation of the ergosterol biosynthetic pathway

treatment of tumors but was shown to have antifungal properties. Flucytosine is transported to the cytoplasm of fungal cells through cytosine permease; in the cytoplasm, cytosine deaminase converts it to 5-fluorouracil, which blocks protein and DNA synthesis. When phosphorylated, 5-fluorouracil is incorporated into RNA, leading to miscoding and inhibition of protein synthesis. Furthermore, phosphorylated 5-fluorouracil can be converted into the deoxynucleoside form by uridine monophosphate pyrophosphorylase; thereafter, it inhibits the enzyme thymidylate synthetase and consequently disrupts DNA synthesis (Vermes et al. 2000).

More recently, the fungal cell wall has become a specific target of antifungal drugs, since it is absent from mammalian cells. Caspofungin was the first compound to target the fungal cell wall and was approved for clinical use in 2001. It is a member of the echinocandin class, which inhibits the enzyme $(1,3)$-β-D-glucan synthase (FKS1 and FKS2), thus preventing the synthesis of $(1,3)$-β-D-glucan and disrupting cell wall biosynthesis. In addition to caspofungin, two other echinocandins, micafungin and anidulafungin, are commercially available. These drugs are only available as intravenous infusions and are indicated to treat invasive aspergillosis and candidiasis. They have fungicidal activity against most *Candida* species and fungistatic activity against *Aspergillus* species. Although most fungal species encode orthologs of FKS1 and FKS2, echinocandins are not effective against Zygomycetes *C. neoformans*, or *Fusarium* spp. (Denning 2003).

Transcriptome analyses have been used to evaluate the responses of pathogenic fungi, such as *C. albicans, A. fumigatus,* and *T. rubrum,* to several antifungal drugs, including azoles, polyenes, terbinafine, and echinocandins (Liu et al. 2005; da Silva Ferreira et al. 2006; Cervelatti et al. 2006; Yu et al. 2007b; Paião et al. 2007; Gautam et al. 2008; Diao et al. 2009; Zhang et al. 2009; Peres et al. 2010b). These studies revealed that the modulation of genes in the ergosterol biosynthetic pathway varies significantly among species and drugs (Table 13.3). For instance, in response to itraconazole, *C. albicans* upregulated the following genes related to ergosterol biosynthesis: *erg1, erg2, erg3, erg4, erg5, erg6, erg9, erg10, erg11,* and *erg25* (De Backer et al. 2001). In contrast, *T. rubrum* only upregulated *erg11, erg24,* and *erg25* (Diao et al. 2009). Similarly, in response to voriconazole *A. fumigatus* only upregulated *erg3, erg24,* and *erg25* (da Silva Ferreira et al. 2006). Although caspofungin and flucytosine do not primarily target the ergosterol biosynthetic pathway, they elicited the upregulation of some ergosterol biosynthetic genes in *C. albicans* (Liu et al. 2005) (Table 13.3).

In response to ketoconazole, *C. albicans* upregulated genes involved in the biosynthesis of ergosterol, lipids, and fatty acids. Ketoconazole also induced the expression of the major transporter genes *cdr1* and *cdr2* (Liu et al. 2005). Similarly, in response to ketoconazole, *T. rubrum* upregulated genes involved in the metabolism of lipids, fatty acids, and sterols, including *erg3, erg4, erg6, erg11, erg24, erg25,* and *erg26* as well as the multidrug-resistance gene encoding ABC1, which is a homolog of *C. albicans* CDR1 (Yu et al. 2007a).

In response to AMB, *C. albicans* downregulated genes related to ergosterol biosynthesis and upregulated genes related to cell stress, including those encoding nitric oxide oxidoreductase (YHB1), catalase 1 (CTA1), aldehyde oxidase 1 (AOX1), and superoxide dismutase 2 (SOD2) (Liu et al. 2005). *A. fumigatus* exposed to AMB upregulated *erg11* and downregulated *erg6*. In addition, it modulated genes involved in cell stress, transport, oxidative phosphorylation, nucleotide metabolism, cell cycle control, and protein metabolism. Moreover, in response to the oxidative damage caused by AMB exposure, *A. fumigatus* overexpressed several genes encoding antioxidant enzymes, such as Mn-SOD, catalase, the thiol-specific antioxidant protein LsfA, glutathione S-transferase (GST), and thioredoxin. *A. fumigatus* downregulated ergosterol biosynthetic genes in response to AMB, possibly in attempt to use alternate sterols or sterol intermediates in the cell membrane (Gautam et al. 2008).

C. albicans exposed to caspofungin induced the expression of genes encoding cell wall maintenance proteins, including a target of caspofungin (the β-1,3-glucan synthase subunit homolog to FKS3), a pH-regulated glucan-remodeling enzyme (PHR1), extracellular matrix proteins (ECM21 and ECM33), and a putative fatty acid elongation enzyme (FEN12). Interestingly, *fen12* was upregulated in response to caspofungin and downregulated in response to AMB. In response to flucytosine, *C. albicans* upregulated the *cdc21* gene, which encodes thymidylate synthetase. This enzyme is the target of flucytosine and is associated with DNA synthesis; therefore, its upregulation may prevent fungal death. Other genes that were upreguvwhich is

Table 13.3 Modulation of genes related to ergosterol biosynthesis in response to drug exposure

Gene	C. albicans					T. rubrum				A. fumigatus	
	ITRA	AMB	KTC	CPF	5-FC	ITRA	AMB	KTC	TRB	AMB	VRCZ
erg1	13.2	–	–	–	–	–	–	–	–	–	0.34
erg2	3.9	1.35	2.06	1.26	0.70	–	–	–	-1.64	–	–
erg3	12.4	0.26	2.10	0.27	1.45	–	–	3.6	-2.1	–	2.34
erg4	11.6	0.73	1.57	0.61	1.00	–	-4.7	6.5	–	–	-1.38
erg5	12	1.05	1.40	3.13	1.39	–	–	–	–	–	–
erg6	50.3	0.68	1.92	0.80	0.98	2.27	–	37.5	–	-2.6	–
erg7	–	0.77	1.45	0.58	1.18	2.35	–	–	–	–	0.21
erg9	4.9	0.54	1.50	0.51	1.16	–	–	–	–	–	–
erg10	10.9	0.70	3.14	0.65	1.58	–	–	–	1.61	–	0.1
erg11	12.5	0.56	2.43	0.64	1.41	13.55	–	3.6	–	2.8	0.34
erg13	–	0.41	1.50	0.44	2.95	–	–	–	1.64	–	–
erg20	–	0.96	1.02	0.70	1.53	–	–	–	–	–	–
erg24	–	–	–	–	–	6.76	–	8.6	-2.01	–	2.44
erg25	4.2	0.43	2.52	0.53	2.73	7.92	–	11.9	-2.24	–	1.65
erg26	4.8	0.59	1.27	0.49	1.12	2.26	–	11.7	–	–	–
erg251	–	0.37	2.74	0.22	2.50	–	–	–	–	–	–

ITRA itraconazole, *AMB* amphotericin B, *KTC* ketoconazole, *CPF* caspofungin, *5-FC* flucytosine, *TRB* terbinafine and *VRCZ* voriconazole

a nucleoside diphosphate kinase, and FUR1 an uracil phosphoribosyltransferase, as well as genes associated with protein synthesis (Liu et al. 2005).

Terbinafine is commonly used to treat dermatophytosis. Exposure of *T. rubrum* to TRB decreased the expression of genes related to ergosterol biosynthesis, such as *erg2*, *erg4*, *erg24*, and *erg25*, and increased the expression of genes involved in lipid metabolism, such as *erg10*, *erg13*, and *ino1*. Although TRB primary target is squalene epoxidase (ERG1), *T. rubrum* did not differentially express *erg1* after exposure to TRB. It did, however, upregulate multidrug-resistance (MDR) genes, including *mdr1* and *mdr2* (Zhang et al. 2009). Indeed, MDR2 is associated with drug susceptibility. Overexpression of *mdr2* likely causes the efflux of TRB, since deletion of *mdr2* increased *T. rubrum* susceptibility to TRB (Fachin et al. 2006).

The emergence of resistant strains is an important obstacle to effective antifungal therapy. Azoles are the first-line treatment for many fungal infections; however, their use may select for azole-resistant mutants. Several mechanisms contribute to drug resistance, including alteration of the drug target, increased drug efflux, and increased cellular stress responses. Both mutations in and overexpression of the ergosterol biosynthesis gene *erg11/cyp51* confer resistance to azoles in *C. albicans* and *A. fumigatus*. For instance, one mutation causes the synthesis of an alternative protein that is insensitive to azoles and diminishes drug efficacy. At least 12 different point mutations in *erg11* have been identified in azole-resistant clinical isolates of *C. albicans* (Shapiro et al. 2011).

Overexpression of efflux pumps are associated with antifungal resistance in *C. albicans*. CDR1 and CDR2 confer resistance to multiple azoles, while MDR1 confers resistance to fluconazole (White et al. 2002). Similarly, azole-resistant clinical isolates of *C. glabrata* have been shown to overexpress genes encoding CDR1 and CDR2 as well as SNQ2, another ATP-binding cassette ABC transporter (Sanguinetti et al. 2005). In response to azoles and other structurally distinct drugs, *T. rubrum* overexpressed *Tru*MDR1 and *Tru*MDR2, which encode ABC transporters. Thus, these genes may participate in drug efflux (Cervelatti et al. 2006; Fachin et al. 2006).

In order to identify genes associated with fluconazole resistance, a laboratory strain of *C. albicans* susceptible to fluconazole was subjected to successive passages in media containing fluconazole to induce resistance, and the transcriptional profile was subsequently analyzed. Some genes were modulated in coordination with the upregulation of CDR1 and CDR2. For instance, there was an upregulation of genes coding the glutathione peroxidase 1 (GPX1) and RTA3, a protein involved in 7-aminocholesterol resistance, and a downregulation of NADPH oxidoreductase (EBP1). Genes that were modulated in coordination with the overexpression of MDR1 included the upregulation of genes encoding aldo-keto reductase family proteins (IFD1, IFD4, IFD5, and IFD7), methylglyoxal reductase (GRP2), pyrophosphate phosphatase (DPP1), and inositol-1-phosphate synthase (INO1) and the downregulation of multi-copper ferroxidase (FET34), phosphatidylethanolamine N-methyltransferase (OPI3), and Cu and Zn-containing superoxide dismutase (IPF1222) (Rogers and Barker 2002, 2003).

Genes encoding the ABC transporters CDR1 and CDR2 in azole-resistant *C. albicans* strains are regulated by the zinc cluster transcription factor TAC1. This transcription factor was initially identified in *C. albicans* in a search for genes containing the *cis*-acting drug-responsive element (DRE) Zn(2)-Cys(6) finger, which is present in the promoter region of the *cdr1* and *cdr2* genes. Further characterization showed that deletion of *tac1* gene prevented the upregulation of *cdr* genes. Moreover, introduction of a *tac1* allele recovered from an azole-resistant strain into an azole-susceptible strain induced overexpression of CDR1 and CDR2 (Coste et al. 2004). In addition, the TAC1 regulon contains 31 upregulated and 12 downregulated genes, including those encoding IFU5, HSP12, phospholipid flippase (RTA3), glutathione peroxidase (GPX1), histidine kinase (CHK1), sphingosine kinase (LCB4), NADH dehydrogenase (NDH2), and sorbose dehydrogenase (SOU1) as well as TAC1, CDR1, and CDR2. Among the downregulated genes in the TAC1 regulon are an iron transporter (FTR1), a putative glycosyl phosphatidyl inositol-anchored protein (IHD1), and an oligopeptide transporter (OPT6), all of them encoding integral membrane proteins, and the superoxide dismutase SOD5, which is a cell wall protein. Furthermore, ChIP-chip experiments demonstrated that TAC1 directly binds to the promoter region of eight of these genes, including CDR1, CDR2, GPX1, LCB4, RTA3, a putative lipase, and two genes with unknown functions (Liu et al. 2007).

A genome-wide expression analysis of resistant clinical isolates of *C. albicans* identified a transcription factor that was upregulated in coordination with MDR1. This gene encodes the multidrug resistance regulator MRR1, which is a zinc cluster transcription factor and the main regulator of MDR1 expression. Gain-of-function mutations in MRR1 are responsible for overexpression of MDR1 and are associated with fluconazole resistance in *C. albicans* (Morschhauser et al. 2007). In addition to regulating MDR1 expression, MRR1 regulated at least 14 other genes that may also contribute to fluconazole resistance. These genes encoded mainly oxidoredutases. Notably, MRR1 does not target CDR1 or CDR2. Overall, large-scale transcriptional analyses have identified several genes associated with drug response and resistance in pathogenic fungi (Table 13.4).

Transcriptional analyses have been performed in order to identify additional genes associated with azole resistance. Recently, RNA-seq analyses were performed on two isogenic *C. albicans* strains that differed only in fluconazole resistance. These studies identified novel genes associated with azole resistance, including the transcription factor CZF1, which is involved in the hyphal transition and the white/opaque switch. Inactivation of CZF1 increased the susceptibility to fluconazole as well as unrelated antifungal drugs, such as TRB and anisomycin. Furthermore, the CZF1 mutant strain displayed increased resistance to the cell-wall-disrupting agent Congo red. The mutant also overexpressed the gene encoding β 1,3-glucan synthase (GLS1), suggesting that CZF1 represses β-glucan synthesis and regulates cell wall integrity (Dhamgaye et al. 2012).

Table 13.4 Putative proteins associated with drug response and resistance in fungal pathogens

Protein description	Gene expression modulation and functional analysis	References
ATP-binding cassette (ABC) multidrug transporter (CDR1)	Up-regulated in *C. albicans* FLU resistant strains	Rogers and Barker (2002, 2003)
ATP-binding cassette (ABC) multidrug transporter (CDR2)	Up-regulated in *C. albicans* FLU resistant strains	Rogers and Barker (2002, 2003)
Transcriptional activator of CDR genes (TAC1)	Gain of function mutations lead to up-regulation of the CDRs genes and FLU resistance in *C. albicans*	Liu et al. (2007)
Lanosterol 14-alpha-demethylase Ergosterol biosynthesis (ERG11)	Mutations or overexpression lead to resistance to FLU in *C. albicans* and *A. fumigatus*	De Backer et al. (2001); Liu et al. (2005); Diao et al. (2009); Gibbons et al. (2012)
	Up-regulated in response to azoles in *C. albicans* and *T. rubrum*	
	Up-regulated in *A. fumigatus* biofilm cells	
Zn2-Cys6 transcription factor (UPC2)	Gain of function mutations lead to up-regulation of the ERG11 and FLU resistance in *C. albicans*	Silver et al. (2004)
MDR/MFS multidrug efflux pump (MDR1)	Up-regulated in *C. albicans* FLU resistant strains	Morschhauser et al. (2007); Zaugg et al. (2009)
	Up-regulated in response to TRB in *T. rubrum*	
Multidrug resistance regulator (MRR1)	Gain of function mutations lead to up-regulation of the MDR1 and FLU resistance in *C. albicans*	Morschhauser et al. (2007)
MDR/MFS multidrug efflux pump (MDR2)	Up-regulated in response to terbinafine in *T. rubrum*	Zaugg et al. (2009); Fachin et al. (2006); Gibbons et al. (2012)
	Up-regulated in *A. fumigatus* biofilm cells	
MDR/MFS multidrug efflux pump (MDR4)	Up-regulated in *A. fumigatus* biofilm cells	Gibbons et al. (2012)
Transcription factor (CZF1)	Up-regulated in *C. albicans* FLU resistant strains	Dhamgaye et al. (2012)

Table 13.4 (continued)

Protein description	Gene expression modulation and functional analysis	References
Thymidylate synthetase associated with DNA synthesis (CDC21)	Up-regulated in response to flucytocine	Liu et al. (2005)
(1,3)-β-D-glucan synthase (FKS3)	Up-regulated in *C. albicans* in response to caspofungin	Liu et al. (2005)
Heat shock protein 90 (HSP90)	Inhibition can improve azole and echinocandins effectiveness against *C. albicans* and *A. fumigatus*	Cowen (2009)
Catalase (CTA1)	Up-regulated in response to AMB in *C. albicans* and in *A. fumigatus*	Liu et al. (2005); Gautam et al. (2008)
Superoxide dismutase 2 (SOD2)	Up-regulated in response to AMB in *C. albicans*	Liu et al. (2005)

In pathogenic fungi, mitochondrial dysfunction has been associated with altered susceptibility to antifungal drugs. In *C. albicans*, inhibition or mutation of the mitochondrial complex I (CI) increased susceptibly to fluconazole even in resistant clinical isolates. Transcriptional analysis was performed on the Δ*goa1* and Δ*ndh51* mutant strains, which are associated with CI-induced susceptibility to fluconazole. GOA1 is required for function of the electron transport chain, and the Δ*goa1* mutant accumulates ROS, undergoes apoptosis, and is avirulent. *Ndh51* encodes a 51-kDa subunit of the NADH dehydrogenase of the electron transport chain, and the Δ*ndh51* mutant exhibits defects in morphogenesis. RNA-seq analyses of these strains demonstrated downregulation of transporters, including the CDR1/CDR2 efflux pumps but not MDR1. Genes related to ergosterol biosynthesis were downregulated in the Δ*ndh51* mutant. In contrast, genes associated with peroxisomes, gluconeogenesis, β-oxidation, and mitochondria were downregulated in the Δ*goa1* mutant (Sun et al. 2013). NDH51 is conserved among eukaryotes, including mammals; nevertheless, GOA1 is only conserved in some *Candida* species. Therefore, fungi-specific mitochondrial genes may be targets for the development of novel antifungal drugs. Indeed, acriflavine, an acridine derivative that has antibacterial, antifungal, antiviral, and antiparasitic properties, induces the overexpression of genes involved in the mitochondrial electron transport chain of *T. rubrum* (Segato et al. 2008).

Interestingly, chemical inhibition of fungal HSP90 improved the activity of azoles and echinocandins against *C. albicans* and echinocandins against *A. fumigatus* (Cowen 2009). Inhibition of HPS90 prevents the stress-response cascade mediated by calcineurin, which is normally activated in response to antifungal drugs. Blunting of the stress-response cascade enhances the fungicidal effects of drugs, leading to cell death (Singh et al. 2009). It will be a challenge to develop an inhibitor selective for fungal HSP90 and inactive against human HSP90. Nevertheless, HSP90 may be a promising target for treatment of resistant fungal diseases and may combat the emergence of drug resistance (Cowen 2009).

In addition to the emergence of drug-resistant strains, another major clinical problem is the formation of microbial biofilms. Biofilms possess specific traits as compared to planktonic cells, such as intrinsic resistance to drugs. In immunocompromised individuals, both *C. albicans* and *A. fumigatus* can form biofilms on implanted medical devices, such as catheters, and cause persistent infections. In particular, biofilms have decreased susceptibility to antifungal drugs, such as polyenes and azoles. In order to understand their mechanisms of resistance, mature biofilms cells were exposed to fluconazole, AMB, and caspofungin. Fluconazole exposure did not significantly alter gene expression, and AMB exposure resulted in only minor alterations in gene expression. On the other hand, biofilms exposed to caspofungin underwent more pronounced alteration in gene expression, including the upregulation of several genes associated with biofilm formation, such as ALS3, a cell-wall adhesin, the transcription factor TEC1, and genes associated with cell-wall remodeling (Vediyappan et al. 2010). Furthermore, AMB and fluconazole bind to the extracellular matrix of the biofilm, which is comprised of β-glucans; such binding inhibits effective drug action (Vediyappan et al. 2010; Nett et al. 2007). An RNA-seq analysis compared the transcriptional profile of an *A. fumigatus* biofilm to that of planktonic cells. Thousands of genes were differentially expressed between the biofilm and planktonic cells. Specifically, the biofilm exhibited an upregulation of secondary metabolism genes, cell-wall-related genes, sterol biosynthetic genes (e.g., *erg11*), transporters associated with antifungal resistance (MDR1, MDR2, and MDR4), and hydrophobins, which are associated with structural organization of biofilms (Gibbons et al. 2012). The complex gene network involved in biofilm formation (Nobile et al. 2008) is consistent with the fact that *C. albicans* can form biofilms in different niches, such as the bloodstream and oral cavity. This highlights the challenge inherent to treating these infections as well as the importance of searching for new antifungal targets.

In conclusion, analyses of the transcriptional changes that occur in response to cytotoxic drugs have identified genes with known mechanisms of action. Moreover, these studies have suggested novel effects of antifungal drugs. In addition, responses shared across multiple classes of antifungal agents were identified in *C. albicans* (Liu et al. 2005) and *T. rubrum* (Fachin et al. 2006; Paião et al. 2007; Peres et al. 2010b). Therefore, there may be non-specific responses to a drug challenge that allow fungi to adapt to stress.

13.5 Concluding Remarks

The pathogenesis of fungal infections involves changes in gene expression and metabolic pathways, which enable fungal invasion and survival. At the same time, fungi elicit host responses aimed at eliminating the pathogen. Genome-wide transcriptional profiling has identified the molecular responses of both host and pathogen during interaction and has provided insight into the adaptive responses that occur during the establishment of infection. The combination of large-scale transcriptomic analysis and systems biology approaches has enabled the development of regulatory molecular models that can assess the dynamic behavior of host-pathogen interactions and elucidate the pathogenesis of human mycoses. These regulatory models have been validated through reverse-genetic approaches by evaluating the physiological behavior of knockout strains *in vitro*, *ex vivo*, and *in vivo*. Furthermore, transcriptomics is a valuable source of data on gene expression regulation, gene structure and function, also providing information regarding the mechanisms of fungal responses and resistance to drugs. These insights will further support the development of novel therapeutic approaches to prevent and control fungal infections.

Acknowledgements This work was supported by grants from Brazilian funding agencies FAPESP (Grant No. 2008/58634-7), CNPq, CAPES and FAEPA. NTAP was supported by postdoctoral fellowships from FAPESP (2009/08411-4) and CNPq (503809/2012-8), GFP was supported by postdoctoral fellowships from FAPESP (2012/22232-8 and 2013/19195-6) and EASL by postdoctoral fellowships from FAPESP (2011/08424-9) and CNPq (150980/2013-2).

References

Aimanianda V, Bayry J, Bozza S et al (2009) Surface hydrophobin prevents immune recognition of airborne fungal spores. Nature 460(7259):1117–1121

Barelle CJ, Priest CL, Maccallum DM et al (2006) Niche-specific regulation of central metabolic pathways in a fungal pathogen. Cell Microbiol 8(6):961–971

Barker KS, Park H, Phan QT et al (2008) Transcriptome profile of the vascular endothelial cell response to *Candida albicans*. J Infect Dis 198(2):193–202

Bedoya SK, Lam B, Lau K et al (2013) Th17 cells in immunity and autoimmunity. Clin Dev Immunol 2013:986789

Biondo C, Midiri A, Gambuzza M et al (2008) IFN-alpha/beta signaling is required for polarization of cytokine responses toward a protective type 1 pattern during experimental cryptococcosis. J Immunol 181(1):566–573

Biondo C, Signorino G, Costa A et al (2011) Recognition of yeast nucleic acids triggers a host-protective type I interferon response. Eur J Immunol 41(7):1969–1979

Brock M (2009) Fungal metabolism in host niches. Curr Opin Microbiol 12(4):371–376

Brown AJ, Haynes K, Quinn J (2009) Nitrosative and oxidative stress responses in fungal pathogenicity. Curr Opin Microbiol 12(4):384–391

Brown GD, Denning DW, Gow NA et al (2012) Hidden killers: human fungal infections. Sci Transl Med 4(165):165rv113

Bruns S, Seidler M, Albrecht D et al (2010) Functional genomic profiling of *Aspergillus fumigatus* biofilm reveals enhanced production of the mycotoxin gliotoxin. Proteomics 10(17):3097–3107

Burmester A, Shelest E, Glockner G et al (2011) Comparative and functional genomics provide insights into the pathogenicity of dermatophytic fungi. Genome Biol 12(1):R7

Cairns T, Minuzzi F, Bignell E (2010) The host-infecting fungal transcriptome. FEMS Microbiol Lett 307(1):1–11

Cambier L, Weatherspoon A, Defaweux V et al (2014) Assessment of the cutaneous immune response during *Arthroderma benhamiae* and *A. vanbreuseghemii* infection using an experimental mouse model. Br J Dermatol 170(3):625–633

Cenci E, Mencacci A, Casagrande A et al (2001) Impaired antifungal effector activity but not inflammatory cell recruitment in interleukin-6-deficient mice with invasive pulmonary aspergillosis. J Infect Dis 184(5):610–617

Cervelatti EP, Fachin AL, Ferreira-Nozawa MS et al (2006) Molecular cloning and characterization of a novel ABC transporter gene in the human pathogen *Trichophyton rubrum*. Med Mycol 44(2):141–147

Chen Y, Toffaletti DL, Tenor JL et al (2014) The *Cryptococcus neoformans* transcriptome at the site of human meningitis. MBio 5(1):e01087–e01013

Conti HR, Shen F, Nayyar N et al (2009) Th17 cells and IL-17 receptor signaling are essential for mucosal host defense against oral candidiasis. J Exp Med 206(2):299–311

Cornet M, Gaillardin C (2014) pH signaling in human fungal pathogens: a new target for antifungal strategies. Eukaryot Cell 13(3):342–352

Cortez KJ, Lyman CA, Kottilil S et al (2006) Functional genomics of innate host defense molecules in normal human monocytes in response to *Aspergillus fumigatus*. Infect Immun 74(4):2353–2365

Coste AT, Karababa M, Ischer F et al (2004) TAC1, transcriptional activator of CDR genes, is a new transcription factor involved in the regulation of *Candida albicans* ABC transporters CDR1 and CDR2. Eukaryot Cell 3(6):1639–1652

Cowen LE (2009) Hsp90 orchestrates stress response signaling governing fungal drug resistance. PLoS Pathog 5(8):e1000471

da Silva Ferreira ME, Malavazi I, Savoldi M et al (2006) Transcriptome analysis of *Aspergillus fumigatus* exposed to voriconazole. Curr Genet 50(1):32–44

Davis DA (2009) How human pathogenic fungi sense and adapt to pH: the link to virulence. Curr Opin Microbiol 12(4):365–370

De Backer MD, Ilyina T, Ma XJ et al (2001) Genomic profiling of the response of *Candida albicans* to itraconazole treatment using a DNA microarray. Antimicrob Agents Chemother 45(6):1660–1670

de Jesús-Berríos M, Liu L, Nussbaum JC et al (2003) Enzymes that counteract nitrosative stress promote fungal virulence. Curr Biol 13(22):1963–1968

Decken K, Kohler G, Palmer-Lehmann K et al (1998) Interleukin-12 is essential for a protective Th1 response in mice infected with *Cryptococcus neoformans*. Infect Immun 66(10):4994–5000

Deepe GS Jr (2000) Immune response to early and late *Histoplasma capsulatum* infections. Curr Opin Microbiol 3(4):359–362

Denning DW (2003) Echinocandin antifungal drugs. Lancet 362(9390):1142–1151

Dhamgaye S, Bernard M, Lelandais G et al (2012) RNA sequencing revealed novel actors of the acquisition of drug resistance in *Candida albicans*. BMC Genomics 13:396

Diao YJ, Zhao R, Deng XM et al (2009) Transcriptional profiles of *Trichophyton rubrum* in response to itraconazole. Med Mycol 47(3):237–247

Fachin AL, Contel EPB, Martinez-Rossi NM (2001) Effect of sub-MICs of antimycotics on expression of intracellular esterase of *Trichophyton rubrum*. Med Mycol 39(1):129–133

Fachin AL, Ferreira-Nozawa MS, Maccheroni W et al (2006) Role of the ABC transporter TruMDR2 in terbinafine, 4-nitroquinoline N-oxide and ethidium bromide susceptibility in *Trichophyton rubrum*. J Med Microbiol 55(8):1093–1099

Fan W, Kraus PR, Boily MJ et al (2005) *Cryptococcus neoformans* gene expression during murine macrophage infection. Eukaryot Cell 4(8):1420–1433

Ferreira-Nozawa MS, Silveira HCS, Ono CJ et al. (2006) The pH signaling transcription factor PacC mediates the growth of *Trichophyton rubrum* on human nail *in vitro*. Med Mycol 44(7):641–645

Fradin C, Kretschmar M, Nichterlein T et al (2003) Stage-specific gene expression of *Candida albicans* in human blood. Mol Microbiol 47(6):1523–1543

Fradin C, De Groot P, MacCallum D et al (2005) Granulocytes govern the transcriptional response, morphology and proliferation of *Candida albicans* in human blood. Mol Microbiol 56(2):397–415

Fradin C, Mavor AL, Weindl G et al (2007) The early transcriptional response of human granulocytes to infection with *Candida albicans* is not essential for killing but reflects cellular communications. Infect Immun 75(3):1493–1501

Gautam P, Shankar J, Madan T et al (2008) Proteomic and transcriptomic analysis of *Aspergillus fumigatus* on exposure to amphotericin B. Antimicrob Agents Chemother 52(12):4220–4227

Gibbons JG, Beauvais A, Beau R et al (2012) Global transcriptome changes underlying colony growth in the opportunistic human pathogen *Aspergillus fumigatus*. Eukaryot Cell 11(1):68–78

Gomez P, Hackett TL, Moore MM et al (2011) Functional genomics of human bronchial epithelial cells directly interacting with conidia of *Aspergillus fumigatus*. BMC Genomics 11:358

Heddergott C, Bruns S, Nietzsche S et al (2012) The *Arthroderma benhamiae* hydrophobin HypA mediates hydrophobicity and influences recognition by human immune effector cells. Eukaryot Cell 11(5):673–682

Hu G, Cheng PY, Sham A et al (2008) Metabolic adaptation in *Cryptococcus neoformans* during early murine pulmonary infection. Mol Microbiol 69(6):1456–1475

Ibrahim-Granet O, Dubourdeau M, Latge JP et al (2008) Methylcitrate synthase from *Aspergillus fumigatus* is essential for manifestation of invasive aspergillosis. Cell Microbiol 10(1):134–148

Idnurm A, Giles SS, Perfect JR et al (2007) Peroxisome function regulates growth on glucose in the basidiomycete fungus *Cryptococcus neoformans*. Eukaryot Cell 6(1):60–72

Idnurm A, Walton FJ, Floyd A et al (2009) Identification of ENA1 as a virulence gene of the human pathogenic fungus *Cryptococcus neoformans* through signature-tagged insertional mutagenesis. Eukaryot Cell 8(3):315–326

Inglis DO, Berkes CA, Hocking Murray DR et al (2010) Conidia but not yeast cells of the fungal pathogen *Histoplasma capsulatum* trigger a type I interferon innate immune response in murine macrophages. Infect Immun 78(9):3871–3882

Jung WH, Hu G, Kuo W et al. (2009) Role of ferroxidases in iron uptake and virulence of *Cryptococcus neoformans*. Eukaryot Cell 8(10):1511–1520

Kane PM (2007) The long physiological reach of the yeast vacuolar H+-ATPase. J Bioenerg Biomembr 39(5–6):415–421

Kim HS, Choi EH, Khan J et al (2005) Expression of genes encoding innate host defense molecules in normal human monocytes in response to *Candida albicans*. Infect Immun 73(6):3714–3724

Kretschmer M, Wang J, Kronstad JW (2012) Peroxisomal and mitochondrial beta-oxidation pathways influence the virulence of the pathogenic fungus *Cryptococcus neoformans*. Eukaryot Cell 11(8):1042–1054

Lambou K, Lamarre C, Beau R et al (2010) Functional analysis of the superoxide dismutase family in *Aspergillus fumigatus*. Mol Microbiol 75(4):910–923

Lim CS, Rosli R, Seow HF et al (2011) Transcriptome profiling of endothelial cells during infections with high and low densities of *C. albicans* cells. Int J Med Microbiol 301(6):536–546

Liu TT, Lee RE, Barker KS et al (2005) Genome-wide expression profiling of the response to azole, polyene, echinocandin, and pyrimidine antifungal agents in *Candida albicans*. Antimicrob Agents Chemother 49(6):2226–2236

Liu TT, Znaidi S, Barker KS et al (2007) Genome-wide expression and location analyses of the *Candida albicans* Tac1p regulon. Eukaryot Cell 6(11):2122–2138

Livonesi MC, Souto JT, Campanelli AP et al (2008) Deficiency of IL–12p40 subunit determines severe paracoccidioidomycosis in mice. Med Mycol 46(7):637–646

Lorenz MC, Fink GR (2001) The glyoxylate cycle is required for fungal virulence. Nature 412 (6842):83–86

Lorenz MC, Bender JA, Fink GR (2004) Transcriptional response of *Candida albicans* upon internalization by macrophages. Eukaryot Cell 3(5):1076–1087

Lupo P, Chang YC, Kelsall BL et al (2008) The presence of capsule in *Cryptococcus neoformans* influences the gene expression profile in dendritic cells during interaction with the fungus. Infect Immun 76(4):1581–1589

Maranhão FCA, Paião FG, Martinez-Rossi NM (2007) Isolation of transcripts over-expressed in human pathogen *Trichophyton rubrum* during growth in keratin. Microbial Pathog 43(4):166–172

Martin R, Moran GP, Jacobsen ID et al (2011) The *Candida albicans*-specific gene EED1 encodes a key regulator of hyphal extension. PLoS One 6(4):e18394

Martinez-Rossi NM, Peres NTA, Rossi A (2008) Antifungal resistance mechanisms in dermatophytes. Mycopathologia 166(5–6):369–383

Martinez-Rossi NM, Persinoti GF, Peres NTA et al (2012) Role of pH in the pathogenesis of dermatophytoses. Mycoses 55(5):381–387

Mayer FL, Wilson D, Hube B (2013) *Candida albicans* pathogenicity mechanisms. Virulence 4(2):119–128

McDonagh A, Fedorova ND, Crabtree J et al (2008) Sub-telomere directed gene expression during initiation of invasive aspergillosis. PLoS Pathog 4(9):e1000154

Missall TA, Lodge JK, McEwen JE (2004) Mechanisms of resistance to oxidative and nitrosative stress: implications for fungal survival in mammalian hosts. Eukaryot Cell 3(4):835–846

Monod M (2008) Secreted proteases from dermatophytes. Mycopathologia 166(5–6):285–294

Morschhauser J, Barker KS, Liu TT et al (2007) The transcription factor Mrr1p controls expression of the MDR1 efflux pump and mediates multidrug resistance in *Candida albicans*. PLoS Pathog 3(11):e164

Morton CO, Varga JJ, Hornbach A et al (2011) The temporal dynamics of differential gene expression in *Aspergillus fumigatus* interacting with human immature dendritic cells *in vitro*. PLoS One 6(1):e16016

Muller V, Viemann D, Schmidt M et al (2007) *Candida albicans* triggers activation of distinct signaling pathways to establish a proinflammatory gene expression program in primary human endothelial cells. J Immunol 179(12):8435–8445

Mullick A, Elias M, Harakidas P et al (2004) Gene expression in HL60 granulocytoids and human polymorphonuclear leukocytes exposed to *Candida albicans*. Infect Immun 72(1):414–429

Nett J, Lincoln L, Marchillo K et al (2007) Beta-1,3 glucan as a test for central venous catheter biofilm infection. J Infect Dis 195(11):1705–1712

Nobile CJ, Solis N, Myers CL et al (2008) *Candida albicans* transcription factor Rim101 mediates pathogenic interactions through cell wall functions. Cell Microbiol 10(11):2180–2196

Nobile CJ, Fox EP, Nett JE et al (2012) A recently evolved transcriptional network controls biofilm development in *Candida albicans*. Cell 148(1–2):126–138.

O'Meara TR, Norton D, Price MS et al (2010) Interaction of *Cryptococcus neoformans* Rim101 and protein kinase A regulates capsule. PLoS Pathog 6(2):e1000776

O'Meara TR, Xu W, Selvig KM et al (2013) The *Cryptococcus neoformans* Rim101 transcription factor directly regulates genes required for adaptation to the host. Mol Cell Biol 34(4):673–684

Ok M, Latge JP, Baeuerlein C et al (2009) Immune responses of human immature dendritic cells can be modulated by the recombinant *Aspergillus fumigatus* antigen Aspf1. Clin Vaccine Immunol 16(10):1485–1492

Oosthuizen JL, Gomez P, Ruan J et al (2011) Dual organism transcriptomics of airway epithelial cells interacting with conidia of *Aspergillus fumigatus*. PLoS One 6(5):e20527

Paião FG, Segato F, Cursino-Santos JR et al (2007) Analysis of *Trichophyton rubrum* gene expression in response to cytotoxic drugs. FEMS Microbiol Lett 271(2):180–186

Peres NTA, Maranhao FCA, Rossi A et al (2010a) Dermatophytes: host-pathogen interaction and antifungal resistance. An Bras Dermatol 85(5):657–667

Peres NTA, Sanches PR, Falcão JP et al (2010b) Transcriptional profiling reveals the expression of novel genes in response to various stimuli in the human dermatophyte *Trichophyton rubrum*. BMC Microbiol 10:39–48

Petzold EW, Himmelreich U, Mylonakis E et al (2006) Characterization and regulation of the trehalose synthesis pathway and its importance in the pathogenicity of *Cryptococcus neoformans*. Infect Immun 74(10):5877–5887

Prigneau O, Porta A, Poudrier JA et al (2003) Genes involved in beta-oxidation, energy metabolism and glyoxylate cycle are induced by *Candida albicans* during macrophage infection. Yeast 20(8):723–730

Ramirez MA, Lorenz MC (2007) Mutations in alternative carbon utilization pathways in *Candida albicans* attenuate virulence and confer pleiotropic phenotypes. Eukaryot Cell 6(2):280–290

Rogers PD, Barker KS (2002) Evaluation of differential gene expression in fluconazole-susceptible and -resistant isolates of *Candida albicans* by cDNA microarray analysis. Antimicrob Agents Chemother 46(11):3412–3417

Rogers PD, Barker KS (2003) Genome-wide expression profile analysis reveals coordinately regulated genes associated with stepwise acquisition of azole resistance in *Candida albicans* clinical isolates. Antimicrob Agents Chemother 47(4):1220–1227

Romani L (2011) Immunity to fungal infections. Nat Rev Immunol 11(4):275–288

Rossi A, Cruz AHS, Santos RS et al (2013) Ambient pH sensing in filamentous fungi: pitfalls in elucidating regulatory hierarchical signaling networks. IUBMB Life 65(11):930–935

Rude TH, Toffaletti DL, Cox GM et al (2002) Relationship of the glyoxylate pathway to the pathogenesis of *Cryptococcus neoformans*. Infect Immun 70(10):5684–5694

Sanguinetti M, Posteraro B, Fiori B et al (2005) Mechanisms of azole resistance in clinical isolates of *Candida glabrata* collected during a hospital survey of antifungal resistance. Antimicrob Agents Chemother 49(2):668–679

Schobel F, Ibrahim-Granet O, Ave P, Latge JP, Brakhage AA, and Brock M (2007) Aspergillus fumigatus does not require fatty acid metabolism via isocitrate lyase for development of invasive aspergillosis. Infect Immun 75:1237–1244

Segato F, Nozawa SR, Rossi A et al (2008) Over-expression of genes coding for proline oxidase, riboflavin kinase, cytochrome c oxidase and an MFS transporter induced by acriflavin in *Trichophyton rubrum*. Med Mycol 46(2):135–139

Shapiro RS, Robbins N, Cowen LE (2011) Regulatory circuitry governing fungal development, drug resistance, and disease. Microbiol Mol Biol Rev 75(2):213–267

Shiraki Y, Ishibashi Y, Hiruma M et al (2006) Cytokine secretion profiles of human keratinocytes during *Trichophyton tonsurans* and *Arthroderma benhamiae* infections. J Med Microbiol 55(Pt 9):1175–1185

Silva SS, Tavares AHFP, Passos-Silva DG et al (2008) Transcriptional response of murine macrophages upon infection with opsonized *Paracoccidioides brasiliensis* yeast cells. Microbes Infect 10(1):12–20

Silva MG, Schrank A, Bailao EF et al (2011) The homeostasis of iron, copper, and zinc in *Paracoccidioides brasiliensis, Cryptococcus neoformans* var. *grubii*, and *Cryptococcus gattii*: a comparative analysis. Front Microbiol 2:49

Silveira HCS, Gras DE, Cazzaniga RA et al (2010) Transcriptional profiling reveals genes in the human pathogen *Trichophyton rubrum* that are expressed in response to pH signaling. Microb Pathog 48(2):91–96

Silver PM, Oliver BG, White TC (2004) Role of *Candida albicans* transcription factor Upc2p in drug resistance and sterol metabolism. Eukaryot Cell 3(6):1391–1397

Singh SD, Robbins N, Zaas AK et al (2009) Hsp90 governs echinocandin resistance in the pathogenic yeast *Candida albicans* via calcineurin. PLoS Pathog 5(7):e1000532

Souto JT, Figueiredo F, Furlanetto A et al (2000) Interferon-gamma and tumor necrosis factor-alpha determine resistance to *Paracoccidioides brasiliensis* infection in mice. Am J Pathol 156(5):1811–1820

Staib P, Zaugg C, Mignon B et al (2010) Differential gene expression in the pathogenic dermatophyte *Arthroderma benhamiae in vitro* versus during infection. Microbiology 156(Pt 3):884–895

Steen BR, Zuyderduyn S, Toffaletti DL et al (2003) *Cryptococcus neoformans* gene expression during experimental cryptococcal meningitis. Eukaryot Cell 2(6):1336–1349

Sugui JA, Kim HS, Zarember KA et al (2008) Genes differentially expressed in conidia and hyphae of *Aspergillus fumigatus* upon exposure to human neutrophils. PLoS One 3(7):e2655

Sun N, Fonzi W, Chen H et al (2013) Azole susceptibility and transcriptome profiling in *Candida albicans* mitochondrial electron transport chain complex I mutants. Antimicrob Agents Chemother 57(1):532–542

Tavares AH, Derengowski LS, Ferreira KS et al (2012) Murine dendritic cells transcriptional modulation upon *Paracoccidioides brasiliensis* infection. PLoS Negl Trop Dis 6(1):e1459

Thewes S, Kretschmar M, Park H et al (2007) *In vivo* and *ex vivo* comparative transcriptional profiling of invasive and non-invasive *Candida albicans* isolates identifies genes associated with tissue invasion. Mol Microbiol 63(6):1606–1628

Tierney L, Linde J, Muller S et al (2012) An Interspecies regulatory network inferred from simultaneous RNA-seq of *Candida albicans* invading innate immune cells. Front Microbiol 3:85

Vediyappan G, Rossignol T, d'Enfert C (2010) Interaction of *Candida albicans* biofilms with antifungals: transcriptional response and binding of antifungals to beta-glucans. Antimicrob Agents Chemother 54(5):2096–2111

Vermes A, Guchelaar HJ, Dankert J (2000) Flucytosine: a review of its pharmacology, clinical indications, pharmacokinetics, toxicity and drug interactions. J Antimicrob Chemother 46(2):171–179

Viriyakosol S, Singhania A, Fierer J et al (2013) Gene expression in human fungal pathogen *Coccidioides immitis* changes as arthroconidia differentiate into spherules and mature. BMC Microbiol 13:121

Wachtler B, Wilson D, Haedicke K et al (2011) From attachment to damage: defined genes of *Candida albicans* mediate adhesion, invasion and damage during interaction with oral epithelial cells. PLoS One 6(2):e17046

Walker LA, Maccallum DM, Bertram G et al (2009) Genome-wide analysis of *Candida albicans* gene expression patterns during infection of the mammalian kidney. Fungal Genet Biol 46(2):210–219

Wang Z, Gerstein M, Snyder M (2009) RNA-Seq: a revolutionary tool for transcriptomics. Nat Rev Genet 10(1):57–63

Westermann AJ, Gorski SA, Vogel J (2012) Dual RNA-seq of pathogen and host. Nat Rev Microbiol 10(9):618–630

White TC, Holleman S, Dy F et al (2002) Resistance mechanisms in clinical isolates of *Candida albicans*. Antimicrob Agents Chemother 46(6):1704–1713

Woodfolk JA, Platts-Mills TA (1998) The immune response to dermatophytes. Res Immunol 149(4–5):436–445

Yu L, Zhang W, Liu T et al (2007a) Global gene expression of *Trichophyton rubrum* in response to PH11B, a novel fatty acid synthase inhibitor. J Appl Microbiol 103(6):2346–2352

Yu L, Zhang W, Wang L et al (2007b) Transcriptional profiles of the response to ketoconazole and amphotericin B in *Trichophyton rubrum*. Antimicrob Agents Chemother 51(1):144–153

Zakikhany K, Naglik JR, Schmidt-Westhausen A et al (2007) *In vivo* transcript profiling of *Candida albicans* identifies a gene essential for interepithelial dissemination. Cell Microbiol 9(12):2938–2954

Zaugg C, Monod M, Weber J, Harshman K, Pradervand S, Thomas J et al. (2009) Gene expression profiling in the human pathogenic dermatophyte Trichophyton rubrum during growth on proteins. Eukaryot Cell 8:241–250

Zhang W, Yu L, Yang J et al (2009) Transcriptional profiles of response to terbinafine in *Trichophyton rubrum*. Appl Microbiol Biotechnol 82(6):1123–1130

Chapter 14
Transcriptomics of the Host–Pathogen Interaction in Paracoccidioidomycosis

Patrícia Albuquerque, Hugo Costa Paes, Aldo Henrique Tavares, Larissa Fernandes, Anamélia Lorenzetti Bocca, Ildinete Silva-Pereira, Maria Sueli Soares Felipe and André Moraes Nicola

Abstract Due to a lack of molecular tools that enable gain- and loss-of-function studies, much research with the fungi of the *Paracoccidioides* genus has consisted of gene expression studies. These have addressed the direct interaction of these fungi with the mammalian host or their response to environmental stimuli of interest to the study of their adaptability to said host, such as the temperature shift that triggers dimorphic transition. In this chapter, we present a review of findings of host–pathogen interaction studies and what evidence they found of mechanisms whereby *Paracoccidioides* is able to overcome differences in environment and establish disease, and of how the host responds to the pathogen. In the first part, which deals with the pathogen response, expression studies have identified metabolic pathways genes thereof are upregulated when the fungi are exposed to different organs, as well as blood and derivatives, of mice and humans. Of note, these studies have suggested an important role, in the adaptation to host tissues, of a metabolic shift away from glycolysis and aerobic respiration and towards fermentative and non-aerobic ways of obtaining energy. With regard to the remarkable preference of the genus for male hosts, studies of the response of *Paracoccidioides* to oestradiol have suggested a role of Rho GTPases in the process. As for the second part, dealing with the host response to the fungus, despite the paucity of data, the few large-scale studies available offer evidence to support the model whereby Th1-driven immune responses are protective and disease is associated with Th2 and Th17 responses, in keeping with small-scale studies. Overall, gene expression studies have supplied a large amount of data that lack direct experimental confirmation but which keep revealing new research avenues.

A. M. Nicola (✉)
Programa de Pós-Graduação em Ciências Genômicas e Biotecnologia, Universidade Católica de Brasília, SGAN 916, módulo B, bloco C, sala 211, Brasília, DF, 70790-160, Brazil
e-mail: nicola@ucb.br

P. Albuquerque · A. H. Tavares · L. Fernandes
Universidade de Brasília, Faculdade de Ceilândia,
Brasília, DF 72220–900, Brazil

H. C. Paes · A. L. Bocca · I. Silva-Pereira · M. S. S. Felipe
Universidade de Brasília, Campus Universitário Darcy Ribeiro,
Brasília, DF 70910–900, Brazil

© Springer International Publishing Switzerland 2014
G. A. Passos (ed.), *Transcriptomics in Health and Disease,*
DOI 10.1007/978-3-319-11985-4_14

14.1 Introduction

Paracoccidioides brasiliensis and its sister clade *P. lutzii* are the causative agents of paracoccidioidomycosis (PCM), one of a handful of illnesses caused by thermally dimorphic fungi that is endemic to Latin America and constitutes the main cause of mortality due to primary, systemic mycoses in that region (Restrepo et al. 2012). As other primary mycoses such as histoplasmosis, blastomycosis and coccidioidomycosis, it affects healthy, immunocompetent individuals, is fatal if untreated and its clinical course consists of a progressive granulomatous infection of the lungs with possible systemic dissemination and involvement of other organs as well as skin and mucosae. Even successful treatment may result in significant lifestyle impairment due to lung scarring (Restrepo et al. 2012).

As all thermally dimorphic fungi, *Paracoccidioides* exist in the environment as saprophytic mycelia the spores or hypha fragments of which are inhaled by mammalian hosts and germinate in the lung as yeast cells that cause the illness; this morphologic transition can be reproduced *in vitro* by shifting the temperature from 25 to 37 °C and is fully reversible (San-Blas et al. 2002). The fungi do not need to pass through a host to complete their life cycle and the ability to cause infection is considered an "evolution accident", as possibility first postulated for *Cryptococcus neoformans* whereby the fungi, possibly in order to survive predation by environmental amoeba and other protozoans, evolved mechanisms that enabled them to resist phagocytosis and proliferate in the interior of phagocytic cells (Steenbergen et al. 2001). This in turn would have allowed them to survive within macrophages in the host, which are one of the first lines of defence against intracellular pathogens.

A hallmark of PCM, however, is its preference for male hosts, who are infected in a proportion that may reach 30:1 relative to women (Shankar et al. 2011a); this proportion becomes more pronounced if women out of reproductive age are factored out, and the underlying mechanism has not been clarified, although there is evidence that oestradiol impairs the onset of the yeast phenotype (Restrepo et al. 1984).

14.2 Pathogen Response to the Host

In contrast with *Histoplasma capsulatum* and *Blastomyces dermatitidis*, thermally dimorphic fungi for which a full molecular toolkit for genetic engineering is available, *Paracoccidioides* are yet to have their genomes successfully manipulated in a consistent, flexible and reproducible way: apart from sparse reports on gene silencing allowing phenotypical studies, gene deletion and introduction remain unfulfilled goals of research on these agents. Thus, much attention has been devoted to high-throughput studies of genetic reprogramming of *Paracoccidioides* in response to environmental cues of interest, such as the different temperatures that allow for each of the two phenotypes (Felipe et al. 2005), the interior of macrophages (Tavares et al. 2007) and internal organs (Bailao et al. 2006) and the presence of oestradiol (Shankar et al. 2011b) and Fernandes et al., unpublished data). It is expected that indirect evidence derived of these gene expression studies will shed light on which genes are essential for adaptation to the mammalian host and thus point to novel therapeutic approaches.

Even before the release of the first full genomes of *P. brasiliensis* and *P. lutzii*, the transcriptome approach had identified the major differences in expression patterns between the steady-state mycelium and yeast phases of both fungi (Felipe et al. 2005), and temporally in the process of transition from one to the other (Goldman et al. 2003). Since then, much work has been carried out to identify genetic programmes involved in the growth of the fungus in conditions that correspond to different niches in the host, such as blood (Bailao et al. 2006), plasma (Bailao et al. 2007), and the liver parenchyma (Bailao et al. 2006; Costa et al. (2007); or that replicate *in vitro* aspects of them, such as the co-cultivation with peritoneal macrophages (Tavares et al. 2007), lung epithelial cells (Oliveira et al., unpublished data) or basal lamina components as collagen and fibronectin (Bailao et al. 2012). In the same approach, the influence of oestradiol on the transcriptome of *Paracoccidioides* has also been investigated (Shankar et al. 2011b) and Fernandes et al., unpublished data). We present here the main discoveries that resulted from this approach and how they fit into the global picture of the pathogenesis process.

14.2.1   The Metabolic Response to the Host Environment

At first sight, data obtained from the differential transcriptome experiments with *Paracoccidioides* conform with work hypotheses and predictions. For example, Expressed Sequence Tag (EST) on *Paracoccidioides* exposed to murine liver revealed upregulation of glycolysis-related transcripts such as those for the enzymes glucokinase, acyl-phosphatase and phosphoglyceratemutase (Costa et al. 2007). This is all in keeping with the fact that the liver is a carbohydrate storage and as such, glucose derivatives would be the major carbon source available to the fungus in that site.

In other instances, though, findings are less straightforward to explain. In the same liver experiment, the authors found an accumulation of the quinoprotein alcohol dehydrogenase transcripts, and postulated that *Paracoccidioides* favours alcoholic fermentation in the liver. This is similar to the metabolic response of the yeast cell to *in vitro* culture, where indeed the fungus adopts a fermentative lifestyle that contrasts with the mycelium, which favours glycolysis coupled with the tricarboxylic acid cycle and the respiratory chain (Felipe et al. 2005). However, in the same hepatic environment *Candida albicans*, a human commensal fungus that causes opportunistic systemic infections, consumes its sugars aerobically rather than fermentatively (Thewes et al. 2007). An explanation for this difference is can be that the commensal fungus evolved a more refined respiration machinery that can cope with the oxygen levels of the liver, whereas the free-living *Paracoccidioides* needs a higher concentration of oxygen than is found in the liver to perform respiration adequately. It must be noted that hepatocytes themselves, as most cells of the body outside of nutritional stress, are primarily aerobic. It would be interesting, in this context, to investigate the transcriptional response of *Paracoccidioides* to the oxygen-rich lung environment.

To get to other organs, *Paracoccidioides* need to circulate in the bloodstream, which prompted investigators to analyse their response to human blood and plasma (Bailao et al. 2006; Bailao et al. 2007). Their observations indicate that the fungus responds to the blood milieu by inducing gluconeogenesis via upregulation of phosphoenolpyruvate carboxykinase, and fatty acid catabolism by upregulation of enzymes from both beta-oxidation (e.g. acyl-CoA dehydrogenase) and methylcitrate cycle (e.g. 2-methycitrate dehydratase) pathways, which enable the fungus to consume even- and odd-carbon compounds. Both responses seem appropriate for an oxygen-rich environment such as the blood, but they are also suggestive of a glucose-poor niche and the blood is known to have high glucose availability. One of the explanations advanced to account for the apparent contradiction is that *Paracoccidioides* cells travel by the blood within macrophages, and the interior of the phagosome and phagolysosome are poor in glucose (Lorenz et al. 2004). However, this does not explain why "naked" yeast cells would respond to the blood or plasma the same way they respond to the interior of phagocytic vesicles. It is more likely, given the evidence, that those observations reflect a decrease in blood glucose during the time course of the experiment. In *C. albicans*, exposure to whole human blood resulted in upregulation of both glycolytic and gluconeogenetic enzymes, which reflects a population of both phagocytosed and free cells (Fradin et al. 2003). *Cryptococcus neoformans*, in contrast, maintains transcripts for both pathways and controls which one will predominate by posttranslational mechanisms (Price et al. 2011). Similar scenarios might be at play in *Paracoccidioides* as well.

As for the macrophage, transcriptional analyses of the response of *Paracoccidioides* to its phagosome was in keeping with expectations for a glucose-poor environment: a decrease of the glycolytic enzyme phosphofructokinase and an upregulation of the glyoxylate cycle enzymes isocitrate lyase and malate synthase, which enable pathogens to use fatty acid carbon chains, once reduced to acetate, to be used in gluconeogenesis (Tavares et al. 2007; Derengowski et al. 2008). Similar observations have been made for other pathogens (Lorenz et al. 2004; Miramon et al. 2012; Fukuda et al. (2013) and thus the glyoxylate cycle is of interest as a source of drug targets, as it appears to be important to fungal adaptation to the host.

Amino acid metabolism becomes more biased towards biosynthetic pathways as *Paracoccidioides* cells are exposed to the liver. There and within murine macrophages, transcripts for proteins implicated in methionine salvage and uptake, such as the *MUP1* permease and cystathionine beta-lyase, are upregulated (Tavares et al. 2007; Costa et al. (2007). The trend is also observed in *C. albicans* (Fradin et al. 2005) and in the systemic fungal pathogen *Aspergillus fumigatus* (Morton et al. 2011), which means that this is a common response of this class of infectious agents to the low availability of methionine in the host.

14.2.2   Access to Metals in the Host

Iron, zinc and copper, and to a lesser extent some other metals, are essential cofactors of several proteins, from respiration enzymes to transcription factors, and their depletion disrupts homeostasis of any cell. Fungal pathogens (Nevitt 2011) such as *Paracoccidioides* are no different, and they must cope with the host's defences, which include iron and zinc sequestration. Thus, it comes as no surprise that *Paracoccidioides* cells respond to the liver environment by upregulating iron/zinc and copper transporter transcription (Bailao et al. 2006), or that they activate iron-independent, fermentative pathways such as glycolysis in response to iron depletion *in vitro* (Parente et al. 2011).

14.2.3   Biosignalling, Stress Responses and Cell Wall Biogenesis

All transcriptional and ultimately phenotypic responses of a cell to changes in its environment are triggered by signal transduction mechanisms and thus, it is to be expected that *Paracoccidioides*, in contact with the host, will show changes in their signalling cascades.

Importantly, the osmotic and oxidative stresses that are met by the fungal cells once in the host must be counteracted by stress response pathways. Indeed, *Paracoccidioides* respond to blood and plasma by inducing the putative osmosensor SHO1, which is related to the Hog1 pathway of oxidative and osmotic stress response in other fungi (Bailao et al. 2007; Bailao et al. 2006). A more consistent observation is the upregulation of the transcript for the Rab GTPase in response to several host niches (Bailao et al. 2006; Bailao et al. 2007; Costa et al. 2007). This monomeric GTPase has been shown to be involved in organelle biogenesis and vesicle trafficking in other fungal pathogens (Johnston et al. 2009), and it may be related to vacuole formation in response to osmotic stress, but its precise role in infection, if any, remains to be elucidated.

In a more hypothesis-driven approach, Fernandes et al. (unpublished data) and others (Shankar et al. 2011b) have investigated whether the presence of oestradiol changed signalling patterns involved with the dimorphic transition, given that previous studies show that *Paracoccidioides* has diminished formation of yeast cells in female hosts (Restrepo et al. 1984). Findings suggest that changes do occur in response to oestradiol, and they mainly involve transcriptional downregulation of another GTPase, Rho, and of the geranylating enzyme GGTase that adds the geranyl to Rho so it can be anchored in the membrane and exert its signalllng functions. Rho is a major regulator of fungal response to the host, and is implicated, in several species studied, in cell shape and polarity, osmotic integrity, cytoskeleton organisation and cell wall biogenesis (Yamochi et al. 1994; Zhang et al. 2013; Richthammer et al. 2012). Of note, it is a regulator of beta-1,3-glucan synthase (Fks1), the enzyme that synthesises the major component of the mycelium cell wall (Liu and Balasubramanian 2001), and it has more generally been implicated in cell wall biogenesis.

Indeed, mycelium treated with oestradiol (the condition that most likely simulates the early events of infection in the female host) showed downregulation of not only the FKS1 transcript, but of other cell wall synthesis enzymes such as chitin synthase and alfa-1,3-glucan synthase (Shankar et al. 2011b) and Fernandes et al., unpublished). These enzymes are respectively responsible for the higher chitin content of the yeast cell wall upon transition (Kanetsuna et al. 1969) and the replacement of beta-1,3-glucan by alfa-1,3-glucan in the yeast cell wall, which is hypothesised to impair recognition of fungal cells by the host innate immunity (Rappleye et al. 2007). The general picture from these experiments is that oestradiol somehow impairs Rho-mediated signalling and cell wall remodelling in response to the host, and ultimately blocks dimorphic transition.

Another important aspect of adaptation to the host is the heat shock response. The growth at higher temperatures than room has long been considered a hallmark of virulent pathogens, and heat shock proteins (Hsps) have been reported to be induced by pathways independent of dimorphic transition in *H. capsulatum* (Nguyen and Sil 2008). They are also implicated in other kinds of stress, which is in keeping with their role as protein folding agents. In *Paracoccidioides* exposed to murine liver and human blood, Hsp70 is induced even relative to yeast cells grown *in vitro* at 37 °C; so is Hsp30, which has been implicated in the osmotic stress response, several components of which such as thioredoxin are also upregulated (Bailao et al. 2006; Bailao et al. 2007). The HSP90 transcript is also induced, and the corresponding protein has already been advanced as a possible drug target given its apparent role in fungal viability and response osmotic and thermal stress, and to survival within host macrophages (Nicola et al. 2008; Tamayo et al. 2013; Goldman et al. 2003). Several chaperones were found to be downregulated in *Paracoccidioides* in response to female serum or oestradiol, which suggests that this hormone may cause a more widespread disruption in signalling and stress response in this fungus (Shankar et al. 2011b) and Fernandes et al., unpublished).

Oxidative and nitrosative stress response genes in *Paracoccidioides*, as one might expect, are upregulated in response to internalization by macrophages, the single most important phagocyte in containing systemic fungal infections. The most induced transcript of phagocytosed yeast cells is the glycosylphosphatidylinositol (GPI-) anchored copper-zinc (Cu-Zn) superoxide dismutase (Tavares et al. 2007), which, by analogy with similar proteins from closely related fungi as *H. capsulatum* (Youseff et al. 2012), is probably involved in detoxification of macrophage-derived superoxide. Another induced gene under the same conditions, the alternative oxidase (AOX), is one of the few that have had their role in virulence analysed directly, by RNA interference (Ruiz et al. 2011). It was found necessary for resistance against macrophage killing and detoxification of peroxide.

Finally, the transcriptional response of *Paracoccidioides* during incubation with the lung epithelial cell immortal lineage A549 (Oliveira et al., unpublished data) and with extracellular matrix proteins has been analysed (Bailao et al. 2012). The logic of these analyses was to study the interaction of the fungus to the first host cell type it meets in the process of infection, as well as to basal lamina components that it needs to adhere to in the process of tissue penetration. The genes for two enzymes with pleiotropic functions, enolase and glyceraldehyde 3-phosphate dehydrogenase, were found to be upregulated in this context. Both enzymes have been described as adhesion proteins when associated with the fungal cell wall (Barbosa et al. 2006; Nogueira et al. 2010), and thus the findings in *Paracoccidioides* add to the existing evidence in literature that, in addition to being key enzymes in sugar metabolism, they are virulence factors necessary for establishment of early infection.

In summary, as it often happens with transcriptional analysis data, much of what is known about the *Paracoccidioides* response to the host environment agrees with evidence from other organisms and work hypotheses. New routes of investigation can be opened by this kind of data upon careful mining, though; in the case of *Paracoccidioides*, the role of GTPase signalling in the establishment of infection and dimorphic transition, which could not be inferred from pre-transcriptome knowledge, is an example of needed confirmation with experiments at the lab bench. The wise application of high-throughput techniques to tangible biological questions regarding the interaction of fungi with their mammalian hosts, as outlined by the above, is a necessity made all the more pressing in organisms that defy genetic manipulation such as *Paracoccidioides*.

14.3 Host Response to the Pathogen

14.3.1 Large Scale Transcriptomics

PCM is known to happen in only about 2% of the people who are infected with *Paracoccidioides spp* (McEwen et al. 1995). The fact that most infected people do not ever develop the disease is a strong evidence that in addition to virulence factors of the pathogen, host determinants are also essential to its onset. However, in contrast with the numerous transcriptomic analyses made with *P. brasiliensis* cells, only a handful of published studies have dealt with the large-scale transcriptional response of the host upon infection with fungi from the genus *Paracoccidioides* (Table 14.1). Two papers using high-throughput microarray techniques have analysed the response of murine macrophages (Silva et al. 2008) and dendritic cells (Tavares et al. 2012) to in vitro infection by *Paracoccidioides spp*.

Silva and colleagues used a nylon microarray containing several hundred cDNAs to obtain a picture of how immune-related genes in macrophages respond to infection by *P. lutzi*. In addition to phagocytosing and killing fungal cells, macrophages are also key players in the regulation of the immune response due to the cytokines they secrete. Upon infection, they increased the expression of the pro-inflammatory cytokine TNF-α and chemokines that recruit more monocytes/macrophages, lymphocytes, neutrophils, NK and dendritic cells (Godiska et al. 1997; Nagira et al. 1997; Moser et al. 1990; Kurth et al. 2001) and decreased expression of chemokines associated with IL-13-mediated inflammation (Ma et al. 2004) and basophil attraction and activation (Bischoff et al. 1992). The transcriptional regulation of transcription factors and macrophage effectors also suggests they could be preparing for augmented cellular immunity, with induction of the MHC II gene H2-Eb1, of the interferon-responsive Stat1 transcriptional factor (Katze et al. 2002) and the C-type lectin receptor CLEC1b involved in macrophage activation (Mourao-Sa et al. 2011).

This pro-inflammatory response of infected murine peritoneal macrophages, however, contrasts with previous reports showing that in vitro these cells are not very efficient in controlling *P. brasiliensis* growth but may actually be niches for faster fungal replication (Brummer et al. 1989) unless the macrophages are activated by IFN-γ (Brummer et al. 1988). The transcriptional response observed by Silva and colleagues is thus representative of a biological situation in which macrophages are not able to kill or restrict fungal growth.

In their recent work, Tavares and colleagues used more advanced high-throughput techniques to analyse the modulation of thousands instead of hundreds of host genes in the dendritic cell response to *Paracoccidioides* infection (Tavares et al. 2012). These cells responded to infection by *P. brasiliensis* by augmenting transcription of the genes encoding pro-inflammatory cytokines TNF-α and IL12B, which are important in activating cell-mediated immunity to *P. brasiliensis* either via macrophage activation or by inducing production of IFN-γ by NK and Th1 cells. Other inflammatory mediators that were also transcriptionally regulated include chemokines Cxcl10, Ccl22 and Ccl27, all involved in chemotaxis of monocytes, NK and T cells (Godiska et al. 1997; Dufour et al. 2002). Another important trend that is noticeable in the transcriptional response to *P. brasiliensis* infection is the down-regulation of phagocytic receptors such as FcγRI, CR3, TLR4 and DC-SIGN, which could indicate a process of dendritic cell maturation; however, there were no concomitant increases in MHC II or T-cell co-stimulatory molecules.

In addition to these two reports, we have also considered as a large scale study the one performed by Castro and colleagues, who studied the transcription of 17 cytokines and transcriptional regulators in cells isolated from PCM patients (de Castro et al. 2013). The techniques used and the number of genes assessed is very different from the two previous microarray studies, but the authors did a very thorough analysis of the cytokines and transcriptional factors that regulate the differentiation of activated Th cells into classical Th1 and Th2, the more recently proposed Th17 and the more controversial Th9 and Th22 subsets (Zhu et al. 2010). The expression of the genes encoding these key proteins was measured in peripheral blood mononuclear cells isolated from patients with paracoccidioidomycosis and controls after exposure to non-specific mitogens, a measurement of the intrinsic tendencies that the immune system of these patients has of polarizing to different Th phenotypes. The results found associations between increased expression of Th1-related genes and lack of clinical disease, increased Th17/Th22 responses and the adult form of the disease and a Th2 response with juvenile PCM. These correlations add to the growing body of evidence that after infection by *P. brasiliensis* an important factor in determining if and what type of disease will occur is the type of Th cells that are formed (reviewed by (Calich et al. 1998; Fortes et al. 2011)).

Table 14.1 Technical details and main results of three large scale transcriptomic studies of host response to *Paracoccidioides* infection

	Silva et al. (2008)	Tavares et al. (2012)	de Castro et al. (2013)
Paracoccidioides spp. Used in the study	Pb01 isolate, described in the manuscript as *P. brasiliensis* but now know to be *P. lutzi* Teixeira et al. (2009)	Pb18 isolate, *P. brasiliensis*	Species not determined. Patients were confirmed to be infected with *Paracoccidioides spp.* by direct observation or serological tests
Host component from which gene transcription was analyzed	Thioglycollate-induced peritoneal macrophages from BALB/c mice	Dendritic cells differentiated from BALB/c murine bone marrow using granulocyte-macrophage colony stimulating factor (GM-CSF)	Peripheral blood mononuclear cells (PBMCs) obtained from patients with paracoccidioidomycosis and controls. PBMCs were stimulated with non-specific mitogens
In vivo or in vitro study	In vitro	In vitro	In vivo
Disease type	Does not apply	Does not apply	Adult and juvenile forms
Control group(s) against which gene expression was compared	Unifected macrophages	Uninfected dendritic cells	Healthy persons with asymptomatic infection or with no infection at all (determined by paracoccidioidin cutaneous test)
Time of host–pathogen interaction tested	6 h, 24 h and 48 h	6 h of infection	Not determined
Transcriptomic technique(s) used	Nylon microarrays and real-time RT-PCR	Glass slide microarrays and real-time RT-PCR	Real-time RT-PCR
Number of genes tested	624	4500	17
Number of differentially expressed genes	118 (105 up-regulated and 13 down-regulated)	299 (81 up-regulated and 218 down-regulated)	–
Experiments used to confirm transcriptomic data	Cytokine ELISA	Cytokine ELISA and receptor blockage assays	Cytokine ELISA, immunohisto-chemistry and flow cytometry immunophenotyping
Microarray cutoff considered for differential expression	Fold-change of 1.5. q value <0.05. FDR <5%	Fold-change of 1.2. q value <0.05. FDR <0.5%	Statistically significant difference using T-test ($p < 0.05$)

Table 14.1 (continued)

	Silva et al. (2008)	Tavares et al. (2012)	de Castro et al. (2013)
Expression of inflammatory mediators (cytokines and chemokines)	Increased: TNF-α, Ccl21, Ccl22, Cxcl1, Cxcl14 Decreased: Ccl6, Ccl2	Increased: TNF-α, IL12b, CXCL10, CCL22, CCL27 Decreased: CCL25	Increased: IFN-γ and IL-17 (adult form), IL-4 and IL-5 (juvenile form), IL-10, TGF-β, IL-2 and TNF-α (all patients)
Pattern recognition receptors	Increased: Clec1b, Cd14	Decreased: DC-SIGN (CD209a), TLR4, CR3 (Itgam + Itgb2)	–
T-cell activation signals	Increased: H2-Eb1, CD28	–	–
Immune effectors	Increased: Complement C2 and C3	Increased: Complement C1	–
Transcriptional regulators	Increased: Ier5, Cnot10, Cnot2, Stat1, Nfκb, Nkrf Decreased: Pias1	Increased: PIAS1, NFκB, ZFHX1 Decreased: NκRF, STAT2, STAT6	Increased: T-bet, RORγt and FoxP3 (adult form), GATA-3 and PU.1 (juvenile form)

14.3.2 Small-Scale Transcriptional Analysis

The number of studies focusing on gene expression of host cells during or after fungal interaction is significantly smaller than the ones dealing with expression changes in *P. brasiliensis*. Despite two genome-wide studies (Silva et al. 2008; Tavares et al. 2012) and a thorough study of cytokine- and immune response-related transcription factor expression in different PCM forms (de Castro et al. 2013), most of the other studies target only a few genes, mainly those for cytokines. For this reason, we made a comprehensive search for all manuscripts presenting data on PCM, even when only one or just a few genes were studied, thus, we have gathered most of the available data on this still poorly studied subject (Table 14.2).

Most of the studies on the host immune response to *P. brasiliensis* suggest that resistance to infection is mainly due to stimulation of a Th1 response with production of Th1 cytokines (INF-γ, IL-2, TNF-α) while susceptibility is linked to a Th2 polarized response (IL-4, IL-5, IL-10, TGF-γ (de Castro et al. 2013).

14.3.2.1 Modulation of Immune Response Genes in Response to the Interaction with High Virulence Relative to Low Virulence *P. Brasiliensis* Strains

Some of the host transcript modulation studies deal with differences in immune response against highly virulent and lowly virulent strains, particularly Pb18 and Pb265, respectively. Kurokawa and colleagues observed that both strains induced

Table 14.2 Small Scale Studies of the Host Response to *Paracoccidioides*

Reference	Gene(s) tested	Technique	Host cell/sample	*P. brasiliensis* strain	Main transcriptional changes described
Kurokawa et al. (2007)	TNF-α, IL-1β IL-6 IL-10	Real-time PCR	Monocytes isolated from human peripheral blood mononuclear cells from healthy individuals	High virulent *P. brasiliensis* (Pb18) and low virulent *P. brasiliensis* (Pb265)	High and similar levels of TNF-α transcriptsfor both Pb18 and Pb265 infected cells
					Pb265 caused an earlier induction in IL-1β transcript levels than Pb265
					Pb18 induced an earlier increase of IL-6 and higher levels of IL-10 than Pb265
Ferreira et al. (2007)	TLR-2 TLR-4 MyD88	Reverse-transcription PCR	Lung dendritic cells of B10-A and A/J mice strains	Pb18	Only DCs from susceptible mice (B10.A) expressed TLR-2 after infection
					Both mouse strains expressed TLR-4 and MyD88
Reis et al. (2008)	TGF-α INF-γ IL-10	Real-time PCR	Lungs of infected BALB/c mice injected with adjuvant or rPb27-immunised mice	Pb18	Immunisation with rPb27 enhanced TGF-β and INF-γ levels
					Infection also increased TGF-β, but not at the same levels
Oarada et al. (2009)	MyeloperoxidaseCathepsin-GElastase-2IL-18CXCL10 iNOS GM-CSF NF-kB	Real-time PCR	Mouse spleen and liver of BALB/c mice submitted to three diets with different protein content (0%, 1.5%, or 20% of casein) after infection with *P. brasiliensis*.	Pb18	Higher levels of myeloperoxidase, cathepsin-G, and elastase-2 transcripts in the liver and spleen of mice fed with the higher protein diet
					Higher levels of IL-18, CXCL10, iNOS, GM-CSF and NF-kB) transcripts in the liver of mice fed with higher protein diet

Table 14.2 (continued)

Reference	Gene(s) tested	Technique	Host cell/sample	P. brasiliensis strain	Main transcriptional changes described
Bonfim et al. (2009)	TLR-1 TLR-2 TLR-4 Dectin-1	Real-time PCR	Human peripheral blood mononuclear cells from healthy individuals	Pb18 and Pb265	Both strains induced higher levels of TLR-2 transcripts in monocytes
					Pb265 induced an earlier increase in the levels of dectin-1 in monocytes than Pb18
					Pb265 induced TLR-2 in neutrophils while Pb18 induced TLR-4
Ferreira et al. (2010)	IL-10 TGFβ	Real-time PCR	PBMC from healthy donors, and PCM patients (with active disease or after treatment)		Higher levels of IL-10 in patients with active PCM compared to healthy individuals or post-treatment patients
					TGFβ is higher in both PCM patients groups compared to healthy individuals
Soares et al. (2010)	phagocytosis CLEC2, TLR-2, CD14, NKRF, NF-κB, TNF-α, IL-1β	Real-time PCR	Alveolar macrophag lineage MH-S infected with P. brasiliensis in the presence of a PLB inhibitor	Pb18 (ATCC 32069)	Lower levels of TLR-2, CD14, IL-1β, NF-κB and TNF-α and higher levels of CLEC2 and NKRF (pro-inflammatory inhibitor) in the presence PLB inhibitor
Loures et al. (2011)	IL-18	Real-time PCR	Infected lungs of MyD88−/− mice on a C57BL/6 background	Pb18	MyD88−/− had decreased IL-18 transcript levels in comparison to WT 8 weeks post infection
Fernandes et al. (2011)	IL-10 IL-4 TNF-a TGFβ IFN-γ iNOS.	Real-time PCR	Lungs of male BALB/c mice infected with P. brasiliensis and treated with fluconazole alone or in the presence of rPb40 and rPb27	Virulent clinical isolate	The two proteins used in immunisation caused increased levels of IFN-γ in the first time point analysed (70 d after infection) Decrease in the levels of IL-10 and iNOS in mice submitted to combined treatment with the two proteins and fluconazole relative to adjuvant or fluconazole

Table 14.2 (continued)

Reference	Gene(s) tested	Technique	Host cell/sample	P. brasiliensis strain	Main transcriptional changes described
Ruiz et al. (2011)	IL-6 IL-10 IL-12p40 TNFα	Real-time PCR	MH-S -mouse alveolar macrophages infected with PbWt and PbAOX-aRNA conidia and yeast cells	PbWt (P. brasiliensis ATCC 60855) and a PbAOX-aRNA (P. brasiliensis strain transformed with an antisense cassette for PbGAOX)	None of the cytokines was expressed in the non-activated or IFN-g-activated macrophages. Higher levels of TNFα post infection with PbWt or PbAOX-aRNA strains
					Conidia of both strains produced higher levels of TNF-α than yeast cells
Konno et al. (2012)	TNF-α IL10 IL6 TLR-2	Real-time PCR	BMM from BALB/c mice, stimulated with zymosan and treated with GP43-derived peptides	Zymosan stimulation in the presence or absence of GP43 derived peptides P4 and P23	P4 and P23 induced higher levels of TNF-α, IL-10 and IL-6
Longhi et al. (2012)	granzyme perforin granulysin	Real-time PCR	Purified NK cells from peripheral venous blood of healthy donors and PCM patients	Pb18 and Pb265	NK cells stimulated with Pb18 or Pb265 showed higher granzyme, perforin and granulysin mRNA levels
Bordon-Graciani et al. (2012)	iNOS	Real-time PCR	Peripheral blood mononuclear cells from healthy individuals primed or not with IFN-γ, TNF-α or GM-CSF	Pb18 e Pb265	Higher levels of iNOS mRNA after cytokine stimulation in control mice cells
					Pb18 caused lower expression of NOS in non-primed cells but the expression increased after cytokine activation
					Pb265 induced higher levels of NOS in all conditions compared to Pb18

Table 14.2 (continued)

Reference	Gene(s) tested	Technique	Host cell/sample	P. brasiliensis strain	Main transcriptional changes described
Pina et al. (2013)	TGF-β	Real-time PCR	DCs from B10.A and A/J mice	Pb18 (highly virulent)	Higher TGF-β mRNA levels in A/J DCs
Tristao et al. (2013)	5-LO Arginase-1 NOS-2 T-bet, IL–12p40, INF-γ	Real-time PCR	Lungs of 5-LO-deficient mice (5-LO−/−) and wild-type B6.129 mice	Pb18	Earlier increase of 5-LO in WT compared to mutant

Higher levels of NOS-2 in WT compared to mutant at 30 days after infection. Higher levels of IL-12p40, INF-γ and T-bet and lower levels of Arg1 in mutant compared to WT |
Feriotti et al. (2013)	SOCS3 SOCS1 ARG1 NOS2 FIZZ Ym1	Real-time PCR	B10.A and A/J macrophages Aftermannan treatment and P. brasiliensis infection	Pb18	After mannan treatment, both uninfected and infected B10.A macrophages presented higher levels of NOS2 and SOCS3 transcripts than A/J macrophages A/J macrophages presented higher transcript levels of ARG1, FIZZ1, YM1, SOCS1 after both P. brasiliensis and mannan treatment
Voltan et al. (2013)	Eea1	Real-time PCR	Macrophages alveolar murine line AMJ2-C11	Pb18	Lower eea1 transcript expression in Pb18 infected macrophages
Torres et al. (2013)	IL-6 IL-10 IL12p40 TNF-α	Real-time PCR	MH-S -mouse alveolar macrophages Activated or not with IFN-γ	P. brasiliensis Pb339 And PbGP43 aRNA1 (P. brasiliensis transformed with an antisense cassete for PbGP43)	Higher TNF-α transcript levels in the absence of WT PbGP43 levels

Table 14.2 (continued)

Reference	Gene(s) tested	Technique	Host cell/sample	P. brasiliensis strain	Main transcriptional changes described
Menino et al. (2013)	TNF-α IL-6 MIP-2 IL-17 IL-10	Real-Time PCR	Liver of C57BL/6 mice and TLR9$^{-/-}$ (C57BL/6 background)	P. brasiliensis ATCC 60855	Higher mRNA levels of all cytokines in TLR9$^{-/-}$ mice compared to wild type Protein levels followed the same differences, although TNF-α levels were not statistically different
Bernardino et al. (2013)	iNOS, ARG1 TGF-β, IL-12 TNF-α, IDO	Real-time PCR	Peritoneal macrophages of C57BL/6 mice iNOS-deficient (iNOS$^{-/-}$)	Pb18 (highly virulent)	iNOS-deficient mouse macrophages had higher levels of IL-12, Arg1, TGF-β and TNF-α transcripts while wild-type macrophages had higher levels of IL-12 and iNOS
Burger et al. (2013)	IL-10 TGF-β TNF-α IFN-γ	Real-time PCR	A/J and B10.A peripancreatic omentum	Pb18	A/J had high levels of TNF-α and IFN-γ while B10.A showed no up regulation of these genes. B10.A presented higher levels of TGF-β
Torres et al. (2014)	IL-6 IL-10 IL12P40 TNF-α	Real-time PCR	MH-S macrophage cell line (activated or not with IFN-γ)	Pb339 and PbP27 aRNA (P. brasiliensis transformed with antisense cassete for Pb27)	They could not detect the expression of those cytokines in the tested conditions
Loures et al. (2014)	TLR2, TLR4 SOCS3, SOCS1, ARG1,NOS2, FIZZ1 Yml	Real-time PCR	C57BL/6 Clec7a$^{-/-}$ (Dectin1$^{-/-}$) and wild type (WT)	Pb18	Dectin1$^{-/-}$ mice had higher levels of Yml, arg1, SOCS1 and TLR-2 and reduced levels of TLR-4 Higher levels of SOCS3

monocyte production of TNF-α, IL-1β, IL-6 and IL-10. However, Pb18 produced earlier and higher levels of IL-1β, IL-6, and IL-10 transcripts than Pb265. They also observed a similar difference in cytokine protein levels between the two strains. Their results indicate that the regulation of pro and anti-inflammatory cytokines in response to *P. brasiliensis* is very complex and Pb18 early and continuous induction of pro- and anti-inflammatory cytokines can create an imbalance in monocyte functions favouring fungal growth and immune system evasion (Kurokawa et al. 2007). A complementary study showed that human monocytes and neutrophils also modulate the expression of pattern recognition receptors (PRRs) in response to *P. brasiliensis* (Bonfim et al. 2009). TLR-2, TLR-4 and dectin-1 transcripts were differentially modulated in response to high or low virulent strains. They observed that stimulation with both strains induced a general decrease in cell surface levels of TLRs and dectin-1 expression on monocytes, but not in neutrophils. However, transcript levels followed an opposite trend. TLR-2 transcripts levels were stimulated in response to both strains in monocytes and Pb265 induced a higher increase in TLR-2 expression in neutrophils, although both strains had positive effects on this receptor expression. The authors did not observe any effects on TLR-4 transcript levels on monocytes, but Pb18 showed stronger positive effects on TLR-4 expression on neutrophils. The expression of dectin-1 was also higher in response to Pb265 in monocytes and at earlier times in the interaction with neutrophils. They suggested that the high content of β-glucan in the cell wall of Pb265 was responsible for its interaction with dectin-1 and TLR-2 and fungal internalization; the stimulation of those receptors would in turn lead to IL-10 secretion, thus avoiding an excessive inflammatory response and favouring a protective immune response to the fungus (Bonfim et al. 2009). In addition, Bondon-Graciani and colleagues showed that Pb18 induces a decrease in the expression of iNOS mRNA was not observed when monocytes had been previously stimulated with IFN-γ, TNF-α or GM-CSF. On the other hand, Pb265 induced higher levels of iNOS mRNA in all tested conditions, although the mRNA levels did not correlate with an actual increase in NO production (Bordon-Graciani et al. 2012). Although production of NO is regarded as an important macrophage microbicidal tool, another study using iNOS-deficient murine macrophages showed that this compound could have deleterious effects in pulmonary PCM by suppressing TNF-α production, T cell immunity and pulmonary granuloma organization (Bernardino et al. 2013). Murine macrophages that lack iNOS presented reduced levels of IL-12 and increased production of arg1 and TGF-b transcripts suggesting an anti-inflammatory profile while WT macrophages expressed high levels of iNOS and IL-12. However, this early anti-inflammatory response was matched by higher levels of TNF-α and the development of an enhanced Th1 response pattern that resulted in a lower overall mortality rates of iNOS mice compared to WT mice (Bernardino et al. 2013).

Comparison between NK cells derived from healthy individuals and PCM patients stimulated by Pb18 and Pb265 showed that NK cells of PCM patients had lower cytotoxic effects in comparison with healthy individuals, although both strains induced higher levels of granzyme, perforin and granulysin transcripts (Longhi et al. 2012). They suggest that the impaired cytotoxic response despite higher amounts of granzyme and perforin could result from problems in NK cell activation, which could be overcome by treatment with IL-15. As an important component of the innate immune response, further studies on NK cells could reveal an important impact of these cells on the development of the immune response to PCM (Longhi et al. 2012).

14.3.2.2   Differences in the Immune Response Modulation in Response to *P. Brasiliensis*: Disease-Resistant and Susceptible Murine Models

There are a few mouse strains with different degrees of susceptibility against infection by *P. brasiliensis*. Most of the studies comparing low and high susceptibility hosts employ A/J and B10.A mice strains, respectively. We found a few studies comparing transcriptional changes in response to fungal stimulation aiming to characterise the importance of host response in the disease outcome. A study of Ferreira and collaborators using pulmonary dendritic cells suggested that the expression of PRRs might have a role in host susceptibility against *P. brasiliensis* infection (Ferreira et al. 2007). The authors observed that only dendritic cells from susceptible mice expressed TLR-2 transcripts after infection. As TLR-2 could induce IL-10 secretion in susceptible mice, they concluded that differential receptor binding of the fungus by dendritic cells has significant effects on cytokine production and host susceptibility to infection.

14.3.2.3   Modulation of Immune Response Genes by Vaccine Immunogen Candidates

A few groups have been investigating potential vaccine targets for *P. brasiliensis* as an alternative to the still considerably high toxicity of currently available antifungal treatments. Two studies in this field evaluated the potential of *P. brasiliensis* antigens in the modulation of immune response against this fungus. In one of these studies a recombinant form of a 27 kDa protein from *P. brasiliensis* called Pb27 which was previously shown to be highly immunogenic was also shown to induce a protective response in immunisation assays using Balb/c mice (Reis et al. 2008). Among the observed effects they found that this protein induced an upregulation of TGFβ and IFN-γ transcript levels in the sera from immunised mice, while infection with the highly virulent Pb18 induced only TGF-β and not at the same levels. In another study using Pb27 in combination with another protein, Pb40, the group found that immunisation with these two proteins induced increased levels of INF-γ and TNF-α when compared to a group of mice that was infected with a virulent clinical isolate. The association of the two proteins and fluconazole treatment also induced the production of IFN-γ transcripts, an smaller increase in TNF-α and a significant decrease in IL-10 mRNA levels. All treated groups (immunization and/or fluconazole) showed lower levels, relative to the infected controls, of iNOS in early (70 days post infection) and later time points (120 days), and of the TGF-β mRNAs in the later one.

14.4   Concluding Remarks

PCM still is as an important cause of morbidity and mortality in Latin America. The treatment remains sub-optimal, relying mostly on toxic, expensive or lengthy regimens. One crucial way to overcome this scenario is to understand better how *Paracoccidioides* causes disease and how the human host responds to infection, so we can find more efficient therapeutic strategies. The studies reviewed in this chapter help provide a picture of how *Paracoccidioides* transcription responds to the interaction with the host. However, the paucity of genome-wide studies on the complex host transcriptional response to infection by *Paracoccidioides* suggests that there is still a lot to be learnt about this host–pathogen interaction before benefits can be reaped by PCM patients and their healthcare professionals.

References

Bailao AM, Schrank A, Borges CL, Dutra V, Walquiria Ines Molinari-Madlum EE, Soares Felipe MS, Soares Mendes-Giannini MJ, Martins WS, Pereira M, Maria de Almeida Soares C (2006) Differential gene expression by *Paracoccidioides brasiliensis* in host interaction conditions: representational difference analysis identifies candidate genes associated with fungal pathogenesis. Microbes Infect 8(12–13):2686–2697. doi:10.1016/j.micinf.2006.07.019

Bailao AM, Shrank A, Borges CL, Parente JA, Dutra V, Felipe MS, Fiuza RB, Pereira M, de Almeida Soares CM (2007) The transcriptional profile of *Paracoccidioides brasiliensis* yeast cells is influenced by human plasma. FEMS Immunol Med Microbiol 51(1):43–57. doi:10.1111/j.1574-695X.2007.00277.x

Bailao AM, Nogueira SV, Rondon Caixeta Bonfim SM, de Castro KP, de Fatima da Silva J, Mendes Giannini MJ, Pereira M, de Almeida Soares CM (2012) Comparative transcriptome analysis of *Paracoccidioides brasiliensis* during in vitro adhesion to type I collagen and fibronectin: identification of potential adhesins. Res Microbiol 163(3):182–191. doi:10.1016/j.resmic.2012.01.004

Barbosa MS, Bao SN, Andreotti PF, de Faria FP, Felipe MS, dos Santos Feitosa L, Mendes-Giannini MJ, Soares CM (2006) Glyceraldehyde-3-phosphate dehydrogenase of *Paracoccidioides brasiliensis* is a cell surface protein involved in fungal adhesion to extracellular matrix proteins and interaction with cells. Infect Immun 74(1):382–389. doi:10.1128/IAI.74.1.382-389.2006

Bernardino S, Pina A, Felonato M, Costa TA, Frank de Araujo E, Feriotti C, Bazan SB, Keller AC, Leite KR, Calich VL (2013) TNF-α and CD8[+] T cells mediate the beneficial effects of nitric oxide synthase-2 deficiency in pulmonary paracoccidioidomycosis. PLoS Negl Trop Dis 7(8):e2325. doi:10.1371/journal.pntd.0002325

Bischoff SC, Krieger M, Brunner T, Dahinden CA (1992) Monocyte chemotactic protein 1 is a potent activator of human basophils. J Exp Med 175(5):1271–1275

Bonfim CV, Mamoni RL, Blotta MH (2009) TLR-2, TLR-4 and dectin-1 expression in human monocytes and neutrophils stimulated by *Paracoccidioides brasiliensis*. Med Mycol 47(7):722–733. doi:10.3109/13693780802641425

Bordon-Graciani AP, Dias-Melicio LA, Acorci-Valerio MJ, Araujo JP, Jr, Soares AM (2012) High expression of human monocyte iNOS mRNA induced by *Paracoccidioides brasiliensis* is not associated with increase in NO production. Microbes Infect/Institut Pasteur 14(12):1049–1053. doi:10.1016/j.micinf.2012.07.009

Brummer E, Hanson LH, Stevens DA (1988) Gamma-interferon activation of macrophages for killing of *Paracoccidioides brasiliensis* and evidence for nonoxidative mechanisms. Int J Immunopharmacol 10(8):945–952

Brummer E, Hanson LH, Restrepo A, Stevens DA (1989) Intracellular multiplication of *Paracoccidioides brasiliensis* in macrophages: killing and restriction of multiplication by activated macrophages. Infect Immun 57(8):2289–2294

Burger E, Nishikaku AS, Gameiro J, Francelin C, Camargo ZP, Verinaud L (2013) Cytokines expressed in the granulomatous lesions in experimental Paracoccidioidomycosis: role in host protective immunity and as fungal virulence factor. Clin Cell Immunol S1:010. doi:10.4172/2155-9899.S1-010

Calich VL, Vaz CA, Burger E (1998) Immunity to *Paracoccidioides brasiliensis* infection. Res Immun 149(4–5):407–417; discussion 499–500

Costa M, Borges CL, Bailao AM, Meirelles GV, Mendonca YA, Dantas SF, de Faria FP, Felipe MS, Molinari-Madlum EE, Mendes-Giannini MJ, Fiuza RB, Martins WS, Pereira M, Soares CM (2007) Transcriptome profiling of *Paracoccidioides brasiliensis* yeast-phase cells recovered from infected mice brings new insights into fungal response upon host interaction. Microbiology 153(Pt 12):4194–4207. doi:10.1099/mic.0.2007/009332-0

de Castro LF Ferreira MC da Silva RM Blotta MH Longhi LN Mamoni RL (2013) Characterization of the immune response in human paracoccidioidomycosis. J Infect 67(5):470–485. doi:10.1016/j.jinf.2013.07.019

Derengowski LS, Tavares AH, Silva S, Procopio LS, Felipe MS, Silva-Pereira I (2008) Upregulation of glyoxylate cycle genes upon *Paracoccidioides brasiliensis* internalization by murine macrophages and in vitro nutritional stress condition. Med Mycol 46(2):125–134. doi:10.1080/13693780701670509

Dufour JH, Dziejman M, Liu MT, Leung JH, Lane TE, Luster AD (2002) IFN-gamma-inducible protein 10 (IP-10; CXCL10)-deficient mice reveal a role for IP-10 in effector T cell generation and trafficking. J Immunol 168(7):3195–3204

Felipe MS, Andrade RV, Arraes FB, Nicola AM, Maranhao AQ, Torres FA, Silva-Pereira I, Pocas-Fonseca MJ, Campos EG, Moraes LM, Andrade PA, Tavares AH, Silva SS, Kyaw CM, Souza DP, Pereira M, Jesuino RS, Andrade EV, Parente JA, Oliveira GS, Barbosa MS, Martins NF, Fachin AL, Cardoso RS, Passos GA, Almeida NF, Walter ME, Soares CM, Carvalho MJ, Brigido MM, PbGenome N (2005) Transcriptional profiles of the human pathogenic fungus *Paracoccidioides brasiliensis* in mycelium and yeast cells. J Biol Chem 280(26):24706–24714. doi:10.1074/jbc.M500625200

Feriotti C, Loures FV, Frank de Araujo E, da Costa TA, Calich VL (2013) Mannosyl-recognizing receptors induce an M1-like phenotype in macrophages of susceptible mice but an M2-like phenotype in mice resistant to a fungal infection. PLoS One 8(1):e54845. doi:10.1371/journal.pone.0054845

Fernandes VC, Martins EM, Boeloni JN, Coitinho JB, Serakides R, Goes AM (2011) The combined use of *Paracoccidioides brasiliensis* Pb40 and Pb27 recombinant proteins enhances chemotherapy effects in experimental paracoccidioidomycosis. Microbes Infect/Institut Pasteur 13(12–13):1062–1072. doi:10.1016/j.micinf.2011.06.004

Ferreira KS, Bastos KR, Russo M, Almeida SR (2007) Interaction between *Paracoccidioides brasiliensis* and pulmonary dendritic cells induces interleukin-10 production and toll-like receptor-2 expression: possible mechanisms of susceptibility. J Infect Dis 196(7):1108–1115

Ferreira MC, de Oliveira RT, da Silva RM, Blotta MH, Mamoni RL (2010) Involvement of regulatory T cells in the immunosuppression characteristic of patients with paracoccidioidomycosis. Infect Immun 78(10):4392–4401. doi:10.1128/IAI.00487-10

Fortes MR, Miot HA, Kurokawa CS, Marques ME, Marques SA (2011) Immunology of paracoccidioidomycosis. An Bras Dermatol 86(3):516–524

Fradin C, Kretschmar M, Nichterlein T, Gaillardin C, d'Enfert C, Hube B (2003) Stage-specific gene expression of *Candida albicans* in human blood. Mol Microbiol 47(6):1523–1543

Fradin C, De Groot P, MacCallum D, Schaller M, Klis F, Odds FC, Hube B (2005) Granulocytes govern the transcriptional response, morphology and proliferation of *Candida albicans* in human blood. Mol Microbiol 56(2):397–415. doi:10.1111/j.1365-2958.2005.04557.x

Fukuda Y, Tsai HF, Myers TG, Bennett JE (2013) Transcriptional profiling of *Candida glabrata* during phagocytosis by neutrophils and in the infected mouse spleen. Infect Immun 81(4):1325–1333. doi:10.1128/IAI.00851-12

Godiska R, Chantry D, Raport CJ, Sozzani S, Allavena P, Leviten D, Mantovani A, Gray PW (1997) Human macrophage-derived chemokine (MDC), a novel chemoattractant for monocytes, monocyte-derived dendritic cells, and natural killer cells. J Exp Med 185(9):1595–1604

Goldman GH, dos Reis Marques E, Duarte Ribeiro DC, de Souza Bernardes LA, Quiapin AC, Vitorelli PM, Savoldi M, Semighini CP, de Oliveira RC, Nunes LR, Travassos LR, Puccia R, Batista WL, Ferreira LE, Moreira JC, Bogossian AP, Tekaia F, Nobrega MP, Nobrega FG, Goldman MH (2003) Expressed sequence tag analysis of the human pathogen *Paracoccidioides brasiliensis* yeast phase: identification of putative homologues of Candida albicans virulence and pathogenicity genes. Eukaryot Cell 2(1):34–48

Johnston DA, Eberle KE, Sturtevant JE, Palmer GE (2009) Role for endosomal and vacuolar GTPases in *Candida albicans* pathogenesis. Infect Immun 77(6):2343–2355. doi:10.1128/IAI.01458-08

Kanetsuna F, Carbonell LM, Moreno RE, Rodriguez J (1969) Cell wall composition of the yeast and mycelial forms of *Paracoccidioides brasiliensis*. J Bacteriol 97(3):1036–1041

Katze MG, He Y, Gale M, Jr (2002) Viruses and interferon: a fight for supremacy. Nat Rev Immunol 2(9):675–687. doi:10.1038/nri888

Konno FT, Maricato J, Konno AY, Guereschi MG, Vivanco BC, Feitosa Ldos S, Mariano M, Lopes JD (2012) *Paracoccidioides brasiliensis* GP43-derived peptides are potent modulators of local and systemic inflammatory response. Microbes Infect/Institut Pasteur 14(6):517–527. doi:10.1016/j.micinf.2011.12.012

Kurokawa CS, Araujo JP, Jr, Soares AM, Sugizaki MF, Peracoli MT (2007) Pro- and anti-inflammatory cytokines produced by human monocytes challenged in vitro with *Paracoccidioides brasiliensis*. Microbiol Immunol 51(4):421–428

Kurth I, Willimann K, Schaerli P, Hunziker T, Clark-Lewis I, Moser B (2001) Monocyte selectivity and tissue localization suggests a role for breast and kidney-expressed chemokine (BRAK) in macrophage development. J Exp Med 194(6):855–861

Liu J, Balasubramanian MK (2001) 1,3-beta-Glucan synthase: a useful target for antifungal drugs. Curr Drug Targets Infect Disord 1(2):159–169

Longhi LN, da Silva RM, Fornazim MC, Spago MC, de Oliveira RT, Nowill AE, Blotta MH, Mamoni RL (2012) Phenotypic and functional characterization of NK cells in human immune response against the dimorphic fungus *Paracoccidioides brasiliensis*. J Immunol 189(2):935–945. doi:10.4049/jimmunol.1102563

Lorenz MC, Bender JA, Fink GR (2004) Transcriptional response of *Candida albicans* upon internalization by macrophages. Eukaryot Cell 3(5):1076–1087. doi:10.1128/EC.3.5.1076-1087.2004

Loures FV, Pina A, Felonato M, Feriotti C, de Araujo EF, Calich VL (2011) MyD88 signaling is required for efficient innate and adaptive immune responses to *Paracoccidioides brasiliensis* infection. Infect Immun 79(6):2470–2480. doi:10.1128/IAI.00375-10

Loures FV, Araujo EF, Feriotti C, Bazan SB, Costa TA, Brown GD, Calich VL (2014) Dectin-1 Induces M1 Macrophages and Prominent Expansion of CD8[+] IL-17[+] Cells in Pulmonary Paracoccidiodomycosis. The J Infect Dis 210:762–773 doi:10.1093/infdis/jiu136

Ma B, Zhu Z, Homer RJ, Gerard C, Strieter R, Elias JA (2004) The C10/CCL6 chemokine and CCR1 play critical roles in the pathogenesis of IL-13-induced inflammation and remodeling. J Immunol 172(3):1872–1881

McEwen JG, Garcia AM, Ortiz BL, Botero S, Restrepo A (1995) In search of the natural habitat of *Paracoccidioides brasiliensis*. Arch Med Res 26(3):305–306

Menino JF, Saraiva M, Gomes-Alves AG, Lobo-Silva D, Sturme M, Gomes-Rezende J, Saraiva AL, Goldman GH, Cunha C, Carvalho A, Romani L, Pedrosa J, Castro AG, Rodrigues F (2013) TLR9 activation dampens the early inflammatory response to *Paracoccidioides brasiliensis*, impacting host survival. PLoS Negl Trop Dis 7(7):e2317. doi:10.1371/journal.pntd.0002317

Miramon P, Dunker C, Windecker H, Bohovych IM, Brown AJ, Kurzai O, Hube B (2012) Cellular responses of *Candida albicans* to phagocytosis and the extracellular activities of neutrophils are critical to counteract carbohydrate starvation, oxidative and nitrosative stress. PLoS One 7(12):e52850. doi:10.1371/journal.pone.0052850

Morton CO, Varga JJ, Hornbach A, Mezger M, Sennefelder H, Kneitz S, Kurzai O, Krappmann S, Einsele H, Nierman WC, Rogers TR, Loeffler J (2011) The temporal dynamics of differential gene expression in *Aspergillus fumigatus* interacting with human immature dendritic cells in vitro. PLoS One 6(1):e16016. doi:10.1371/journal.pone.0016016

Moser B, Clark-Lewis I, Zwahlen R, Baggiolini M (1990) Neutrophil-activating properties of the melanoma growth-stimulatory activity. J Exp Med 171(5):1797–1802

Mourao-Sa D, Robinson MJ, Zelenay S, Sancho D, Chakravarty P, Larsen R, Plantinga M, Van Rooijen N, Soares MP, Lambrecht B, Reis e Sousa C (2011) CLEC-2 signaling via Syk in myeloid cells can regulate inflammatory responses. Eur J Immunol 41(10):3040–3053. doi:10.1002/eji.201141641

Nagira M, Imai T, Hieshima K, Kusuda J, Ridanpaa M, Takagi S, Nishimura M, Kakizaki M, Nomiyama H, Yoshie O (1997) Molecular cloning of a novel human CC chemokine secondary lymphoid-tissue chemokine that is a potent chemoattractant for lymphocytes and mapped to chromosome 9p13. J Biol Chem 272(31):19518–19524

Nevitt T (2011) War-Fe-re: iron at the core of fungal virulence and host immunity. Biometals 24(3):547–558. doi:10.1007/s10534-011-9431-8

Nguyen VQ, Sil A (2008) Temperature-induced switch to the pathogenic yeast form of *Histoplasma capsulatum* requires Ryp1, a conserved transcriptional regulator. Proc Natl Acad Sci U S A 105(12):4880–4885. doi:10.1073/pnas.0710448105

Nicola AM, Andrade RV, Dantas AS, Andrade PA, Arraes FB, Fernandes L, Silva-Pereira I, Felipe MS (2008) The stress responsive and morphologically regulated hsp90 gene from *Paracoccidioides brasiliensis* is essential to cell viability. BMC Microbiol 8:158. doi:10.1186/1471-2180-8-158

Nogueira SV, Fonseca FL, Rodrigues ML, Mundodi V, Abi-Chacra EA, Winters MS, Alderete JF, de Almeida Soares CM (2010) *Paracoccidioides brasiliensis* enolase is a surface protein that binds plasminogen and mediates interaction of yeast forms with host cells. Infect Immun 78(9):4040–4050. doi:10.1128/IAI.00221-10

Oarada M, Kamei K, Gonoi T, Tsuzuki T, Toyotome T, Hirasaka K, Nikawa T, Sato A, Kurita N (2009) Beneficial effects of a low-protein diet on host resistance to *Paracoccidioides brasiliensis* in mice. Nutrition 25(9):954–963. doi:10.1016/j.nut.2009.02.004

Parente AF, Bailao AM, Borges CL, Parente JA, Magalhaes AD, Ricart CA, Soares CM (2011) Proteomic analysis reveals that iron availability alters the metabolic status of the pathogenic fungus *Paracoccidioides brasiliensis*. PLoS One 6(7):e22810. doi:10.1371/journal.pone.0022810

Pina A, de Araujo EF, Felonato M, Loures FV, Feriotti C, Bernardino S, Barbuto JA, Calich VL (2013) Myeloid dendritic cells (DCs) of mice susceptible to paracoccidioidomycosis suppress T cell responses whereas myeloid and plasmacytoid DCs from resistant mice induce effector and regulatory T cells. Infect Immun 81 (4):1064–1077. doi:10.1128/IAI.00736-12

Price MS, Betancourt-Quiroz M, Price JL, Toffaletti DL, Vora H, Hu G, Kronstad JW, Perfect JR (2011) *Cryptococcus neoformans* requires a functional glycolytic pathway for disease but not persistence in the host. MBio 2(3):e00103–00111. doi:10.1128/mBio.00103-11

Rappleye CA, Eissenberg LG, Goldman WE (2007) *Histoplasma capsulatum* alpha-(1,3)-glucan blocks innate immune recognition by the beta-glucan receptor. Proc Natl Acad Sci U S A 104(4):1366–1370. doi:10.1073/pnas.0609848104

Reis BS, Fernandes VC, Martins EM, Serakides R, Goes AM (2008) Protective immunity induced by rPb27 of *Paracoccidioides brasiliensis*. Vaccine 26(43):5461–5469. doi:10.1016/j.vaccine.2008.07.097

Restrepo A, Salazar ME, Cano LE, Stover EP, Feldman D, Stevens DA (1984) Estrogens inhibit mycelium-to-yeast transformation in the fungus *Paracoccidioides brasiliensis*: implications for resistance of females to paracoccidioidomycosis. Infect Immun 46(2):346–353

Restrepo A, Gómez BL, Tobón A (2012) Paracoccidioidomycosis: Latin America's own fungal disorder. Curr Fungal Infect Rep 6(4):9. doi:10.1007/s12281-012-0114-x

Richthammer C, Enseleit M, Sanchez-Leon E, Marz S, Heilig Y, Riquelme M, Seiler S (2012) RHO1 and RHO2 share partially overlapping functions in the regulation of cell wall integrity and hyphal polarity in *Neurospora crassa*. Mol Microbiol 85(4):716–733. doi:10.1111/j.1365-2958.2012.08133.x

Ruiz OH, Gonzalez A, Almeida AJ, Tamayo D, Garcia AM, Restrepo A, McEwen JG (2011) Alternative oxidase mediates pathogen resistance in *Paracoccidioides brasiliensis* infection. PLoS Negl Trop Dis 5(10):e1353. doi:10.1371/journal.pntd.0001353

San-Blas G, Nino-Vega G, Iturriaga T (2002) *Paracoccidioides brasiliensis* and paracoccidioidomycosis: molecular approaches to morphogenesis, diagnosis, epidemiology, taxonomy and genetics. Med Mycol 40(3):225–242

Shankar J, Restrepo A, Clemons KV, Stevens DA (2011a) Hormones and the resistance of women to paracoccidioidomycosis. Clin Microbiol Rev 24(2):296–313. doi:10.1128/CMR.00062-10

Shankar J, Wu TD, Clemons KV, Monteiro JP, Mirels LF, Stevens DA (2011b) Influence of 17beta-estradiol on gene expression of *Paracoccidioides* during mycelia-to-yeast transition. PLoS One 6(12):e28402. doi:10.1371/journal.pone.0028402

Silva SS, Tavares AH, Passos-Silva DG, Fachin AL, Teixeira SM, Soares CM, Carvalho MJ, Bocca AL, Silva-Pereira I, Passos GA, Felipe MS (2008) Transcriptional response of murine macrophages upon infection with opsonized *Paracoccidioides brasiliensis* yeast cells. Microbes Infect 10(1):12–20

Soares DA, de Andrade RV, Silva SS, Bocca AL, Soares Felipe SM, Petrofeza S (2010) Extracellular *Paracoccidioides brasiliensis* phospholipase B involvement in alveolar macrophage interaction. BMC Microbiol 10:241. doi:10.1186/1471-2180-10-241

Steenbergen JN, Shuman HA, Casadevall A (2001) *Cryptococcus neoformans* interactions with amoebae suggest an explanation for its virulence and intracellular pathogenic strategy in macrophages. Proc Natl Acad Sci U S A 98 (26):15245–15250. doi:10.1073/pnas.261418798

Tamayo D, Munoz JF, Torres I, Almeida AJ, Restrepo A, McEwen JG, Hernandez O (2013) Involvement of the 90 kDa heat shock protein during adaptation of *Paracoccidioides brasiliensis* to different environmental conditions. Fungal Genet Biol 51:34–41. doi:10.1016/j.fgb.2012.11.005

Tavares AH, Silva SS, Dantas A, Campos EG, Andrade RV, Maranhao AQ, Brigido MM, Passos-Silva DG, Fachin AL, Teixeira SM, Passos GA, Soares CM, Bocca AL, Carvalho MJ, Silva-Pereira I, Felipe MS (2007) Early transcriptional response of *Paracoccidioides brasiliensis* upon internalization by murine macrophages. Microbes Infect 9(5):583–590. doi:10.1016/j.micinf.2007.01.024

Tavares AH, Derengowski LS, Ferreira KS, Silva SS, Macedo C, Bocca AL, Passos GA, Almeida SR, Silva-Pereira I (2012) Murine dendritic cells transcriptional modulation upon *Paracoccidioides brasiliensis* infection. PLoS Negl Trop Dis 6(1):e1459. doi:10.1371/journal.pntd.0001459

Teixeira MM, Theodoro RC, de Carvalho MJ, Fernandes L, Paes HC, Hahn RC, Mendoza L, Bagagli E, San-Blas G, Felipe MS (2009) Phylogenetic analysis reveals a high level of speciation in the *Paracoccidioides* genus. Mol Phylogenet Evol 52(2):273–283. doi:S1055-7903(09)00135-3 [pii] 10.1016/j.ympev.2009.04.005

Thewes S, Kretschmar M, Park H, Schaller M, Filler SG, Hube B (2007) In vivo and ex vivo comparative transcriptional profiling of invasive and non-invasive *Candida albicans* isolates identifies genes associated with tissue invasion. Mol Microbiol 63(6):1606–1628. doi:10.1111/j.1365-2958.2007.05614.x

Torres I, Hernandez O, Tamayo D, Munoz JF, Leitao NP, Jr, Garcia AM, Restrepo A, Puccia R, McEwen JG (2013) Inhibition of PbGP43 expression may suggest that gp43 is a virulence factor in *Paracoccidioides brasiliensis*. PloS One 8(7):e68434. doi:10.1371/journal.pone.0068434

Torres I, Hernandez O, Tamayo D, Munoz JF, Garcia AM, Gomez BL, Restrepo A, McEwen JG (2014) *Paracoccidioides brasiliensis* PbP27 gene: knockdown procedures and functional characterization. FEMS Yeast Res 14(2):270–280. doi:10.1111/1567-1364.12099

Tristao FS, Rocha FA, Moreira AP, Cunha FQ, Rossi MA, Silva JS (2013) 5-Lipoxygenase activity increases susceptibility to experimental *Paracoccidioides brasiliensis* infection. Infect Immun 81(4):1256–1266. doi:10.1128/IAI.01209-12

Voltan AR, Sardi Jde C, Soares CP, Pelajo Machado M, Fusco Almeida AM, Mendes-Giannini MJ (2013) Early Endosome Antigen 1 (EEA1) decreases in macrophages infected with *Paracoccidioides brasiliensis*. Med Mycol 51(7):759–764. doi:10.3109/13693786.2013.777859

Yamochi W, Tanaka K, Nonaka H, Maeda A, Musha T, Takai Y (1994) Growth site localization of Rho1 small GTP-binding protein and its involvement in bud formation in *Saccharomyces cerevisiae*. J Cell Biol 125(5):1077–1093

Youseff BH, Holbrook ED, Smolnycki KA, Rappleye CA (2012) Extracellular superoxide dismutase protects *Histoplasma* yeast cells from host-derived oxidative stress. PLoS Pathog 8(5):e1002713. doi:10.1371/journal.ppat.1002713

Zhang C, Wang Y, Wang J, Zhai Z, Zhang L, Zheng W, Zheng W, Yu W, Zhou J, Lu G, Shim WB, Wang Z (2013) Functional characterization of Rho family small GTPases in *Fusarium graminearum*. Fungal Genet Biol 61:90–99. doi:10.1016/j.fgb.2013.09.001

Zhu J, Yamane H, Paul WE (2010) Differentiation of effector CD4 T cell populations (*). Annu Rev Immunol 28:445–489. doi:10.1146/annurev-immunol-030409-101212

Chapter 15
Dissecting Tuberculosis Through Transcriptomic Studies

Rodrigo Ferracine Rodrigues, Rogério Silva Rosada, Thiago Malardo,
Wendy Martin Rios and Celio Lopes Silva

Abstract Tuberculosis (TB) remains one of the biggest treats in public health, infecting approximately one-third of the human population and killing almost two million people per year. Prophylaxis and treatment methods present some weakness, and HIV co-infection and resistant strains add complexity to the situation, leading WHO to declare the disease a global emergence. To change this situation, new diagnostics, therapies and prevention strategies are urgently needed, but their development relies on biomarker availably. TB biomarker studies have focused especially on diagnostics, treatment efficacy and prophylaxis success by vaccination. Given the multifactorial complexity of this disease, biosignatures are considered more adequate in TB than isolated markers. Standardizations (including assays, definitions and protocols) could accelerate biomarkers research once it would reduce heterogeneous datasets. Thus, specific databanks and integrated platforms of studies are precious resources to conduct broad research. Consequently, to reach the objective of defining TB biosignatures will demand tools from other areas, such as bioinformatics.

15.1 The Problematic of TB

15.1.1 Epidemiologic Aspects

TB is caused by a group of phylogenetically closely related bacteria known as the Mycobacterium tuberculosis complex (MTBC) (Cole et al. 1998). In humans TB is mainly caused by *Mycobacterium tuberculosis* (Mtb) and *M. africanum*, a phylogenetic variant of MTBC restricted to West Africa (de Jong et al. 2010). It is believed that one-third of the human population is infected with the TB bacillus, and about 1.8 million individuals die each year from this disease (Lönnroth et al. 2010; WHO 2010). In 2012 the incidence of the TB was 8.6 million cases (WHO 2013).

R. F. Rodrigues (✉) · R. S. Rosada · T. Malardo · W. M. Rios · C. L. Silva
Center for Tuberculosis Research, Ribeirão Preto Medical School, University of São Paulo,
14049-900 Ribeirão Preto, São Paulo, Brazil
e-mail: ferracine.rodrigues@gmail.com

© Springer International Publishing Switzerland 2014   289
G. A. Passos (ed.), *Transcriptomics in Health and Disease*,
DOI 10.1007/978-3-319-11985-4_15

TB is acquired by inhalation of aerosol particles containing bacilli, expelled by an individual with active disease, and it mainly affects the lungs (Wallis et al. 2010; Walzl et al. 2011). Much of the population that comes into contact with Mtb can eliminate or contain the infection in a latent form due to the immune response (Wallis et al. 2010). However, in a small group of people, bacilli overcome the immune system defenses, resulting in the progression of infection (active disease) (Wallis et al. 2010). This progression can occur soon after infection (1–5% of cases) or years later with the reactivation of the disease (5–9% of cases). In total, about 10% of infected individuals will develop the disease during their lifetime (Sudre et al. 1992; WHO 2013).

The diagnosis is carried out by medical history, physical examination, sputum smear microscopy, thorax radiography and microbiological cultures, taking days up to months (Parida and Kaufmann 2010). Generally, treatment is performed with four drugs for 6 to 9 months, depending on the severity of the disease and response to therapy (Parida and Kaufmann 2010). The long and rigorous treatment favors its abandonment, which in turn cooperates to select resistant strains (multidrug resistant (MDR) and extensive drug resistant (XDR)) (Parida and Kaufmann 2010). Moreover, according to WHO, the treatment regimen is considered effective only when patients are cured and remain without disease after 2 years of observation. The unique prophylactic measure available is the BCG vaccine, which is effective in the protection of severe disease in children (miliary TB and meningococcal TB) (Parida and Kaufmann 2010). However, it fails to protect against pulmonary TB in adults, the most prevalent form of the disease (Parida and Kaufmann 2010).

15.1.2 Social Aspects

TB kills people in the most economically productive period of their lives (Lopez et al. 2006) and can leave sequelae that reduce life quality (Miller et al. 2009). The economic damage caused by mortality and morbidity deriving from the disease is 12 billion dollars annually (WHO 2008). Thus, the social consequences of this disease, which causes great loss in productivity, are often disastrous (Hanson et al. 2006).

TB treatment delay is a common problem that happens mainly due to knowledge lack about TB, poor access to health systems and failures of diagnosis (John and John 2009). Many studies have observed that approximately 50% of individuals with pulmonary TB confirmed do not exhibit symptoms commonly assessed in the disease diagnosis (Hoa et al. 2010; Ayles et al. 2009). This scenario is principally caused by inadequate and inaccurate diagnostic methods (Wallis et al. 2010).

15.1.3 "To-dos" of TB

TB is a multifaceted disease. Therefore, it is not surprising that the immune response to TB is equally complex and variable, involving a plethora of elements from innate and adaptative immune systems, whose number and roles has been constantly enlarged and reviewed (O'Garra et al. 2013). In addition to contributing to the understanding of the pathophysiology of TB, information about the immune system can also be exploited to define biomarkers of disease.

Identifying biomarkers is a mandatory step to TB control because this will help to development of better diagnostics, effective vaccines, new and shorter therapies (Wallis et al. 2010; Abu-Raddad et al. 2009) and interventions to prevent the progression of latent infection to active disease (Lönnroth et al. 2010). Biomarkers are described as characteristics that define a host state with respect to any disease, process and therapeutic or prophylactic intervention (Biomakers Definition Working Group 2001; Parida and Kaufmann 2010; Zarate-Blades et al. 2011). In this way, biomarkers may provide a rational basis for TB studies (Ottenhoff et al. 2012b). In TB, biomarkers are mainly wanted to discriminate latent and active TB (diagnosis and classification), assess the reactivation/relapse risk and indicate vaccine protection and therapeutic results (Parida and Kaufmann 2010; Wallis et al. 2010).

Biomarkers could replace clinical parameters and assess the effectiveness of vaccine and drug at early stages of development, decreasing the risk of initiating clinical trials with candidates that will not confer further advantages (Ottenhoff et al. 2012b) and decreasing time and costs of their licensing process (Wallis et al. 2010).

In drug development, biomarkers can help in the clinical effects evaluation, drug selection, patient stratification and dose-response prediction and toxicity (Parida and Kaufmann 2010). In prophylactic studies, the biomarkers would assist in preclinical and clinical trials defining the best antigen, adjuvant, dose, route, delivery system, the type of vaccine or combination of them (prime boost) and animal model (Ottenhoff et al. 2012b). And although any type of element or molecule can be a biomarker, RNA studies have received considerable attention.

Thus, over the past decade, the literature about transcriptomic studies in TB research field has grown considerably. Taking into account the different profiles and backgrounds of readers interested in this chapter, we sought to address the most consistent studies dealing with transcriptomic studies on TB, covering such relevant topics as bacilli virulence, characterization of different stages of the disease, therapeutic effects of drugs and vaccines and the interaction between the bacterium and host, including mouse models and non-human primates and also studies with human subjects, finally leading the reader to the most modern trends of the area.

15.2   Experimental Tuberculosis Biomarkers

15.2.1   The Mice Model

Mice are the most common *in vivo* model to study TB due to advantageous features compared to other models, like cheaper and more practical manipulation, well-characterized genetics and abundance of reagents to research (Smith 2003). For these reasons, there are many transcriptomic studies in murine TB, often using lung samples. Usually, these articles list functionally categorized genes and some studies tried to set a limited list of genes that could work as biomarkers in preclinical studies. Herein, we contemplate some examples of studies exploring the effects of BCG vaccination and Mtb infection in different models of murine TB.

15.2.1.1 Correlates of Protection

Thinking about the necessities of trustworthy parameters in preclinical tests, the differences between BCG-vaccinated and Mtb-exposed animals are important. One study using microarrays compared these groups at two time points. The lung transcriptome from Balb/c mice revealed that gene expression alterations were significantly higher in infected animals than in BGC-vaccinated mice, highlighting changes in the expression of genes related to granulocytes. Mtb infection stimulated yet stronger expression of genes related to antigen processing and presentation and to interferon gamma (IFN-γ) expression than BCG vaccination did (Mollenkopf et al. 2006). Farther, this study highlighted the importance of expression and downstream effects of IFN-γ, which is an intriguing point in TB research because it is considered important but not sufficient to drive the combat the infection. The relevance of induction of IFN-γ related transcripts as a indicative of protection was also discussed in a study that investigated transcriptional changes in lungs from aerosol-challenged mice previously vaccinated with BCG (Rodgers et al. 2006), strengthening the importance of the induction of these molecules and their effects during the establishment of infection and over the latter stages of the disease.

Another study used Balb/c mice to investigate transcriptomic lung changes in BCG-vaccinated animals before and after infection. Six weeks after BCG administration, a clear pattern of expression differentiated vaccinated uninfected group from others (vaccinated infected; unvaccinated infected; unvaccinated uninfected). Functional analyses showed that these genes were highly related to connective tissue and to the dynamic and diversified cell activities. Next, transcriptional lung profiles of mice vaccinated and infected were compared to mice just infected at different time points. There was a strong difference between groups 14 days after Mtb challenge, and these genes were related to the expression and signaling pathways of IL-17 and IFN-γ cytokines (Aranday Cortes et al. 2010). Thus, the study aimed to identify biosignatures to predict vaccine success before challenge and to identify biomarkers reflecting protective response after exposition to virulent bacillus, although in this case *M. bovis* was used instead of Mtb. Even betting on a signature composed by IL-17 and Th1 profile as an indication of protection, no definitive conclusion was possible since confirmation experiments are needed.

Thus, transcriptomic studies have increased our understanding about the mechanisms underlying BCG protection or failure. Many groups are working to use this knowledge in rational approaches to develop new vaccines to boost or substitute BGC (Rosada et al. 2014). DNA-based vaccines are considered an interesting alternative because they can be engineered to evoke a specific type of immune response against a specific target and they can be employed as prophylactic or therapeutic interventions (Khan 2013). As an example, we briefly expose results of transcriptomic studies about DNA-hsp65, a DNA-based vaccine constructed with the 65 kDa heat-shock protein of *M. leprae* (Box 15.1).

Box 15.1—DNA-hsp65   Vaccine

During its history of development, the DNA-hsp65 vaccine proved to be capable of preventing or treating TB in mice (Lowrie et al. 1997; Lowrie et al. 1999). Moreover, as an immunotherapeutic intervention, DNA-hsp65 shortened the duration of conventional therapy, conferring an outstanding adjuvant effect (Silva et al. 2005). Although DNA-hsp65 effects were related to induction of Th1 CD4$^+$ helper-T cells as well as CD8$^+$ cytotoxic T-lympho-cytes, further characterization of the underlying immunotherapy mechanisms of this vaccine was performed in murine model by means of microarray and real-time PCR assays. The results showed that 98 genes could distinguish the DNA-hsp65-treated group from other groups that received saline or empty vector. Functional analyses of these genes suggested the enhancement of Th1 immunity, inhibition of Th2 cytokines, induction of regulatory media-tors and IL-17-mediated response, and control of excessive inflammation and lung damage, which was confirmed by histopathologic analysis. For these reasons, the authors claimed that such differential gene expression could rep-resent transcriptional biomarkers of DNA-hsp65 immunotherapy against TB (Zárate-Bladés et al. 2009). Taken together, the body of research about DNA-hsp65 highlights the importance and encourages the use of DNA vaccine technology to prevent and treat infectious diseases such as TB.

15.2.1.2   Development of Mtb Infection

Some studies investigated more comprehensively the evolution of Mtb infection us-ing C57BL/6 mice, one of the resistant mouse strains. A transcriptional study from lung of C57BL/6 mice infected with Mtb was performed at early stages (12, 15 and 21 days) after aerosol Mtb infection. Differential expression showed lung tran-scripts related to heat shock proteins at day 12. Recognition of pathogens, immuno-globulin receptors, activation of antigen presenting cells (APCs), inflammatory cy-tokines, like tumor necrosis factor alpha (TNF-α), and Th1 response induction were observed mainly 21 days after infection. At that time, the authors also observed many differentially expressed genes. This time point also revealed the induction of genes that confer advantage to pathogen survival, showing that Mtb could modulate the immune response. Different metalloproteases transcripts were detected at dif-ferent time points, highlighting the importance of these proteins to tissue and cell organization. Interestingly, this study cleverly showed the occurrence of differen-tial expression of transcripts representing neutrophils and monocytes before Th1-related transcripts could be detected. Confirmatory assays showed that depletion of neutrophils is detrimental to the posterior establishment of a Th1 immune response.

However, surprisingly, there was no difference in the outcome of animals depleted or not of neutrophils (Kang et al. 2011). This work added the important feature of the events after infection and prior Th1 immune response that could be useful in the design of new vaccines, but it did not define an early signature of TB.

If the knowledge about the early events in the immune response to TB is lacking in many aspects, as we have just seen, the same is true about later stages of the disease. Exploring the development of chronic TB, a study evaluated transcriptional lung profiles from C57BL/6 mice after 20, 40 and 100 days after aerosol Mtb infection. The transcriptional response observed was robust and progressive. Principal component analysis (PCA) indicated that gene expression profiles from 40 and 100 days are closer to each other than they are to the response at 20 days. Functional analysis suggested a huge induction of transcripts related to humoral and cellular immune responses. Over the entire period, there was a dynamic change in expression of recognition receptor, which could indicate some differences between early and late immune response to TB. This study also found IFN-γ-related genes (Th1 type response) induced at 20 days, but importantly, a type of saturation was detected. That means that although the stimulus is present, the signaling pathway is limited. Here also the induction of transcripts related to proteins that favor Mtb survival was observed. Many cytokines and chemokines were induced over the 100 days, supporting the inflammatory landscape. CXCL9 showed a huge induction again highlighting the importance of cells like neutrophils to the establishment and evolution of the immune response to TB. Finally, to find markers of the disease state and progression, tandem self-organizing mapping (SOM) analysis was applied to examine the distribution of genes. The final result was the segregation of 712 genes into five discriminative subgroups according to their trends of expression and time point (Gonzalez-Juarrero et al. 2009). Even so, there is not at this point a final group of genes that can be used as biomarkers in preclinical mice studies.

However, it is important to be aware that the genetic background of mice can directly influence the results and, consequently, the final conclusions. A comparison between susceptible (I/St) and resistant (A/Sn) mouse strains, 2 weeks after Mtb (H37Rv) challenge, revealed that in the resistant strain the host response is more complex, versatile and strong, as reflected in the number of genes whose expression was altered. These overrepresented genes include those related to innate (mast cells, γδ T cell, NK cells and immunoglobulin) as well as to the adaptive immune response (transcripts related to B and T cells). In contrast, genes related to neutrophils and nitric oxide synthesis were overrepresented in susceptible strains. Confirmatory experiments showed that highly susceptible animals presented a more intense granulocyte response, but the bacterial burden was the same in both strains (Shepelkova et al. 2013).

Taken together, these studies show the enormous plasticity of the mouse immune response to TB, the importance and influence of granulocytes at the early stages and their influence in the posterior Th1 type immune response. At same time, such plasticity poses a great challenge to define a biosignature of TB mice infection, as well as of vaccine protection in these models.

15.2.2 Non-human Primate Experimental Model

Despite requiring large physical structures and high operational costs, the use of non-human primates (NHP) as a preclinical model for studies on TB has aroused attention. Monkeys are close to humans from the phylogenetic point of view, and their immune system is well characterized, making them an excellent model to assess the immunogenicity and safety of new vaccines and drugs (Gupta and Katoch 2009; McMurray 2000). In addition, non-human primates are susceptible to Mtb infection and develop most of the clinical and pathological features observed in human pulmonary TB, such as the organization of granuloma, considered a hallmark of the disease (Capuano et al. 2003; Flynn et al. 2003; McMurray 2000; Kita et al. 2005; Gupta and Katoch 2005; Gupta and Katoch 2009).

15.2.2.1 Characterizing Mtb Infection in Non-human Primates

Using microarrays, a study compared the transcriptome profiles of early (4 weeks) and late stage (13 weeks) of disease from granulomas of rhesus macaques (Mehra et al. 2010). In this work it was observed that around 1200 rhesus genes, from excised granulomas, presented differential expression between 4 and 13 weeks post infection, exhibiting an intense proinflammatory profile at early stage that is reversed at the late weeks of infection. This rapid and profound change in transcriptional scenario was associated with the previous knowledge that Mtb presents several mechanisms for persistence in the host that are increased in late stages of infection, forcing the host to reprogram the immune response dramatically (Mehra et al. 2010).

A stress response factor named Sigma H (sigH) is the major factor expressed by Mtb in response to stress conditions, phagocytosis, cell wall damage and hypoxia, and modulates the host immune response (Kaushal et al. 2002; Graham and Clark-Curtiss 1999; Mehra and Kaushal 2009; Dutta et al. 2012). A mutant Mtb deleted for sigH region (Δ-sigH) was generated, infected in NHP and compared to wild-type Mtb infection (Mehra et al. 2012). Using global transcriptomics, the study found that the expression of MMP9, CCL5, LTA and FOSB was higher in lesions from wild-type-Mtb-infected animals and at normal levels in mutant-Mtb-infected animals; nonetheless, SOCS3, FOXJ1, BAX and CCL14 were increased in mutant compared to the wild-type-Mtb-infected group. The overall scenario indicated that proinflammatory genes were induced at higher levels in wild-type compared to mutant-Mtb-infected animals (Mehra et al. 2012).

Taken together, these are great examples how transcriptomics could refine and accelerate the description of a given phenomenon, allowing biomarker discovery and ultimately providing data to extrapolate to humans.

15.2.2.2 Correlates of Protection

Considering that the NHP model infected with Mtb resembles what occurs in human TB, it is desirable to describe in this model biomarkers indicating the efficacy or inefficacy of new vaccines or therapeutic agents. These studies could drive future work to develop new vaccines/therapy against TB for the human population. In BCG-vaccinated cynomolgus macaques, Mtb post-infection results in reductions in bacterial load, clinical signs of TB and immunopathology when compared to non-vaccinated but infected macaques (Mehra et al. 2013). Transcriptional data from lung lesions showed that this was accompanied by similar expression levels of α-chemokines as CXCL1, CXCL2, CXCL3, CXCL6, and CXCL8, which attract neutrophils, and higher expression levels of β-chemokines related to macrophage and lymphocyte recruitment, named CCL2, CCL3, CCL4, CCL5, in vaccinated animals. This study confirmed suggestions from some authors that the recruitment of more neutrophils compared to macrophage numbers leads to intense immunopathology (Lowe et al. 2012; Eum et al. 2010). Furthermore, this work showed that indoleamine 2,3-dioxygenase (IDO), which exerts a strong inhibitory effect on T cells (Blumenthal et al. 2012), was highly expressed on the ring walls of granulomas from non-vaccinated animals, supporting the worst scenario seen, most likely related to the suppression of T cell enrollment.

Because BCG vaccination is effective in macaques challenged with Mtb, a study employing large-scale real-time quantitative PCR to test 138 genes associated with immune response on circulating lymphocytes showed up to a 600-fold increased network activity, with at least 78 up-regulated genes, which maintained the high expression level for more than 6 weeks. These up-regulated genes were grouped by immune function including: cytokines and their receptors associated with T cell responses; chemokines and their receptors for tissue migration and T cell activation; signal costimulators related to T cells elicited by vaccination; transcription and transactivation; T helper assignments; cytotoxic effectors; and innate and other immune factors (Huang et al. 2007).

In this sense, the high cost and extensive resources necessary to perform an entire experiment with NHP are still worth the effort because this animal model of infection could generate a reliable basis to search for biomarkers, shortening the pathway to address human TB disease mechanisms.

15.3 Human TB

15.3.1 Assessing the Complexity of TB in Humans

Although mice and NHPs are valuable models to study TB, it is important to consider some gaps between those models and human disease, which could delay the transference of knowledge to the clinical use. For this reason, many research groups have opted to work directly with human samples. Peripheral blood mononuclear cells (PBMCs) have been chosen as the preferred sample for several studies due to their easy accessibility compared to the lung and their sufficient complexity to reflect the changes caused by the infection. Quantitation of gene expression by high-throughput methods is one of the best tools available to evaluate and discriminate between the diversity of TB patients' statuses (Maertzdorf et al. 2012). Several studies have shown that TB patients exhibit chronic immune system activation with increased pro-inflammatory signaling, expression of Fc-γ receptors and related downstream activation, and apoptosis involvement (Mistry et al. 2007; Jacobsen et al. 2008; Berry et al. 2010; Maertzdorf et al. 2011a; Maertzdorf et al. 2011b; Lesho et al. 2011).

Therefore, the current potential to investigate TB human samples through transcriptomics is vast and some of the most relevant questions will be addressed below.

15.3.2 Disease Biomarkers and Diagnosis Development

It is clear that there is an urgent need to classify patients presenting active TB, latent Mtb infection (LTBI) or without disease, ensuring the correct treatment according to patient status (McNerney and Daley 2011; Wallis et al. 2010). However, there is growing knowledge in the field that these compartmentalized profiles do not exist in humans, resulting in a plastic range of several states presented by TB patient (Walzl et al. 2011). In this sense, only very accurate biomarkers that correlate with the clinical condition could prove if it is possible to discriminate between those three states (McNerney et al. 2012).

In that context, a study compared the gene expression of patients with active TB, LTBI and non-infected individuals using blood samples (Jacobsen et al. 2007). Using microarrays, the authors detected a set of genes that could discriminate between the three groups, including lactoferrin, Rab33A and CD64. Additionally, this work provided evidence to allocate and discriminate 4 groups—LTBI, active TB, recurrent disease and cured TB patients—through a set of 9 genes (NOLA3, ATP5G1, ASNA1, KIAA2013, SOCS3, C14orf2, TEX264, LY6G6D and RIN3) (Mistry et al. 2007).

In an elegant and complex work, a 393-transcript signature was identified in the blood of patients using unsupervised analysis and statistical filter that could discriminate between active disease from latent and healthy patients (Berry et al. 2010). The active TB signature was dominated by genes related to type I interferon produced by neutrophils. Furthermore, exploring transcriptomic data from blood by applying analyses of significance and a modular data mining approach (Chaussabel et al. 2005; Chaussabel et al. 2008), the authors could separate patients with active TB from those with streptococcal and staphylococcal infections and autoimmune diseases (Berry et al. 2010). Other authors also validated those transcriptional patterns independently (Cliff et al. 2013; Ottenhoff et al. 2012a; Maertzdorf et al. 2011b).

Through literature review, a study pointed the immunologic markers interferon-inducible protein (IP-10), interleukin (IL)-6, IL-10, IL-4, forkhead box P3 (FOXP3) and IL-12, as the main candidates of TB biomarkers (John et al. 2012). However, it was not clear whether these markers have the potential to distinguish TB from several other pulmonary infectious diseases or merely between individuals infected or not with Mtb.

The transcriptomic approach is increasing and accelerating the knowledge of human TB, allowing the description of reliable biomarkers for disease condition and/or diagnosis. Further studies focusing on integration of those data will able to describe biosignatures of TB.

15.3.3 Therapy and Vaccine Efficacy

Searching for early biomarkers of therapeutic efficacy is of interest for assessment of therapeutic success as well as to accelerate the evaluation of new drugs in clinical trials. Changes in the peripheral blood transcriptome detected during the treatment of successfully cured patients could be correlated to bacterial clearance and used as biomarkers or biosignatures to indicate the efficacy of therapies shortly after their initiation, saving time and money. Analyzing blood samples of effectively treated patients at different time points, a study found 780 genes differentially expressed just 1 week after regular TB therapy, which remained significantly different over time, and 373 genes whose expression changed in late stages of therapy. Based on these sets, the application of neural network modeling showed efficiency at classifying patients according their status and to predict treatment response based on a combined list of 62 genes (Cliff et al. 2013).

Analyzing blood samples from South African TB patients, differential transcripts profiles were detected between 2 weeks, 2 months, 6 months and 12 months after antituberculosis treatment start (Bloom et al. 2012). Through expression profiles of 15837 transcripts, the authors compiled a TB signature with 664 transcripts representing a fast transcriptional change that was observed at 2 weeks of treatment, maintaining similar pattern during 2 and 6 months of treatment. Furthermore, a comparison between latent TB with active TB patients 6 and 12 months after the beginning of treatment showed that were no transcriptional differences among these groups, most likely due to the efficacy of treatment and cure of TB-active patients (demonstrated by other routine techniques). More importantly, the authors pointed that 2 weeks was sufficient to evaluate if a treatment would be successful or not because in the later time points the transcriptional profile was maintained (Bloom et al. 2012). These findings are extremely important since in some TB patients, the classical treatment is not effective, and a prompt and strategic modification in the therapeutic scheme could be pivotal to a successful treatment.

Using microarray analysis, the gene-expression profile from blood of patients with active disease, in the treatment phase (four drugs, standard protocol), after treatment finished ("cured"), and in healthy controls were compared. Interestingly, using a total of 875 obtained transcripts, they observed that while the gene expression pattern from individuals with active disease and under treatment were similar, the expression of healthy and cured patients exhibited similar pattern (Ottenhoff et al. 2012a). In this sense, transcriptomic pattern comparison studies like that one could be helpful to interrogate and follow up the length of new therapeutic strategies on a time course manner.

Regarding the evaluation of vaccination strategies by transcriptomic profiling, a study involving different strains of BCG showed that, from 16 immune system genes evaluated by qRT-PCR, vaccinated children exhibited in blood samples abundant Th1 and Treg profile–related genes. Additionally when the transcriptomic profiles induced by BCG in neonates from Denmark or Brazil were compared whit those from Japan, the result was a Th1 and an inflammatory profile of immune response, respectively (Wu et al. 2007). In this sense, these results show that transcriptomics is a relevant tool to investigate the effect of a given vaccine on humans, allowing a feasible interrogation of the desired strategy. More importantly, this work could help to understand the benefits of BCG vaccination in children, which is correlated with TB-meningitis protection, and the mechanism remains somewhat unanswered (Roth et al. 2006; Sierra 2006).

15.4 Mtb: Revisiting the Causative Agent from New Perspectives

Since the publication of the Mtb genome, many studies investigated gene expression to understand drug effects on bacilli in the search for new drug targets as well as the mechanisms behind the development of drug resistance (Boshoff et al. 2004; Wilson et al. 1999; Waddell et al. 2004; Keren et al. 2011; Tudó et al. 2010; Wei et al. 2013). Although our comprehension of the host response is becoming clearer, the necessity of exploring bacillus gene expression in the context of modern approaches is evident. Transcriptional changes promoted by different concentrations and exposure times to rifampicin in susceptible and mutant resistant H37Rv Mtb were investigated using microarrays. The results showed that the cluster Rv0559c-Rv0560c was up-regulated in the resistant strain, suggesting the significant participation of these genes in resistance to rifampicin, but their role remains to be clarified (de Knegt et al. 2013).

Microarrays also became a valuable tool to study transcriptional changes in Mtb after macrophage internalization (Domenech et al. 2001) and under diverse phagosomal conditions (Schnappinger et al. 2003). In a temporal serial approach, the importance of acidification of the phagocytic vacuole to the expression of Mtb genes and its difference from non-pathogenic strains were observed, opening new perspectives for understanding the host-pathogen interaction (Rohde et al. 2007). Another study comparing MTBC clinical isolates identified strain-specific transcriptomes *in vitro* and also identified a highly conserved transcriptome among the strains composed of a set of 280 genes related to adaptation to intracellular environment and reduction of growth and metabolism after 24 h of infection in murine macrophages (Homolka et al. 2010). These similarities and differences can be relevant to trials of new vaccines and drugs.

The origin of resistance was investigated by microarray transcriptomic analysis of longitudinal samples of Mtb from patients under supervised therapy that developed resistance during the treatment. Resistant strains underexpressed genes related to protein synthesis, including transcription factors, cell wall biosynthesis and other pivotal metabolic pathways, while genes encoding trans-membrane proteins, transporters, stress response transcriptional factors and drug efflux pumps were overexpressed in the MDR isolates compared to their sensitive counterparts (Chatterjee et al. 2013). Interestingly, mutations were not related to drug target genes, suggesting that metabolic alterations associated with lower DNA repair could be a combined phenomena favoring the survival of resistant strains. Investigation of H37Rv transcriptome at different time points during host infection by deep sequencing showed that genes whose expression products are also related to bacillus metabolism and to adaptation to immune host response were upregulated in susceptible and resistant mice strains (Skvortsov et al. 2013).

As observed, Mtb transcriptomic studies have revealed a remarkable plasticity of gene expression representing the broad range of mechanisms necessary for pathogen adaptation to live inside the host and also metabolic responses to the actions of drugs. Although studies can be more or less divergent in their approaches and, consequently, in their results, it is clear that this enormous amount of data can lead to identify new potential target to drugs, antigens to vaccines as well as prospection of biomarkers, accelerating the path to new solution to fight TB.

15.5   New Perspectives

As we have explored here, the interpretation of results from transcriptomic studies about TB and the prospection and definition of different types of biomarkers are hard tasks, although impressive progress has been achieved. However, to reach the ambitious propositions presented here, it will be necessary to develop new and sophisticated ways to interrogate the transcriptome. Important innovations include considering microRNAs as elements of the transcriptome that could improve our understanding of the disease (Box 15.2). Additionally, it is becoming clearer that developing and applying tools to integrate different types of studies is a fundamental approach and a strategy that should not be abandoned (Box 15.3).

Box 15.2—microRNAs

As microRNAs regulate the expression of a variety of genes in different contexts, including infectious diseases such as TB (Singh et al. 2013a; Guo et al. 2010), the hypothesis that differences in their expression levels could be used as biomarkers gained traction and motivated a series of transcriptomic studies.

In a murine model of TB, from the differential expression investigated by microarray, it was found that the suppression of microRNA (miR)-99b increased TNF-α production in Mtb-infected dendritic cells, with consequent bacterial growth reduction (Singh et al. 2013b). Using the Mtb-infected murine cell line RAW264.7 and bone marrow-derived macrophages, another study revealed that the ESAT-6 antigen secreted by Mtb upregulates miR-155, which represses Bach1 and Ship1 translation, responsible for bacillus "dormancy" and survival, respectively, contributing to abrogate the infection (Kumar et al. 2012). Another study demonstrated that deletion of miR-223 in bacillus-infected mice caused susceptibility to TB, due to excessive poly-morphonuclear cell migration and exacerbated lung inflammation, resulting in the death of these animals. Notably, the increased of miR-223 expression was also observed in blood and lung parenchyma of TB patients, and the correspondence among markers to be used in pre-clinical and clinical studies is highly desirable to accelerate new findings (Dorhoi et al. 2013).

Regarding human infection, some works used microarrays to analyze the expression levels of microRNAs presented in blood patients with active pulmonary TB compared to healthy controls. A study detected 92 differentially expressed microRNAs, from which miR-29a was detected in plasma and sputum of patients and pointed as a possible marker of disease (Fu et al. 2011). In serum of TB patients 97 differentially expressed microRNAs were identified and the authors claimed that miR-361-5p, miR-889 and miR-576-3p could be considered markers for a rapid diagnosis (Qi et al. 2012). Comparing PMBCs from pulmonary TB patients and healthy controls, other study pointed miR-144* as one of the most differentially expressed, which is associated with the decrease of TNF-α and IFN-γ production and T cell proliferation, highlighting miR-144* as an interesting target for further investigations (Liu et al. 2011).

Finally, researching the relationship between virulence and microRNA expression, a study found that human macrophages infected or stimulated with molecules from Mtb present miR-125b upregulation with a consequent reduction in the TNF-α production, while *M. smegmatis*, a less virulent strain, caused the opposite effects (Rajaram et al. 2011).

Consequently, microRNAs have the potential to contribute as biomarkers as well as for understanding the establishment and evolution of infection and mechanisms of pathogen virulence.

Box 15.3—The power of integrative studies

Integrative studies, meaning here the use of transcriptomic data associated with another approach, has been productive to explore underappreciated aspects of Mtb. Studying mechanisms of virulence and persistence of Mtb, an algorithm called Differential Producibility Analysis (DPA) was used to infer alteration in bacilli metabolism based on available data from transcriptomic microarrays studies (Bonde et al. 2011). The resultant analysis indicated that alterations in bacilli gene expression are heavy related to remodeling of cell envelope. This study brought an interesting way to fill the gap of knowledge about metabolites of Mtb under diverse conditions using an abundant source of data, like transcriptomic studies.

Also investigating virulence in Mtb complex, a study used previously available data from microarray studies to confirm the importance of polymorphic genes from predicted genomic islands in virulent strains. Data mining was used to explore transcriptional profiles of Mtb that were exposed to different antibiotics to identify potential ribosomal proteins with extraribosomal functions, acting as regulatory proteins or taking part in other cellular process that could help pathogen adaptation (Fan et al. 2013). If confirmed, this set of genes could be useful in the screening of drugs and to explore their mechanisms of action.

Seeking to confirm and to find new drug targets, gene co-expression network analysis of microarray studies of Mtb in log-phase growth showed overlapped clustering of transcripts related to diverse metabolic process from the pathogen. Further extraction of co-expressed connections resulted in the confirmation of known drug targets and in the identification of potential new ones (Puniya et al. 2013).

Text mining brings the exciting prospect of using automated methods for exploration of large amounts of information in texts, as available in the scientific literature, to provide a conceptual network of relationships that have not been established by more conventional research methods even though the information is already available. Plato (http://platao.fmrp.usp.br) is an example of this approach. Figure 15.1 illustrates an example of a conceptual network constructed with this platform to explore the TB literature using important findings from studies explored in this chapter.

Moreover, the TB research community is investing in integrative platforms, which include databanks covering any study and data related to the gene expression of TB bacilli and/or hosts, as well as tools useful to their analysis, like the TB Database (TBDB), as well as new approaches to investigate Mtb in the most comprehensive way, like SysteMTb. Briefly, TBDB (http://www.tbdb.org) offers access to "microarray and RT-PCR data as well as genomic data for *Mycobacterium tuberculosis* organism along with several other relevant host organisms and similar model organisms. Such data can be visualized, annotated, analyzed and shared using TBDB tools." The platform has interesting resources related to systems biology (Reddy et al. 2009). The SysteMTb initiative (http://www.systemtb.org) intends go deep into the research about Mtb, supporting a complex and integrated platform of projects, designed in a complementary way. Its aim is produce the more complete possible Systems Biology analysis of Mtb under the same set of standard conditions.

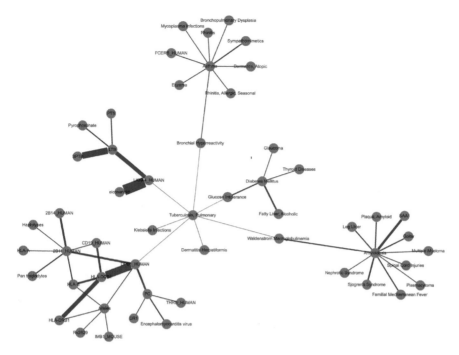

Fig. 15.1 Conceptual network constructed with Plato data mining platform. This network was constructed from the central node "pulmonary tuberculosis". Additional layers were added, searching for the terms "lymphocytes", "macrophages", and "neutrophils", in that order. To avoid an excessive number of nodes and to obtain a readable picture, we limited the outcomes to nodes with high edge strengths and to the following semantic types: organisms, diseases, macromolecules and phenomena and processes. By clicking on the lines, the platform retrieves the articles where the relationships between the terms or concepts were described. The result is a practical and insightful way to search the tuberculosis literature, especially for the correlations between terms representing disease, genes, proteins, cells, etc. This tool could help to save time and increase the accuracy of looking for TB articles describing related events

Mtb will be studied in diverse experimental conditions using a broad range of technologies, including transcriptomics. One interesting study from this consortium reported a surprisingly elevated number of leaderless transcripts whose expression was elevated when Mtb was submitted to starvation, indicating that in this condition leaderless transcripts could have important functions (Cortes et al. 2013).

15.6   Concluding Remarks

As we have seen, although TB transcriptomic signatures remain an open question, undeniable progress has been made. Moreover, defining biomarkers that do not overlap with other diseases is a hard task, and such biomarkers should most likely be adjusted to regional characteristics because studies usually are conducted with limited-size cohorts, even though some studies are becoming more ambitious.

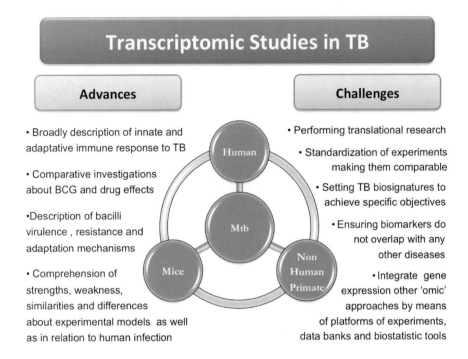

Fig. 15.2 Advances and challenges in TB transcriptomic studies. Over the past few years, transcriptomic studies about TB, exploring the pathogen and its interactions with experimental hosts (such as mice and non-human primates) and humans, have been conducted to advances in the knowledge about this pathology, such as comprehension of disease establishment, its development and also the different stages of immune response. Transcriptomic research has shed light also on molecular mechanisms behind virulence, resistance to drugs and effectiveness of therapeutical and prophylactic interventions. However, the interpretation and exploration of this huge amount of transcriptomic data available to define biosignatures are still great challenges. Specifically, these challenges include determining biomarkers that can be used in pre-clinical and clinical studies and also translated to humans, standardizations of experiments and studies to enable direct comparisons and distinguish candidate biosignatures in TB from other diseases with similar characteristics. Data banks and integrated platforms of studies, which include transcriptomic experiments associated with other 'omic' approaches and biostatistics tools, promise to open a pathway to solve these questions

Understanding the infectious nature of Mtb, almost a human commensal, and its interactions with hosts has shed light on its biology. In Fig. 15.2, we summarize the major progress made with the transcriptomic studies in TB as well as the challenges that still need to be overcome. Deep-sequencing RNA technologies as well as the development of new bioinformatic tools hold great promise for future advances because they could be used to describe the complexities of the transcriptomes of the host and pathogen at the same time during infection. The next decade surely will be revolutionary in the study of TB infection.

References

Abu-Raddad LJ, Sabatelli L, Achterberg JT, Sugimoto JD, Longini IM, Dye C, Halloran ME (2009) Epidemiological benefits of more-effective tuberculosis vaccines, drugs, and diagnostics. Proc Natl Acad Sci USA 106:13980–13985

Aranday Cortes E Kaveh D Nunez-Garcia J Hogarth PJ Vordermeier HM (2010) Mycobacterium bovis-BCG vaccination induces specific pulmonary transcriptome biosignatures in mice. PLoS One 5:e11319

Ayles H, Schaap A, Nota A, Sismanidis C, Tembwe R, De Haas P, Muyoyeta M, Beyers N, Peter Godfrey-Faussett for the ZAMSTAR Study Team (2009) Prevalence of tuberculosis, HIV and respiratory symptoms in two Zambian communities: implications for tuberculosis control in the era of HIV. PLoS One 4:e5602

Berry MP, Graham CM, Mcnab FW, Xu Z, Bloch SA, Oni T, Wilkinson KA, Banchereau R, Skinner J, Wilkinson RJ, Quinn C, Blankenship D, Dhawan R, Cush JJ, Mejias A, Ramilo O, Kon OM, Pascual V, Banchereau J, Chaussabel D, O'garra A (2010) An interferon-inducible neutrophil-driven blood transcriptional signature in human tuberculosis. Nature 466:973–977

Biomakers Definitions Working Group (2001) Biomarkers and surrogate endpoints: preferred definitions and conceptual framework. Clin Pharmacol Ther 69:89–95

Bloom CI, Graham CM, Berry MP, Wilkinson KA, Oni T, Rozakeas F, Xu Z, Rossello-Urgell J, Chaussabel D, Banchereau J, Pascual V, Lipman M, Wilkinson RJ, O'garra A (2012) Detectable changes in the blood transcriptome are present after two weeks of antituberculosis therapy. PLoS One 7:e46191

Blumenthal A, Nagalingam G, Huch JH, Walker L, Guillemin GJ, Smythe GA, Ehrt S, Britton WJ, Saunders BM (2012) M. tuberculosis induces potent activation of IDO–1, but this is not essential for the immunological control of infection. PLoS One 7:e37314

Bonde BK, Beste DJ, Laing E, Kierzek AM, Mcfadden J (2011) Differential producibility analysis (DPA) of transcriptomic data with metabolic networks: deconstructing the metabolic response of M. tuberculosis. PLoS Comput Biol 7:e1002060

Boshoff HI, Myers TG, Copp BR, MCNEIL MR, WILSON MA, BARRY CE (2004) The transcriptional responses of Mycobacterium tuberculosis to inhibitors of metabolism: novel insights into drug mechanisms of action. J Biol Chem 279:40174–40184

Capuano SV 3rd, Croix DA, Pawar S, Zinovik A, Myers A, Lin PL, Bissel S, Fuhrman C, Klein E, Flynn JL (2003) Experimental Mycobacterium tuberculosis infection of cynomolgus macaques closely resembles the various manifestations of human M. tuberculosis infection. Infect Immun 71:5831–5844

Chatterjee A, Saranath D, Bhatter P, Mistry N (2013) Global transcriptional profiling of longitudinal clinical isolates of Mycobacterium tuberculosis exhibiting rapid accumulation of drug resistance. PLoS One 8:e54717

Chaussabel D, Allman W, Mejias A, Chung W, Bennett L, Ramilo O, Pascual V, Palucka AK, Banchereau J (2005) Analysis of significance patterns identifies ubiquitous and disease-specific gene-expression signatures in patient peripheral blood leukocytes. Ann N Y Acad Sci 1062:146–154

Chaussabel D, Quinn C, Shen J, Patel P, Glaser C, Baldwin N, Stichweh D, Blankenship D, Li L, Munagala I, Bennett L, Allantaz F, Mejias A, Ardura M, Kaizer E, Monnet L, Allman W, Randall H, Johnson D, Lanier A, Punaro M, Wittkowski KM, White P, Fay J, Klintmalm G, Ramilo O, Palucka AK, Banchereau J, Pascual V (2008) A modular analysis framework for blood genomics studies: application to systemic lupus erythematosus. Immunity 29:150–164

Cliff JM, Lee JS, Constantinou N, Cho JE, Clark TG, Ronacher K, King EC, Lukey PT, DuncaN K, Van Helden PD, Walzl G, Dockrell HM (2013) Distinct phases of blood gene expression pattern through tuberculosis treatment reflect modulation of the humoral immune response. J Infect Dis 207:18–29

Cole ST, Brosch R, Parkhill J, GARNIER T, Churcher C, Harris D, Gordon SV, Eiglmeier K, Gas S, Barry CE, Tekaia F, Badcock K, Basham D, Brown D, Chillingworth T, Connor R, Davies R,

Devlin K, Feltwell T, Gentles S, Hamlin N, Holroyd S, Hornsby T, Jagels K, krogh A, Mclean J, Moule S, Murphy L, Oliver K, Osborne J, Quail MA, Rajandream MA, Rogers J, Rutter S, Seeger K, Skelton J, Squares R, Squares S, Sulston JE, Taylor K, Whitehead S, Barrell BG (1998) Deciphering the biology of Mycobacterium tuberculosis from the complete genome sequence. Nature 393:537–544

Cortes T, Schubert OT, Rose G, Arnvig KB, Comas I, Aebersold R, Young DB (2013) Genome-wide mapping of transcriptional start sites defines an extensive leaderless transcriptome in Mycobacterium tuberculosis. Cell Rep 5(4):1121–1131

de Jong BC Antonio M Gagneux S (2010) Mycobacterium africanum–review of an important cause of human tuberculosis in West Africa. PLoS Negl Trop Dis 4:e744

de Knegt GJ Bruning O ten Kate MT de Jong M van Belkum A Endtz HP Breit TM Bakker-Woudenberg IA de Steenwinkel JE (2013) Rifampicin-induced transcriptome response in rifampicin-resistant Mycobacterium tuberculosis. Tuberculosis (Edinb) 93:96–101

Domenech P, Barry CE, Cole ST (2001) Mycobacterium tuberculosis in the post-genomic age. Curr Opin Microbiol 4:28–34

Dorhoi A, Iannaccone M, Farinacci M, Faé KC, Schreiber J, Moura-alves P, Nouailles G, Mollen-kopf HJ, Oberbeck-Müller D, Jörg S, Heinemann E, Hahnke K, Löwe D, Del Nonno F, Goletti D, Capparelli R, Kaufmann SH (2013) MicroRNA–223 controls susceptibility to tuberculosis by regulating lung neutrophil recruitment. J Clin Invest 123:4836–4848

Dutta NK, Mehra S, Martinez AN, Alvarez X, Renner NA, Morici LA, Pahar B, Maclean AG, Lackner AA, Kaushal D (2012) The stress-response factor SigH modulates the interaction between Mycobacterium tuberculosis and host phagocytes. PLoS One 7:e28958

Eum SY, Kong JH, Hong MS, Lee YJ, Kim JH, Hwang SH, Cho SN, Via LE, Barry CE 3rd (2010) Neutrophils are the predominant infected phagocytic cells in the airways of patients with active pulmonary TB. Chest 137:122–128

Fan X, Tang X, Yan J, Xie J (2013) Identification of idiosyncratic Mycobacterium tuberculosis ribosomal protein subunits with implications in extraribosomal function, persistence, and drug resistance based on transcriptome data. J Biomol Struct Dyn 32(10):1546–1551

Flynn JL, Capuano SV, Croix D, Pawar S, Myers A, Zinovik A, Klein E (2003) Non-human primates: a model for tuberculosis research. Tuberculosis (Edinb) 83:116–118

Fu Y, Yi Z, Wu X, Li J, Xu F (2011) Circulating microRNAs in patients with active pulmonary tuberculosis. J Clin Microbiol 49:4246–4251

Gonzalez-Juarrero M, Kingry LC, Ordway DJ, Henao-Tamayo M, Harton M, Basaraba RJ, Hanneman WH, Orme IM, Slayden RA (2009) Immune response to Mycobacterium tuberculosis and identification of molecular markers of disease. Am J Respir Cell Mol Biol 40:398–409

Graham JE, Clark-Curtiss JE (1999) Identification of Mycobacterium tuberculosis RNAs synthesized in response to phagocytosis by human macrophages by selective capture of transcribed sequences (SCOTS). Proc Natl Acad Sci U S A 96:11554–11559

Guo W, Li JT, Pan X, Wei L, Wu JY (2010) Candidate Mycobacterium tuberculosis genes targeted by human microRNAs. Protein Cell 1:419–421

Gupta UD, Katoch VM (2005) Animal models of tuberculosis. Tuberculosis (Edinb) 85:277–293

Gupta UD, Katoch VM (2009) Animal models of tuberculosis for vaccine development. Indian J Med Res 129:11–18

Hanson C, Floyd K, Weil D (2006) Tuberculosis in the poverty alleviation agenda. In: Raviglione M (ed) TB: a comprehensive international approach. Informa Healthcare, New York

Hoa NB, Sy DN, Nhung NV, Tiemersma EW, Borgdorff MW, Cobelens FG (2010) National survey of tuberculosis prevalence in Viet Nam. Bull World Health Organ 88:273–280

Homolka S, Niemann S, Russell DG, Rohde KH (2010) Functional genetic diversity among Mycobacterium tuberculosis complex clinical isolates: delineation of conserved core and lineage-specific transcriptomes during intracellular survival. PLoS Pathog 6:e1000988

Huang D, Qiu L, Wang R, Lai X, Du G, Seghal P, Shen Y, Shao L, Halliday L, Fortman J, Shen L, Letvin NL, Chen ZW (2007) Immune gene networks of mycobacterial vaccine-elicited cellular responses and immunity. J Infect Dis 195:55–69

Jacobsen M, Repsilber D, Gutschmidt A, Neher A, Feldmann K, Mollenkopf HJ, Ziegler A, Kaufmann SH (2007) Candidate biomarkers for discrimination between infection and disease caused by Mycobacterium tuberculosis. J Mol Med (Berl) 85:613–621

Jacobsen M, Mattow J, Repsilber D, Kaufmann SH (2008) Novel strategies to identify biomarkers in tuberculosis. Biol Chem 389:487–495

John TJ, John SM (2009) Paradigm shift for tuberculosis control in high prevalence countries. Trop Med Int Health 14:1428–1430

John SH, Kenneth J, Gandhe AS (2012) Host biomarkers of clinical relevance in tuberculosis: review of gene and protein expression studies. Biomarkers 17:1–8

Kang DD, Lin Y, Moreno JR, Randall TD, Khader SA (2011) Profiling early lung immune responses in the mouse model of tuberculosis. PLoS One 6:e16161

Kaushal D, Schroeder BG, Tyagi S, Yoshimatsu T, Scott C, Ko C, Carpenter L, Mehrotra J, Manabe YC, Fleischmann RD, Bishai WR (2002) Reduced immunopathology and mortality despite tissue persistence in a Mycobacterium tuberculosis mutant lacking alternative sigma factor, SigH. Proc Natl Acad Sci U S A 99:8330–8335

Keren I, Minami S, Rubin E, Lewis K (2011) Characterization and transcriptome analysis of Mycobacterium tuberculosis persisters. MBio 2:e00100–11

Khan KH (2013) DNA vaccines: roles against diseases. Germs 3:26–35

Kita Y, Tanaka T, Yoshida S, Ohara N, Kaneda Y, Kuwayama S, Muraki Y, Kanamaru N, Hashimoto S, Takai H, Okada C, Fukunaga Y, Sakaguchi Y, Furukawa I, Yamada K, inoue Y, Takemoto Y, Naito M, Yamada T, Matsumoto M, Mcmurray DN, Cruz EC, Tan EV, Abalos RM, Burgos JA, Gelber R, Skeiky Y, Reed S, Sakatani M, Okada M (2005) Novel recombinant BCG and DNA-vaccination against tuberculosis in a cynomolgus monkey model. Vaccine 23:2132–2135

Kumar R, Halder P, Sahu SK, Kumar M, Kumari M, Jana K, Ghosh Z, Sharma P, Kundu M, Basu J (2012) Identification of a novel role of ESAT–6-dependent miR–155 induction during infection of macrophages with Mycobacterium tuberculosis. Cell Microbiol 14:1620–1631

Lesho E, Forestiero FJ, Hirata MH, Hirata RD, Cecon L, Melo FF, Paik SH, Murata Y, Ferguson EW, Wang Z, Ooi GT (2011) Transcriptional responses of host peripheral blood cells to tuberculosis infection. Tuberculosis (Edinb) 91:390–399

Liu Y, Wang X, Jiang J, Cao Z, Yang B, Cheng X (2011) Modulation of T cell cytokine production by miR–144* with elevated expression in patients with pulmonary tuberculosis. Mol Immunol 48:1084–1090

Lönnroth K, Castro KG, Chakaya JM, Chauhan LS, Floyd K, Glaziou P, Raviglione MC (2010) Tuberculosis control and elimination 2010–50: cure, care, and social development. Lancet 375:1814–1829

Lopez AD, Mathers CD, Ezzati M, Murray CJL, Jamison DT (2006) Global burden of disease and risk factors. Oxford University Press and The World Bank, New York

Lowe DM, Redford PS, Wilkinson RJ, O'garra A, Martineau AR (2012) Neutrophils in tuberculosis: friend or foe? Trends Immunol 33:14–25

Lowrie DB, Silva CL, Colston MJ, Ragno S, Tascon RE (1997) Protection against tuberculosis by a plasmid DNA vaccine. Vaccine 15:834–838

Lowrie DB, Tascon RE, Bonato VL, Lima VM, Faccioli LH, Stavropoulos E, Colston MJ, Hewinson RG, Moelling K, Silva CL (1999) Therapy of tuberculosis in mice by DNA vaccination. Nature 400:269–271

Maertzdorf J, Ota M, Repsilber D, Mollenkopf HJ, Weiner J, Hill PC, Kaufmann SH (2011a) Functional correlations of pathogenesis-driven gene expression signatures in tuberculosis. PLoS One 6:e26938

Maertzdorf J, Repsilber D, Parida SK, Stanley K, Roberts T, Black G, Walzl G, Kaufmann SH (2011b) Human gene expression profiles of susceptibility and resistance in tuberculosis. Genes Immun 12:15–22

Maertzdorf J, Weiner J 3rd, Kaufmann SH (2012) Enabling biomarkers for tuberculosis control. Int J Tuberc Lung Dis 16:1140–8

Mcmurray DN (2000) A nonhuman primate model for preclinical testing of new tuberculosis vaccines. Clin Infect Dis 30(3):210–212

Mcnerney R, Daley P (2011) Towards a point-of-care test for active tuberculosis: obstacles and opportunities. Nat Rev Microbiol 9:204–213

Mcnerney R, Maeurer M, Abubakar I, Marais B, Mchugh TD, Ford N, Weyer K, Lawn S, Grobusch MP, Memish Z, Squire SB, Pantaleo G, Chakaya J, Casenghi M, Migliori GB, Mwaba P, Zijenah L, Hoelscher M, Cox H, Swaminathan S, Kim PS, Schito M, Harari A, Bates M, Schwank S, O'grady J, Pletschette M, Ditui L, Atun R, Zumla A (2012) Tuberculosis diagnostics and biomarkers: needs, challenges, recent advances, and opportunities. J Infect Dis 205(2):147–158

Mehra, S, Kaushal D (2009) Functional genomics reveals extended roles of the Mycobacterium tuberculosis stress response factor sigmaH. J Bacteriol 191:3965–3980

Mehra S, Pahar B, Dutta NK, Conerly CN, Philippi-Falkenstein K, Alvarez X, Kaushal D (2010) Transcriptional reprogramming in nonhuman primate (rhesus macaque) tuberculosis granulomas. PLoS One 5:e12266

Mehra S, Golden NA, Stuckey K, Didier PJ, Doyle LA, Russell-Lodrigue KE, Sugimoto C, Hasegawa A, Sivasubramani SK, Roy CJ, Alvarez X, Kuroda MJ, Blanchard JL, Lackner AA, Kaushal D (2012) The Mycobacterium tuberculosis stress response factor SigH is required for bacterial burden as well as immunopathology in primate lungs. J Infect Dis 205:1203–1213

Mehra S, Alvarez X, Didier PJ, Doyle LA, Blanchard JL, Lackner AA, Kaushal D (2013) Granuloma correlates of protection against tuberculosis and mechanisms of immune modulation by Mycobacterium tuberculosis. J Infect Dis 207:1115–1127

Miller TL, Mcnabb SJ, Hilsenrath P, Pasipanodya J, Weis SE (2009) Personal and societal health quality lost to tuberculosis. PLoS One 4:e5080

Mistry R, Cliff JM, Clayton CL, Beyers N, Mohamed YS, Wilson PA, Dockrell HM, Wallace DM, van Helden PD, Duncan K, Lukey PT (2007) Gene-expression patterns in whole blood identify subjects at risk for recurrent tuberculosis. J Infect Dis 195:357–365

Mollenkopf HJ, Hahnke K, Kaufmann SH (2006) Transcriptional responses in mouse lungs induced by vaccination with Mycobacterium bovis BCG and infection with Mycobacterium tuberculosis. Microbes Infect 8:136–144

O'garra A, Redford PS, Mcnab FW, Bloom CI, Wilkinson RJ, Berry MP (2013) The immune response in tuberculosis. Annu Rev Immunol 31:475–527

Ottenhoff TH, Dass RH, Yang N, Zhang MM, Wong HE, Sahiratmadja E, Khor CC, Alisjahbana B, van Crevel R, Marzuki S, Seielstad M, van de Vosse E, Hibberd ML (2012a) Genome-wide expression profiling identifies type 1 interferon response pathways in active tuberculosis. PLoS One 7:e45839

Ottenhoff TH, Ellner JJ, Kaufmann SH (2012b) Ten challenges for TB biomarkers. Tuberculosis (Edinb) 92(1):17–20

Parida SK, Kaufmann SH (2010) The quest for biomarkers in tuberculosis. Drug Discov Today 15:148–157

Puniya BL, Kulshreshtha D, Verma SP, Kumar S, Ramachandran S (2013) Integrated gene co-expression network analysis in the growth phase of Mycobacterium tuberculosis reveals new potential drug targets. Mol Biosyst 9:2798–2815

Qi Y, Cui L, Ge Y, Shi Z, Zhao K, Guo X, Yang D, Yu H, Shan Y, Zhou M, Wang H, Lu Z (2012) Altered serum microRNAs as biomarkers for the early diagnosis of pulmonary tuberculosis infection. BMC Infect Dis 12:384

Rajaram MV, Ni B, Morris JD, Brooks MN, Carlson TK, Bakthavachalu B, Schoenberg DR, Torrelles JB, Schlesinger LS (2011) Mycobacterium tuberculosis lipomannan blocks TNF biosynthesis by regulating macrophage MAPK-activated protein kinase 2 (MK2) and microRNA miR-125b. Proc Natl Acad Sci U S A 108:17408–17413

Reddy TB, Riley R, Wymore F, Montgomery P, Decaprio D, Engels R, Gellesch M, Hubble J, Jen D, Jin H, Koehrsen M, Larson L, Mao M, Nitzberg M, Sisk P, Stolte C, Weiner B, White J, Zachariah ZK, Sherlock G, Galagan JE, Ball CA, Schoolnik GK (2009) TB database: an integrated platform for tuberculosis research. Nucleic Acids Res 37:499–508

Rodgers A, Whitmore KM, Walker KB (2006) Potential correlates of BCG induced protection against tuberculosis detected in a mouse aerosol model using gene expression profiling. Tuberculosis (Edinb) 86:255–262

Rohde KH, Abramovitch RB, Russell DG (2007) Mycobacterium tuberculosis invasion of macrophages: linking bacterial gene expression to environmental cues. Cell Host Microbe 2:352–364

Rosada RS, Rodrigues RF, Frantz FG, Arnoldi FGC, Torre LG, Silva CL (2014) TB Vaccines: state of the Art and Progresses. In: Giese M (ed) Molecular vaccines: from prophylaxis to therapy. Springer: Vienna

Roth AE, Stensballe LG, Garly ML, Aaby P (2006) Beneficial non-targeted effects of BCG–ethical implications for the coming introduction of new TB vaccines. Tuberculosis (Edinb) 86:397–403

Schnappinger D, Ehrt S, Voskuil MI, Liu Y, Mangan JA, Monahan IM, Dolganov G, Efron B, Butcher PD, Nathan C, Schoolnik GK (2003) Transcriptional Adaptation of Mycobacterium tuberculosis within Macrophages: Insights into the Phagosomal Environment. J Exp Med 198:693–704

Shepelkova G, Pommerenke C, Alberts R, Geffers R, Evstifeev V, Apt A, Schughart K, Wilk E (2013) Analysis of the lung transcriptome in Mycobacterium tuberculosis-infected mice reveals major differences in immune response pathways between TB-susceptible and resistant hosts. Tuberculosis (Edinb) 93:263–269

Sierra VG (2006) Is a new tuberculosis vaccine necessary and feasible? A Cuban opinion. Tuberculosis (Edinb) 86:169–178

Silva CL, Bonato VL, Coelho-Castelo AA, De Souza AO, Santos SA, Lima KM, Faccioli LH, Rodrigues JM (2005) Immunotherapy with plasmid DNA encoding mycobacterial hsp65 in association with chemotherapy is a more rapid and efficient form of treatment for tuberculosis in mice. Gene Ther 12:281–287

Singh PK, Singh AV, Chauhan DS (2013a) Current understanding on micro RNAs and its regulation in response to Mycobacterial infections. J Biomed Sci 20:14

Singh Y, Kaul V, Mehra A, Chatterjee S, Tousif S, Dwivedi VP, Suar M, Van Kaer L, Bishai WR, Das G (2013b) Mycobacterium tuberculosis controls microRNA–99b (miR–99b) expression in infected murine dendritic cells to modulate host immunity. J Biol Chem 288:5056–5061

Skvortsov TA, Ignatov DV, Majorov KB, Apt AS, Azhikina TL (2013) Mycobacterium tuberculosis transcriptome profiling in mice with genetically different susceptibility to tuberculosis. Acta Naturae 5:62–69

Smith I (2003) Mycobacterium tuberculosis pathogenesis and molecular determinants of virulence. Clin Microbiol Rev 16:463–496

Sudre P, ten Dam G, Kochi A (1992) Tuberculosis: a global overview of the situation today. Bull World Health Organ 70:149–159

Tudó G, Laing K, Mitchison DA, Butcher PD, Waddell SJ (2010) Examining the basis of isoniazid tolerance in nonreplicating Mycobacterium tuberculosis using transcriptional profiling. Future Med Chem 2:1371–1383

Waddell SJ, Stabler RA, Laing K, Kremer L, Reynolds RC, Besra GS (2004) The use of microarray analysis to determine the gene expression profiles of Mycobacterium tuberculosis in response to anti-bacterial compounds. Tuberculosis (Edinb) 84:263–274

Wallis RS, Pai M, Menzies D, Doherty TM, Walzl G, Perkins MD, Zumla A (2010) Biomarkers and diagnostics for tuberculosis: progress, needs, and translation into practice. Lancet 375:1920–1937

Walzl G, Ronacher K, Hanekom W, Scriba TJ, Zumla A (2011) Immunological biomarkers of tuberculosis. Nat Rev Immunol 11:343–354

Wei J, Guo N, Liang J, Yuan P, Shi Q, Tang X, Yu L (2013) DNA microarray gene expression profile of Mycobacterium tuberculosis when exposed to osthole. Pol J Microbiol 62:23–30

WHO (2008) In tuberculosis control, the burden of tuberculosis: economic burden. http://www.who.int/trade/distance_learning/gpgh/gpgh3/en/index7.html. Accessed Jan 2014

WHO (2010) Multidrug and extensively drug-resistant TB (M/XDR-TB): 2010 global report on surveillance and response. http://www.who.int/tb/features_archive/m_xdrtb_facts/en/. Accessed Jan 2014

WHO (2013) Global tuberculosis report 2013. http://www.who.int/tb/publications/global_report/en/index.html. Accessed Jan 2014

Wilson M, DeRisi J, Kristensen HH, Imboden P, Rane S, Brown PO, Schoolnik GK (1999) Exploring drug-induced alterations in gene expression in *Mycobacterium tuberculosis* by microarray hybridization. Proc Natl Acad Sci U S A 96:12833–12838

Wu B, Huang C, Garcia L, Ponce de Leon A, Osornio JS, Bobadilla-del-Valle M, Ferreira L, Canizales S, Small P, Kato-Maeda M, Krensky AM, Clayberger C (2007) Unique gene expression profiles in infants vaccinated with different strains of *Mycobacterium bovis* Bacille Calmette-Guerin. Infect Immun 75:3658–3664

Zárate-Bladés CR, Bonato VL, da Silveira EL, Oliveira e Paula M, Junta CM, Sandrin-Garcia P, Fachin AL, Mello SS, Cardoso RS, Galetti FC, Coelho-Castelo AA, Ramos SG, Donadi EA, Sakamoto-Hojo ET, Passos GA, Silva CL (2009) Comprehensive gene expression profiling in lungs of mice infected with *Mycobacterium tuberculosis* following DNAhsp65 immunotherapy. J Gene Med 11:66–78

Zarate-Blades CR, Silva CL, Passos GA (2011) The impact of transcriptomics on the fight against tuberculosis: focus on biomarkers, BCG vaccination, and immunotherapy. Clin Dev Immunol 2011:192630

Chapter 16
Understanding Chagas Disease by Genome and Transcriptome Exploration

Ludmila Rodrigues P. Ferreira and Edecio Cunha-Neto

Abstract The most fundamental level at which the genome information gives rise to the phenotype is by the expression of its genes. Recent results from the ENCODE project, a 10-year effort by hundreds of scientists to characterize the human genome in depth, have indicated that a much larger proportion of our DNA is likely to be expressed and functional than previously estimated. This has put the focus back on RNA as a key component of organism development, meaning that the measurement of gene expression continues to be a critical tool employed across drug discovery, and life science research. Microarray technology has yielded much important information about the 'transcriptome' (or the entire profile of transcripts in a species) and as such has been invaluable in providing the link between information encoded in the genome and phenotype. The great benefit of this approach is that it allows a researcher to investigate the expression of every gene in the genome in a single experiment. This technology has helped scientists to understand biological mechanisms of complex diseases, as Chagas disease or American trypanosomiasis, caused by an intracellular parasite, *Trypanosoma cruzi*, and it is a leading cause of heart failure in Latin America. In this chapter, we will explore the basic aspects of Chagas disease and how research based on genome and trasncriptome exploration has been helping our understanding about different aspects and clinical outcomes of the disease.

16.1 Epidemiology of Chagas Disease

Chagas disease, also called Human American trypanosomiasis, was named after the Brazilian medical doctor Carlos Chagas (Fig. 16.1a), who discovered the disease in 1909 during a campaign to fight malaria in Brazil (Moncayo 2010). Carlos Chagas identified, associated with diseased individuals living in poor dwellings (Fig. 16.1b), a triatomine blood-sucking insect (Fig. 16.1c). He found in the intestine of the bug,

L. R. P. Ferreira (✉) · E. Cunha-Neto
Laboratory of Immunology, Heart Institute (Incor), Avenida Dr. Enéas de Carvalho Aguiar, 44, Bloco II, 9. Andar, 05403-900 São Paulo, Brazil
e-mail: luferreira@lim60.usp.br

© Springer International Publishing Switzerland 2014
G. A. Passos (ed.), *Transcriptomics in Health and Disease,*
DOI 10.1007/978-3-319-11985-4_16

Fig. 16.1 Chagas disease, also called human American trypanosomiasis, was named after the Brazilian medical doctor Carlos Chagas (**a**), he identified, associated with diseased individuals living in poor dwellings (**b**), a triatomine blood-sucking insect (**c**), he found flagellated parasites in the intestine of the bug, which he named *Trypanosoma cruzi (T. cruzi* trypomastigote in a thin blood smear stained with Giemsa (**d**). (Image credits: **a** Public domain. **b** José Eduardo R. Camargo. **c** and **d** Public Health Image Libray—Centers for Disease Control and Prevetion—CDC and Laboratory Identification of Parasitic Diseases of Public Health Concern—DPDx)

flagellated parasites, which he named *Trypanosoma cruzi (T. cruzi)* (Fig. 16.1d). He also found *T. cruzi* parasites in the blood of sick people, and soon correlated the parasitaemia (level of parasites in the blood) with some symptoms of the disease, such as fever, anemia, lymphadenopathy, splenomegaly and a cardiac form of the disease (Kropf 2011; Kropf and Sa 2009; Pays 2009).The disease, begins with a short acute phase characterized by high parasitemia followed by a life-long chronic phase maintained with scarce parasites (Golgher and Gazzinelli 2004). *T. cruzi* is responsible for the infection of ~ 10 million individuals worldwide. The World Health Organization (WHO) estimates that over 25 million people are at risk of the disease, and that more than 10,000 people died from it in 2008 alone (Hotez et al. 2012). Natural transmission of Chagas disease has been controlled in many countries by insecticide targeting of haematophagus bug populations, as well as improved socioeconomic status and quality of dwelling in Latin America. The list of possible infection routes of Chagas disease includes vectorial, transfusional (through *T. cruzi* infected blood), congenital, through organ transplantation, oral transmission and accidental, through laboratory accidents. In 2006 WHO certified Brazil as being free of transmission through *Triatoma infestans*, the main intradomicilliary vector of Chagas disease (Committee 2002). However, there were new reports of oral transmission and oral outbreaks in the Amazon region showing that this victory was only partial (Dias 2009). Chagas disease has even spread outside endemic countries and it has been estimated that 700,000 infected people are living outside of Latin America (Hotez et al. 2012). It has been estimated that in 2007 there were > 300,000 individuals infected with *T. cruzi* in the United States, >5500 in Canada, >80,000 in Europe and in the western Pacific region, >3000 in Japan and >1500 in Australia, thus, transmission by blood transfusion and organ transplants are becoming a new threat for *T. cruzi* infection in non-endemic countries (Hotez et al. 2012; Lescure et al. 2010; Diaz 2007). The pathogenesis of Chagas disease remains largely unknown, and there are still no effective vaccines or drugs to prevent or treat chronic infection with *T. cruzi*.

16.2 *T. cruzi* Life Cycle and Triatomine Vectors

T. cruzi is known to infect eight different mammalian orders including humans and it is transmitted by insect vectors of the Reduviidae family and the subfamily of Triatomines (Committee 2002). There are many popular names for the vector. In Brazil the common name for the vector is *Barbeiro*—"the barber" and the English name is the kissing bug. Around 100 different triatomine species are susceptible to infection with the *T. cruzi* parasite but the principal vector specie has been *Triatoma Infestans* and in Brazil, the species *Triatoma sordida* and *Panstrongylus megistus* are also prevalent (Moncayo and Ortiz Yanine 2006; Siqueira-Batista et al. 2011; Dias 2009). *T. cruzi* has different developmental stages in its life cycle: Epimastigotes are the form stage that proliferates by cell division in the stomach of the triatomine bugs, they migrate to the distal part of the bug's intestine, and by a process called metacyclogenesis, they transform into metacyclic trypomastigotes, the infective form for the vertebrate host. The insects feed on mammals by sucking blood, and *T. cruzi* is transferred via their faeces, deposited on the skin of the host after feeding. The metacyclic trypomastigotes are able to penetrate through mucous membranes as well as skin injuries, when the host scratches the skin after being bitten or rub their eye. The parasites then invade host cells, transforming into amastigotes which replicate and differentiate into trypomastigotes, disrupting host cells and infecting various cell types with a particular tropism for cardiac, skeletal and smooth muscle cells (de Souza et al. 2010). Finally, the bugs are infected by ingesting trypomastigotes in the blood from infected hosts, thus completing the *T. cruzi* life cycle (Fig. 16.2).

16.3 Clinical Features of Chagas Disease: Acute and Chronic Phases of Infection

The parasite *T. cruzi* produces pathological processes in mammals that can occur in various organs and tissues. When *T. cruzi* is transmitted, it invades the victim's bloodstream and the lymphatic system. Hereafter it nestles in many tissues including the skeletal muscle and cardiac tissue, which causes immune responses and inflammation. Chagas disease has an acute as well as a chronic phase. Morbidity and

Trypanosomiasis, American (Chagas disease)
(Trypanosoma cruzi)

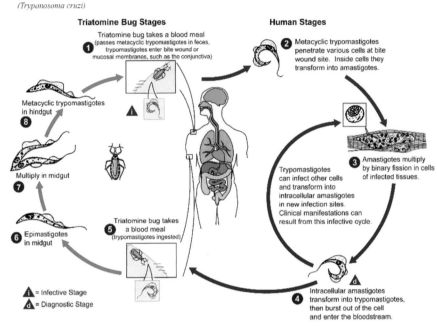

Fig. 16.2 *T.cruzi* life cycle: An infected triatomine insect vector takes a blood meal and releases trypomastigotes in its feces near the site of the bite wound. Trypomastigotes enter the host through the wound or through intact mucosal membranes, such as the conjunctiva *1*. Inside the host, the trypomastigotes invade cells near the site of inoculation, where they differentiate into intracellular amastigotes *2*. The amastigotes multiply by binary fission *3* and differentiate into trypomastigotes, and then are released into the circulation as bloodstream trypomastigotes *4*. The "kissing" bug becomes infected by feeding on human or animal blood that contains circulating parasites *5*. The ingested trypomastigotes transform into epimastigotes in the vector's midgut *6*. The parasites multiply and differentiate in the midgut *7* and differentiate into infective metacyclic trypomastigotes in the hindgut *8*. (Life cycle image and information credit: Laboratory Identification of Parasitic Diseases of Public Health Concern—DPDx (http://www.cdc.gov/dpdx))

mortality are higher in the acute phase for children under five, immune-suppressed people or people with high parasitemia as in patients from outbreaks of food-borne Chagas disease. The acute phase can occur at any age in disease endemic areas, however, the highest frequency is before the age of 15, typically starting in the age group 1–5 years. The acute phase of Chagas disease usually lasts 6–8 weeks, and most frequently is oligo- or asymptomatic and after these phase, most patients appear to be healthy (Moncayo and Ortiz Yanine 2006). The infection by *T. cruzi* can then only be detected by serological or parasitological tests. In the acute phase, if the transmission is vectorial, visible port of entry can be identified, such as the chagoma, a skin lesion in exposed areas of the body, or the Romaña's sign, a purplish edema on the lids of one eye (Fig. 16.3). The sign occurs only in about 10% of infected persons, and can easily be misdiagnosed with conjunctivitis, for example, which is common in rural areas (Dias 1997; Delaporte 1997; Roveda 1967). Other

Fig. 16.3 Romaña's sign, a purplish edema on the lids of one eye that is formed during *T.cruzi* infection. (The illustrations of chagasic patient was obtained from: Public Health Image Libray—Centers for Disease Control and Prevetion—CDC/Dr. Mae Melvin)

clinical features of the acute phase is an excessive activation of the immune system that includes cytokinaemia (high plasma levels of cytokines), intense activation of B and T cells. Generic and unspecific symptoms include diarrhea, vomiting, headache, muscle pain, loss of appetite and extreme fatigue. These symptoms are not very specific, and can easily be confused with other disease etiologies (Coura and Borges-Pereira 2010). Shortly after the acute infection starts, *T. cruzi* components—including its DNA and membrane glycoconjugates—trigger innate immunity via Toll-like receptors in macrophages and dendritic cells, among other cell types. Upon activation, cells from monocytic lineage produce high levels of proinflammatory cytokines like interferon gamma (IFN-γ), interleukin 12 (IL-12) and Tumor necrosis factor alpha (TNF-α). The high level of IFN-γ-induced chemokines and adhesion molecules play an important role in promoting the inflammatory environment in the heart of animals infected with *T. cruzi*. In fact, mice lacking the functional IFN-γ gene display major changes in the CD4+ T and CD8+ T lymphocytes composition of inflammatory infiltrates, as well as enhanced tissue parasitism in the heart (Campos et al. 2004). The essential role of some of these cytokines (e.g. IL-12 and TNF-α) and reactive nitrogen intermediates (RNI) in the control of parasitemia and tissue parasitism is evidenced during the early stages of infection in the murine model (Junqueira et al. 2010).

More precisely, the cells from the macrophage lineage exposed to *T. cruzi* will produce IL-12 that is responsible for initiating IFN-γ synthesis by Natural Killer (NK) cell. IFN-γ plays a major role in resistance through the activation of macrophages to produce high levels of RNI that will effectively control parasite replication (Fig. 16.4). If not controlled by the innate immune system of the host, the infection is fatal as shown in experimental models employing mice lacking functional genes for the IL-12, IFN-γ, IFN-γ receptor, TNF-α receptor or inducible nitric oxide

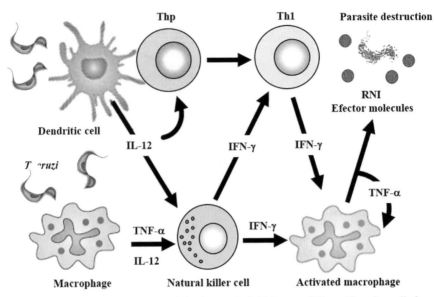

Fig. 16.4 Immune response to *T.cruzi* infection. In the initial stage of *T. cruzi* invasion cells from the innate immune system [dendritic cell, macrophages and natural Killer cells (NK cells)] produce cytokines (IL-12, TNF-α and IFN-γ) and effector molecules [reactive nitrogen intermediates (RNIs)] that lead to parasite destruction. At the same time, innate immune cells, particularly dendritic cells, make the bridge between the innate and acquired immunity, producing cytokines (IL-12) necessary for differentiation and clonal expansion of T helper 1 (Th1) CD4+. IFN-γ produced by CD4+ activates effector mechanisms in macrophages to destroy both amastigotes and phagocytosed trypomastigotes. *IFN* interferon, *IL* interleukin, *Thp* Th precursor cell, *TNF* tumour necrosis factor

(NO) synthase (iNOS) genes (Junqueira et al. 2010; Gazzinelli and Denkers 2006; Golgher and Gazzinelli 2004).

The chronic phase starts with an effective acquired immunity leading to parasitemia drop to a level where it is undetectable with direct parasitological tests, and when symptoms and clinical manifestations typically disappear. However, depending on different factors 10–40% of patients in the chronic phase will develop lesions in target organs, like the intestine (intestinal mega syndrome), esophagus (mega esophagus) and heart (cardiomyopathy), however, up to 70% of infected people remain in an indeterminate asymptomatic form (ASY) for their whole life. The most important clinical consequence of chronic Chagas disease is the chronic Chagas disease cardiomyopathy (CCC), an inflammatory cardiomyopathy that develops in up to 30% of infected individuals. A significant proportion of those patients subsequently develop dilated cardiomyopathy with a fatal outcome. Heart failure of Chagasic aetiology has a worse prognosis and 50% lower survival rate than cardiomyopathies of non-inflammatory aetiology, like ischemic and idiopathic dilated cardiomyopathy (Machado et al. 2012; Bilate and Cunha-Neto 2008). The pathogenesis of CCC is still matter of intense debate. The susceptibility factors that lead to 30% of individuals to develop CCC after *T. cruzi* infection remain unknown.

However, there are three main pathogenetic mechanisms to explain CCC development: cardiac dysautonomy, disorders of the microvascular circulation and inflammatory/immunological tissue damage. Regardless of the mechanisms underlying the initiation and maintenance of the myocarditis, the bulk of the evidence indicates that the inflammatory infiltrate is a significant effector of heart tissue damage. Inflammatory cytokines are produced during the chronic phase of Chagas disease. Mononuclear cells increase their cytokine production, leading to increased plasma levels of TNF-α and IFN-γ, and are even detected in infected ASY individuals, probably in response to parasite persistence. The subset of patients that develop CCC displays an array of immunological alterations consistent with an exacerbated Th1 immune response; The predominance of production of IFN-γ and TNF-α (Abel et al. 2001) associated to the increased expression of the Th1 transcription factor T-bet in the heart, which is not controlled by regulatory T cells in situ is evidence corroborating that the Th1 response is involved in tissue damage in CCC. Chagas thus remains a neglected disease, with no vaccines or anti-parasitic drugs proven efficient in chronically infected adults, when most patients are diagnosed. Development of effective drugs for CCC is hampered by the limited knowledge of its pathogenesis. T cell migration to the myocardium and inflammation, cytokine/chemokine-induced modulation of myocardial gene and protein expression, and genetic components controlling such processes are clearly key events (Nogueira et al. 2012). Available data suggests that investigation of the affected heart will yield the most complete insights on pathogenetic events that are crucial for CCC development.

16.4 Understanding Chagas Disease by Trascriptome Analysis in Patients, in Vitro and in Animal Models

Fundamental questions regarding the pathogenesis of Chagas disease remain unexplained like why different patients develop the cardiac, digestive, cardiodigestive or asymptomatic clinical forms of the disease. A powerful approach to pursue these questions is using different tools to compare and identify genes and/or pathways implicated in the establishment of the infection and pathogenesis. The first gene expression analyses were performed primarily based on observations from immunoblotting, polymerase chain reaction and/or Northern blotting. These techniques were limited only allowing the evaluation of a few pre-selected genes at one time. Another limitation was the access to human heart samples from the acute phase of the disease, so the majority of data available is based on murine models and/or using cells from *in vitro T. cruzi* infection. Several reports have been published with these approaches with a high variability in parasite strains, host cells, mammalian species and times of infection generating a complex picture and few general conclusions. Transcriptome analysis and other high throughput technologies have revolutionized the field of molecular biology and afforded the opportunity to profile the expression of thousands of genes.

16.5   Transcriptome Analysis in Vitro Models

In 2002, Burleigh et al., have performed the first microarray analysis to identify differences in gene expression using an *in vitro* model of human fibroblasts infected with *T. cruzi* (Vaena de Avalos et al. 2002). For this experiment, they used a glass slides high-density microarrays consisting of ~27,000 human cDNAs that were hybridized with fluorescent probes generated from *T. cruzi*-infected human fibroblasts (HFF) at early time points following infection (2–24 h). Surprisingly, they observed that no genes were induced ≥ 2-fold in HFF cDNA between 2 and 6 h post-infection (hpi). A significant increase in transcript abundance for 106 host cell genes was observed only at 24 hpi. Among the most highly induced was a set of interferon-stimulated genes, indicative of a type I interferon (IFN) response to *T. cruzi*. The authors concluded that the delay of *T. cruzi to* induce host cell transcriptional responses is indicative that changes in host cell gene expression may correlate with a particular parasite-dependent event such as differentiation or replication. These events are performed by *T. cruzi* silently without eliciting major changes in the host fibroblasts gene expression.

Because cardiac myocytes are important targets of initial infection with *T. cruzi*, in 2009 another study compared gene profiling of primary cultures of cardiac myocytes infected for 48 h with *T. cruzi* (Goldenberg et al. 2009). They employed microarray analysis with glass slides containing a total of 31,769 70mer oligonucleotide probes. As expected, the results are diverse from the study done using fibroblasts and show a substantial alteration in expression of more than 5% of the sampled genome with major alterations in genes related to inflammation, immunological responses and cell adhesion. Among the pathways most affected from the list of up-regulated genes were those involved in enzymatic activity, immune and stress responses, apoptosis and activation of the proteasome, and calcium-activated potassium channel activity. Down-regulated pathways included calcium and second messenger signaling, cytoskeleton elements (actin filaments, stress fibers, myosin), enzymatic degradation (lysozyme, trypsin, metallopeptidases) and extracellular matrix. This study showed that the cardiac myocytes themselves contribute to the remodeling process even in the absence of other confounding factors, even though in vivo models show contributions by fibroblasts and heart-infiltrating inflammatory cells.

16.6   Transcriptome and Proteome Analysis in Rodent
Models

The study of chagasic heart disease has been aided by the use of the mouse model which recapitulates many of the functional and pathological alterations of the human disease. In 2003 two different groups have performed microarray analysis to detect differences in gene expression in the heart of mice experimentally infected with *T. cruzi*. They have used different microarray platforms i.e., Garg et al., have used commercial nylon membranes microarrays containing a repertoire of 1176 mouse genes printed on the arrays to evaluate the gene expression in whole heart of mice infected with SylvioX10/4 strain of *T. cruzi* for 3, 37 and 110 days post infection (dpi) Garg et al. 2003 and Mukherjee et al., have used glass microarrays containing ~27,400 mouse cDNAs clones to evaluate gene expression also in whole heart from C57BL/6 129sv mice infected for 100 days with the Brazil strain of *T. cruzi* (Mukherjee et al. 2003).

Garg et al., showed that out of a total of 1176 genes printed on the arrays, 31, 89, and 66 genes were differentially regulated in the context of their expression trends at 3, 37, and 110 dpi, respectively. They showed that all of the differentially expressed genes in the myocardium at 3 dpi were up-regulated and encoded immune-related or host defense/stress proteins. During the acute phase (37 dpi), mRNA species for 77 of the 89 differentially regulated genes were increased by at least twofold. Of these, 27 transcripts were increased by > 10-fold, and 18 of the 27 transcripts encoded the immune-related proteins. Out of the 12 transcripts that were reproducibly repressed at 37 dpi, eight were characterized to encode proteins involved in mitochondrial energy metabolism. Surprisingly, a majority of the differentially expressed genes (>63%) in the myocardium of infected mice at 110 dpi were repressed relative to normal controls. From the 66 differentially expressed gene at 110 dpi, 42 were repressed and of these, 26 (60%) transcripts have implications in sustaining the mitochondrial energy metabolism and maintaining the cytoskeletal and extracellular matrix (ECM) structure and function. The study performed by Murkherjee et al. also demonstrated the induction of several genes important to cardiac remodeling, like cytokines and growth factor genes, including growth differentiation factor 3 and insulin-like growth factor-binding proteins, a family of structurally homologous secreted proteins that specifically bind and modulate the activities of insulin-like growth factors (IGF-1 and IGF-2), enhance cellular differentiation and stimulate cell proliferation and muscle cell differentiation. Results from both studies are in accordance showing changes in oxidative phosphorylation and depressed energy metabolism. Soares et al., in 2010 also analysed gene expression profiling in total heart from C57Bl/6 mice chronically infected (8 months of infection) with *T. cruzi* (Colombiana strain) (Soares et al. 2010). They used, for their analysis, glass slides microarrays spotted with 32,620 mouse 70mer oligonucleotides. Their results showed some similarities to the previous studies. As expected, mice chronically infected with *T. cruzi* have intense myocarditis, with a inflammatory infiltrate mainly composed by mononuclear cells, including CD4+ and CD8+ T lymphocytes and macrophages. So the arrays showed alterations in a great number of genes related to inflammation and immune responses. Genes coding for the macrophage cell surface marker CD68 and lymphocytes antigens CD38 and CD 52 had their expression increased, a finding compatible with the presence of these cells in the inflammatory infiltrate. The expression of genes coding for adhesion molecules, such as galectin-3, P-selectin ligand (CD162), integrin β3 (CD61), and ICAM-1 (CD54), was increased in hearts of chagasic mice. Cytokine-associated genes were differentially expressed in hearts of chagasic mice like IFNγ and TNF-α. Another characteristic of hearts in chronically chagasic mice is fibrosis. The results showed up-regulation of genes related to synthesis of extracellular matrix components and a increased expression of lysyl oxidase, an enzyme that promotes the cross-linking of collagen fibers. The tissue inhibitor of metalloproteinase 1 (TIMP-1), an inhibitor of collagen degradation, was also up-regulated in chronic chagasic hearts. Bilate et al. have performed a proteomic analysis in hearts of acutely *T. cruzi* infected Syrian hamsters and have shown that severe acute infection is associated to differential expression of structural/contractile and stress response proteins that may be associated with alterations in the cardiomyocyte cytoskeleton (Bilate et al. 2008).

16.7   Gene Expression Profiling Using CCC Patient Heart Samples

In 2005, Cunha-Neto et al. showed the first gene expression profiling study in human heart samples from Chagas patients and controls, obtained at transplantation (Cunha-Neto et al. 2005). They used a 10,386-element cDNA microarray, built from cardiovascular cDNA libraries, in combination with real-time reverse transcriptase polymerase chain reaction analysis to compare the gene expression fingerprint of five patients with CCC (serological diagnosis, positive epidemiology), seven with DCM (dilated cardiomyopathy in the absence of ischemic disease, and negative epidemiology) and four normal adult heart tissue (obtained from four non failing donor hearts not used for cardiac transplantation due to size mismatch with available recipients). They found that gene expression patterns are markedly different in CCC and DCM, with significant activity of IFN-inducible genes in CCC patients. Indeed, it showed that immune response, lipid metabolism, and mitochondrial oxidative phosphorylation genes were selectively up-regulated in myocardial tissue of the tested Chagas' cardiomyopathy patients. Interferon (IFN)-γ-inducible genes represented 15% of genes specifically up-regulated in Chagas' cardiomyopathy myocardial tissue, indicating the importance of IFN-γ signaling also in the human model. They also tested whether IFN-γ can directly modulate cardio-myocyte gene expression by exposing fetal murine cardiomyocytes to IFN-γ and the IFN-γ-inducible chemokine monocyte chemoattractant protein-1. Atrial natriuretic factor expression increased 15-fold in response to IFN-γ whereas combined IFN-γ and monocyte chemoattractant protein-1 increased atrial natriuretic factor expression 400-fold. The authors concluded that IFN-γ and chemokine signaling may directly up-regulate cardiomyocyte expression of genes involved in pathological hypertrophy, which may lead to heart failure. Another important result was similar to what was observed in the gene expression analysis in the murine model of *T. cruzi* infection: They saw that IFN-γ and *T. cruzi* infection can depress energy metabolism, thus reducing myocardial ATP generation, which has potential consequences for myocardial contractility, electric conduction and rhythm.

16.8   Important Factors That Influence Gene Expression in CCC: Genetic Polymorphisms and MicroRNAs

16.8.1   Genetic Polymorphisms

Various studies have attempted to identify the factors that cause CCC to develop in only a fraction of the population exposed to parasites. Much attention has been given to the environment because parasite transmission depends markedly on environmental factors including vector density, vector distribution, and parasite virulence. However, host genetic polymorphisms are also factors that determine infection and disease phenotypes. Familial aggregation of CCC has been described, suggesting that there might be a genetic component to disease susceptibility. This is also supported by the fact that only one third of *T. cruzi*-infected individuals develop CCC. Despite these difficulties, major advances have been made in the field of genetics of some of the parasitic diseases, and provided important insights into the mechanisms of pathogenesis. Genetic polymorphisms can determine specific gene expression profiles and activate specific disease pathways which are fundamental factors to the increased aggressiveness of CCC. So far, several susceptibility loci were mapped on the human genome and some susceptibility genes were identified. It is likely that these studies will yield extremely useful information for drug and vaccine development. Studies has been performed in Chagas disease in order to identify single- nucleotide polymorphisms showing differential rates of *T. cruzi* infection and disease progression correlating susceptibility to infection (presence of antibody to *T. cruzi*) with genetic factors. Moreover, previous case-control studies have already identified several genes associated to human susceptibility to CCC. These studies have compared polymorphism frequencies in patients with CCC and ASY individuals. Due to the obvious importance of the Th1 T cell-rich myocarditis in the pathogenesis of CCC, the focus has been on genes involved in the innate and adaptive immune responses However, these studies were usually small and led to conflicting results when populations of different ethnicity were studied. Deng et al., in 2013 performed a GWAS study in Chagas disease, obtaining genome-wide genotypes for 580 Chagas seropositives e donors and cases from Brazil (Deng et al. 2013). More than 675,000 SNPs were directly genotyped. They detected important SNPs associated with cardiomyopathy.One of them is the SNP rs4149018 located in the SLCO1B1 gene, a membrane transporter that belongs to a solute carrier family and plays a role in drug metabolism. It is expressed in the liver, brain, heart and kidney, and transports organic anions, such as digoxin, bilirubin, methothrexate and statins. Another important SNP detected was one located in the gene HSPB8, a small heat shock protein whose heart specific overexpression in transgenic mice induces myocardial hypertrophy.

16.8.2   MicroRNAs

Small, non-coding RNA, known as microRNA (miRNA), play a key role in determining which genes are expressed. MiRNA regulate tissue-specific protein expression and are involved in virtually all cellular processes; up to one-third of mammalian mRNAs are susceptible to miRNA-mediated regulation. So far, 2042 human miRNAs have been registered at miRBase in release 19.0 (http://microrna. sanger.ac.uk/). MiRNAs bind to partially complementary sequences present in the 3' untranslated regions (UTR) of specific "target" mRNA. This pairing between the miRNA and its target mRNA leads to cleavage of the target mRNA or translation inhibition, resulting in silencing of gene expression. It has been shown that miRNA are determinants of the physiology and pathophysiology of the cardiovascular system and altered expression of muscle- and/or cardiac-specific miRNAs such as the miRNAs named miR-1, miR-208 and miR-133 in myocardial tissue is involved in heart development and cardiovascular diseases (CD), including myocardial hypertrophy, heart failure and fibrosis (Bostjancic et al. 2010; Chen et al. 2006; Divakaran 2010; Oliveira-Carvalho et al. 2012). Several targets of these three miRNAs are related to CD, among them RhoA and Thrap1, which are involved in cardiac hypertrophy, and connective tissue growth factor (CTGF), related to the development of fibrosis and cardiac remodeling. In 2014, Ferreira et al. published the first description of miRNA expression dysregulation in diseased myocardium of CCC patients (Ferreira et al. 2014). The most important finding was that five muscle specific miRNAs, miR-1, miR-133a-2, miR-133b, and the myocardial-specific miR-208a and miR-208b were downregulated in CCC myocardium as compared to control myocardium. Importantly, this study identified putative targets of the differentially expressed microRNA using an *in silico* analysis. They identified 2226 mRNA transcripts as putative targets of these five miRNAs tested, of which 221 had already been experimentally validated as targets. In order to have a preliminary assessment whether myocardial expression patterns of the 5 miRNAs was associated with concordant (i.e.inverse) expression of target miRNAs in the same tissue, they tested mRNA target matches from the gene expression microarray profiling done by Cunha-neto et al. in 2005. Among 91 mRNAs whose expression was upregulated in CCC myocardium, 11 were targets of the concordantly downregulated miRNAs tested; they also found 3 mRNA targets that were upregulated only in DCM (out of 47) and also 3 target mRNAs up-regulated simultaneously both in CCC and DCM (out of 31 genes). From the gene targets regulated by theses miRNAs there are one transcription factor, i.e., the inflammatory transcription factor and a known mediator in cardiac dysfunction, NF-κB and protein kinases, i.e., mitogen-activated protein kinases (MAPK) including p38MAPK, ERK1/2, c-Jun N-terminal kinases (JNK), Phosphatidylinositide 3-kinases (PI3 K), and the Protein Kinase B (AKT), enzymes that play important roles in signaling pathways leading to cardiac hypertrophy. Another important gene, direct target of miR-1 is Cyclin D 1 (CDND1). This protein, along with other D-type cyclins (D2 and D3) is a positive cell cycle regulator that plays an important role in controlling proliferation of cardiomyocytes during normal heart development. Importantly, the expression of D-type cyclins is generally low in the adult heart and is increased in the diseased heart, where their upregulation may promote cardiac hypertrophy instead of cell proliferation (Hotchkiss et al. 2012). Accordingly, a previous study has shown that CCND1 expression is upregulated during *T. cruzi* acute infection in mice and that the expression of CDND1 and other types of cyclins like A1, B1 and E1 are increased in heart lysates of mice acutely infected with *T. cruzi* compared with uninfected controls (Nagajyothi et al. 2006). The study showed that miR-1 controlled CDND1 might also be a key element in CCC.

Even with all this information generated little progress was made to determine efficient markers that can predict the population most at risk to develop CCC. Nowadays, studies based on genome-wide approaches such as microarrays and RNA sequencing technologies have been markedly improved. The measurement of gene expression continues to be a critical tool employed across drug discovery, life science research. The study of genes differentially expressed during CCC will be improved and different insights linking the information encoded in the genome, and expressed in the transcriptome of the disease phenotype will be achieved in the near future.

References

Abel LC, Rizzo LV, Ianni B, Albuquerque F, Bacal F, Carrara D, Bocchi EA, Teixeira HC, Mady C, Kalil J, Cunha-Neto E (2001) Chronic Chagas' disease cardiomyopathy patients display an increased IFN-gamma response to Trypanosoma cruzi infection. J Autoimmun 17(1):99–107. doi:10.1006/jaut.2001.0523

Bilate AM, Cunha-Neto E (2008) Chagas disease cardiomyopathy: current concepts of an old disease. Rev Inst Med Trop Sao Paulo 50(2):67–74

Bilate AM, Teixeira PC, Ribeiro SP, Brito T, Silva AM, Russo M, Kalil J, Cunha-Neto E (2008) Distinct outcomes of Trypanosoma cruzi infection in hamsters are related to myocardial parasitism, cytokine/chemokine gene expression, and protein expression profile. J Infect Dis 198(4):614–623. doi:10.1086/590347

Bostjancic E, Zidar N, Stajner D, Glavac D (2010) MicroRNA miR-1 is up-regulated in remote myocardium in patients with myocardial infarction. Folia Biol (Praha) 56(1):27–31

Campos MA, Closel M, Valente EP, Cardoso JE, Akira S, Alvarez-Leite JI, Ropert C, Gazzinelli RT (2004) Impaired production of proinflammatory cytokines and host resistance to acute infection with Trypanosoma cruzi in mice lacking functional myeloid differentiation factor 88. J Immunol 172(3):1711–1718

Chen JF, Mandel EM, Thomson JM, Wu Q, Callis TE, Hammond SM, Conlon FL, Wang DZ (2006) The role of microRNA-1 and microRNA-133 in skeletal muscle proliferation and differentiation. Nat Genet 38(2):228–233. doi:10.1038/ng1725

Committee WHOE (2002) Control of Chagas disease. World Health Organ Tech Rep Ser 905:i–vi, 1–109 (back cover)

Coura JR, Borges-Pereira J (2010) Chagas disease: 100 years after its discovery. A systemic review. Acta Trop 115(1–2):5–13. doi:10.1016/j.actatropica.2010.03.008

Cunha-Neto E, Dzau VJ, Allen PD, Stamatiou D, Benvenutti L, Higuchi ML, Koyama NS, Silva JS, Kalil J, Liew CC (2005) Cardiac gene expression profiling provides evidence for cytokinopathy as a molecular mechanism in Chagas' disease cardiomyopathy. Am J Pathol 167(2):305–313. doi:10.1016/S0002-9440(10)62976-8

Delaporte F (1997) Romana's sign. J Hist Biol 30(3):357–366

Deng X, Sabino EC, Cunha-Neto E, Ribeiro AL, Ianni B, Mady C, Busch MP, Seielstad M, International RCSGftNREDS-IC (2013) Genome wide association study (GWAS) of Chagas cardiomyopathy in Trypanosoma cruzi seropositive subjects. PLoS One 8(11):e79629. doi:10.1371/journal.pone.0079629

de Souza W de Carvalho TM Barrias ES (2010) Review on Trypanosoma cruzi: host cell interaction. Int J Cell Biol pii:295394. doi:10.1155/2010/295394

Dias JC (1997) Cecilio Romana, Romana's sign and Chagas' disease. Rev Soc Bras Med Trop 30(5):407–413

Dias JC (2009) Elimination of Chagas disease transmission: perspectives. Mem Inst Oswaldo Cruz 104(Suppl 1):41–45

Diaz JH (2007) Chagas disease in the United States: a cause for concern in Louisiana? J La State Med Soc159(1):21–23, 25–29

Divakaran VG (2010) MicroRNAs miR-1, -133 and -208: same faces, new roles. Cardiology 115(3):172–173. doi:10.1159/000272540

Ferreira LR, Frade AF, Santos RH, Teixeira PC, Baron MA, Navarro IC, Benvenuti LA, Fiorelli AI, Bocchi EA, Stolf NA, Chevillard C, Kalil J, Cunha-Neto E (2014) MicroRNAs miR-1, miR-133a, miR-133b, miR-208a and miR-208b are dysregulated in Chronic Chagas disease Cardiomyopathy. Int J Cardiol 175(3):400–417. doi:10.1016/j.ijcard.2014.05.019

Garg N, Popov VL, Papaconstantinou J (2003) Profiling gene transcription reveals a deficiency of mitochondrial oxidative phosphorylation in Trypanosoma cruzi-infected murine hearts: implications in chagasic myocarditis development. Biochim Biophys Acta 1638(2):106–120

Gazzinelli RT, Denkers EY (2006) Protozoan encounters with Toll-like receptor signalling pathways: implications for host parasitism. Nat Rev Immunol 6(12):895–906. doi:10.1038/nri1978

Goldenberg RC, Iacobas DA, Iacobas S, Rocha LL, da Silva de Azevedo Fortes F, Vairo L, Nagajyothi F, Campos de Carvalho AC, Tanowitz HB, Spray DC (2009) Transcriptomic alterations in Trypanosoma cruzi-infected cardiac myocytes. Microbes Infect 11(14–15):1140–1149. doi:10.1016/j.micinf.2009.08.009

Golgher D, Gazzinelli RT (2004) Innate and acquired immunity in the pathogenesis of Chagas disease. Autoimmunity 37(5):399–409. doi:10.1080/08916930410001713115

Hotchkiss A, Robinson J, MacLean J, Feridooni T, Wafa K, Pasumarthi KB (2012) Role of D-type cyclins in heart development and disease. Can J Physiol Pharmacol 90(9):1197–1207. doi:10.1139/y2012-037

Hotez PJ, Dumonteil E, Woc-Colburn L, Serpa JA, Bezek S, Edwards MS, Hallmark CJ, Musselwhite LW, Flink BJ, Bottazzi ME (2012) Chagas disease: "the new HIV/AIDS of the Americas". PLoS Negl Trop Dis 6(5):e1498. doi:10.1371/journal.pntd.0001498

Junqueira C, Caetano B, Bartholomeu DC, Melo MB, Ropert C, Rodrigues MM, Gazzinelli RT (2010) The endless race between Trypanosoma cruzi and host immunity: lessons for and beyond Chagas disease. Expert Rev Mol Med 12:e29. doi:10.1017/S1462399410001560

Kropf SP (2011) Carlos Chagas: science, health, and national debate in Brazil. Lancet 377(9779):1740–1741

Kropf SP, Sa MR (2009) The discovery of Trypanosoma cruzi and Chagas disease (1908–1909): tropical medicine in Brazil. Hist Cienc Saude Manguinhos 16(Suppl 1):13–34

Lescure FX, Le Loup G, Freilij H, Develoux M, Paris L, Brutus L, Pialoux G (2010) Chagas disease: changes in knowledge and management. Lancet Infect Dis 10(8):556–570. doi:10.1016/S1473-3099(10)70098-0

Machado FS, Jelicks LA, Kirchhoff LV, Shirani J, Nagajyothi F, Mukherjee S, Nelson R, Coyle CM, Spray DC, de Carvalho AC, Guan F, Prado CM, Lisanti MP, Weiss LM, Montgomery SP, Tanowitz HB (2012) Chagas heart disease: report on recent developments. Cardiol Rev 20(2):53–65. doi:10.1097/CRD.0b013e31823efde2

Moncayo A (2010) Carlos Chagas: biographical sketch. Acta Trop 115(1–2):1–4. doi:10.1016/j.actatropica.2009.10.022

Moncayo A, Ortiz Yanine MI (2006) An update on Chagas disease (human American trypanoso-miasis). Ann Trop Med Parasitol 100(8):663–677. doi:10.1179/136485906X112248

Mukherjee S, Belbin TJ, Spray DC, Iacobas DA, Weiss LM, Kitsis RN, Wittner M, Jelicks LA, Scherer PE, Ding A, Tanowitz HB (2003) Microarray analysis of changes in gene expres-sion in a murine model of chronic chagasic cardiomyopathy. Parasitol Res 91(3):187–196. doi:10.1007/s00436-003-0937-z

Nagajyothi F, Desruisseaux M, Bouzahzah B, Weiss LM, Andrade Ddos S, Factor SM, Scherer PE, Albanese C, Lisanti MP, Tanowitz HB (2006) Cyclin and caveolin expression in an acute model of murine Chagasic myocarditis. Cell Cycle 5(1):107–112

Nogueira LG, Santos RH, Ianni BM, Fiorelli AI, Mairena EC, Benvenuti LA, Frade A, Donadi E, Dias F, Saba B, Wang HT, Fragata A, Sampaio M, Hirata MH, Buck P, Mady C, Bocchi EA, Stolf NA, Kalil J, Cunha-Neto E (2012) Myocardial chemokine expression and intensity of myocarditis in Chagas cardiomyopathy are controlled by polymorphisms in CXCL9 and CXCL10. PLoS Negl Trop Dis 6(10):e1867. doi:10.1371/journal.pntd.0001867

Oliveira-Carvalho V, Carvalho VO, Silva MM, Guimaraes GV, Bocchi EA (2012) MicroRNAs: a new paradigm in the treatment and diagnosis of heart failure? Arq Bras Cardiol 98(4):362–369

Pays JF (2009) Chagas Carlos Justiniano Ribeiro (1879–1934). Bull Soc Pathol Exot 102(5):276–279

Roveda JM (1967) Romana's sign. Cole's unilateral trypanosomiasic ophthalmia. Arch Oftalmol B Aires 42(1):1–4

Siqueira-Batista R, Gomes AP, Rocas G, Cotta RM, Rubiao EC, Pissinatti A (2011) Chagas's dis-ease and deep ecology: the anti-vectorial fight in question. Cien Saude Colet 16(2):677–687

Soares MB, de Lima RS, Rocha LL, Vasconcelos JF, Rogatto SR, dos Santos RR, Iacobas S, Goldenberg RC, Iacobas DA, Tanowitz HB, de Carvalho AC, Spray DC (2010) Gene expres-sion changes associated with myocarditis and fibrosis in hearts of mice with chronic chagasic cardiomyopathy. J Infect Dis 202(3):416–426. doi:10.1086/653481

Vaena de Avalos S Blader IJ Fisher M Boothroyd JC Burleigh BA (2002) Immediate/early re-sponse to Trypanosoma cruzi infection involves minimal modulation of host cell transcription. J Biol Chem 277(1):639–644. doi:10.1074/jbc.M109037200

Chapter 17
Expression Tests in Actual Clinical Practice: How Medically Useful is the Transcriptome?

Bertrand R. Jordan

Abstract Expression profiling has proved itself as a discovery tool, and has generated great expectations for use in molecular diagnostics. Microarray technology, available in robust industrial-strength implementation since the late 1990s, appeared well adapted to assess properties of tumours through expression profiling, and received much attention in the last decade. It was expected to provide prognostic information (how is the condition likely to develop) as well as predictive indications on the therapy most likely to succeed. However, requirements for a clinical test are quite different from those applying to a research tool. The test must demonstrate *analytical validity* (technical quality of the assay, reproducibility, robustness…) as well as *clinical validity,* strong correlation between the result of the test and clinical outcomes such as progression-free survival. The test also needs approval by regulatory authorities, and proven *clinical utility*: demonstrated improvement of the outcome for the patient in terms of survival, or reduction of toxic side effects. Cost considerations also enter the picture, since these are expensive tests. Thus the road from a scientific result to a successful diagnostic tool is long and arduous, and the number of expression signatures actually used in medical practice is limited. Different approaches can be implemented: for small sets of genes, RT-PCR is the method of choice. For larger numbers, hundreds or thousands of genes, microarrays are the preferred tools. In addition, approaches based on new-generation sequencing will undoubtedly play a role in the future.

17.1 Introduction

17.1.1 A Very Successful Research Tool

Expression profiling has abundantly proved itself as a discovery tool, to delineate genes potentially involved in a process by virtue of their differential expression and

B. R. Jordan (✉)
Marseille Medical School, Aix-Marseille Université/EFS/CNRS, UMR 7268 ADES,
Marseille, France
e-mail: Bertrand.jordan@univ-amu.fr

© Springer International Publishing Switzerland 2014
G. A. Passos (ed.), *Transcriptomics in Health and Disease,*
DOI 10.1007/978-3-319-11985-4_17

to investigate the mechanisms at work in physiological or pathological situations. The power of this technology has also generated great expectations concerning its use as a molecular diagnostic tool: an expression profile can be expected to provide important information on the aggressiveness of a given tumour or on the state of a patient's immune system, and the microarray technology, available in robust industrial-strength implementation since the late 1990s, appeared very well adapted to assess this at the level of a few tens or hundreds of genes. Thus the use of expression signatures as clinically valuable diagnostic tools has been the subject of much attention since the beginning of the current century, with the hope of providing significant medical information on the prognosis (how is the condition likely to develop) as well as prediction on which therapy is most likely to be effective. To take one of the best known examples, expression analysis of resected tissue from a breast tumour can be expected to indicate how aggressive this particular tumour is (prognostic information) as well as whether or not chemotherapy is likely to be necessary (predictive information).

17.1.2 Specific Requirements for Clinical Tests

However, the requirements for a clinical test are quite different from those applying to a research tool. A test based on measurement of an expression signature must demonstrate *analytical validity* (technical quality of the assay, reproducibility, robustness...) as well as *clinical validity,* that is, conclusive demonstration of the correlation between the result of the test and clinical outcomes such as progression-free survival or objective response to treatment. To be approved in most regulatory systems (such as the Food and Drug Administration (FDA) in the USA), and to be actually used in practice, the test must also meet the more difficult requirement of *clinical utility* (Simon 2008). It must be shown that its use in clinical practice actually improves the outcome for the patient, providing a better cure rate, longer survival, and/or a reduction of toxic side effects. This demonstration usually necessitates a prospective clinical trial, a long and expensive undertaking. Cost considerations also enter the picture, since these are expensive tests and there has to be some kind of economic justification to their use. Thus the road from a scientific result to a widely used molecular diagnostic tool is long and arduous (Koscielny 2010), and accordingly, the number of expression signatures actually used in medical practice is limited (Jordan 2010, for a list of tests used in oncology see Raman et al. 2013).

17.1.3 Different Technical Approaches

In terms of technology, three different approaches can be used to obtain an expression profile: if the number of genes to be assessed is limited, RT-PCR, an industry gold standard, is the method of choice. For larger sets, hundreds or thousands of genes, microarrays are the preferred tools. Finally, sequencing approaches based on new-generation sequencing represent a promising newcomer: they have not yet achieved much penetration in this application, but they will undoubtedly play an increasingly important role. In the following parts I will use examples from each of these categories to show the issues involved and then conclude on the outlook for the future.

17.2   A Successful Low-Complexity Expression test: Oncotype Dx—and Others

17.2.1   An Unmet Need in Breast Cancer Management

This example, as well as the following one, concerns diagnostic tools used in breast cancer management (Arango et al. 2013; Kittaneh et al. 2013). Breast cancer is a frequent affection (lifetime risk ~13% for women), fairly serious (5-years survival rate of the order of 85%, depending on subtype and stage), and in need of better prognostic and predictive biomarkers. One important issue, in the case of early-stage breast cancer[1] is whether or not to perform chemotherapy after resection of the tumour. Retrospective studies show that chemotherapy improves (moderately) the 5-years survival rate of women that undergo it, at the expense of fairly severe side effects; they also demonstrate that this is only effective in approx. 25% of the patients, so that most are treated unnecessarily. Thus a test that would provide strong prognostic information could be used to avoid chemotherapy for those women whose prognosis is very good and, if it was predictive of the effect of the treatment, could lead to treat only those in which chemotherapy can be expected to be useful.

17.2.2   An RT-PCR Test that Works with Fixed Samples

A number of laboratories have accordingly studied expression profiles in breast tumour samples, hoping to derive expression signatures with prognostic and possibly predictive values—in fact, more than 50 expression signatures have been published for breast cancer (see Venet et al. 2011 for a list), and several of them are offered as clinical tests (Raman et al. 2013). The Oncotype Dx breast cancer test marketed by the company Genomic Health is based on the study of 250 candidate genes in samples from 447 patients, resulting in the selection of a set of 16 genes (plus 5 reference genes) whose expression levels combined according to an algebraic formula define a "recurrence score" for each sample (Paik et al. 2004). This signature demonstrated clinical validity in an additional set of 668 tumours and allowed classification of patients into low-risk (51% of patients, recurrence rate 6.8%), intermediate risk (22%, 14.3% recurrence rate) and high-risk (27%, 30.5%

---

[1] Defined as a small tumour (<2 cm) with no axillary node involvement.

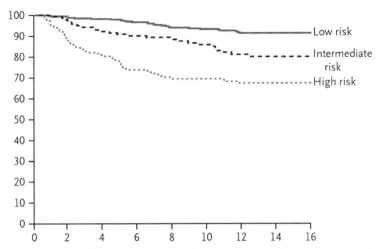

Fig. 17.1 Performance of the Oncotype test developed by Genomic Health: percentage of patients remaining free from distant recurrence of breast cancer (ordinate) as a function of time post-treatment (years, abscissa) for patients whose recurrence score according to the test is less than 18 (*low risk*), between 18 and 31 (*intermediate risk*) and above 31 (*high risk*). The difference between the groups is significant at $p < 0.001$. (Data from Paik et al. 2004)

recurrence rate) (Fig. 17.1). It is important to note that expression was measured by real-time RT-PCR and that the method had been adapted to work on formalin-fixed, paraffin-embedded ("FFPE") tumour tissue. This allowed the authors to use FFPE samples for retrospective studies, a very significant advantage since this is by far the most widely used storage method for tumour samples—only a small fraction are preserved as fresh-frozen tissue, especially in the USA.

17.2.3 Successful Clinical and Commercial Development of Oncotype Dx

The test has been marketed by Genomic Health since 2004 as a "laboratory-developed test" (i.e. performed by the company on samples sent to it, not marketed in kit form), thus not requiring formal FDA registration, although the laboratory itself had to be certified. It has been shown to be predictive for the effect of chemotherapy in low-risk patients, allowing avoidance of this treatment for that category. It has also been shown to have prognostic value in some categories of more advanced breast cancer. An extensive prospective clinical trial called TAILORx (Trial Assigning IndividuaLized Options for Treatment (Rx)) is underway to more precisely evaluate its predictive value in several situations, with final results expected in 2017 (Zujewski et al. 2008).

In spite of its cost (over US$ 3000), Oncotype has been quite successful, with annual sales now approaching 100,000 per year. It is actually, by far, the most successful gene expression test. This is because it answers a real medical need (avoiding unnecessary chemotherapy) in a frequent type of cancer. It was also first on the market for this application, and the use of samples embedded in paraffin blocks (FFPE, Formalin-Fixed Paraffin-Embedded) enabled early validation on retrospective samples and facilitates performance of the test in the normal clinical environment. The test has not been approved by the FDA, but since it is a "laboratory-developed test" (LDT) performed in-house by the company on samples received from hospitals, it does not need such registration, although the facilities had to be approved under the Clinical Laboratory Improvement Amendments (CLIA) regulations. Specialised expert panels such as, in 2011, the 12th St Gallen International Breast Cancer Conference Expert Panel have agreed that Oncotype DX could be used where available to predict chemotherapy responsiveness in certain situations (Goldhirsch et al. 2011; Goldhirsch et al. 2013). According to the company, essentially all insurance companies in the USA have agreed to reimburse the test, resulting in a covered population exceeding 200 million. Interestingly, the National Institute for Health and Care Excellence (NICE) in the UK, an official body charged with evaluating tests for their cost-to-effectiveness ratio before inclusion in the national health service, agreed to approve Oncotype only after its price was reduced by a considerable (but undisclosed) amount. This was necessary as NICE requires treatment to cost no more than £ 20,000–30,000 per "quality-adjusted life year" ("QUALY") saved—an illustration of how cost considerations can be taken into account for such expensive tests…

17.2.4 Other Low-Complexity (RT-PCR) Expression Tests

A number of other expression tests using RT-PCR technology to assess a relatively small number of genes are marketed for clinical use, although none of them has (so far) achieved really wide acceptance. In the breast cancer area, the PAM50 test uses expression values from 50 genes to classify breast cancer into four subtypes that respond differently to standard therapies (Parker at al. 2009), and is marketed as ProSigna by Nanostring (Seattle, USA). Oncotype Dx prostate, another expression assay from Genomic Health, combines the expression of 17 genes measured on needle biopsies with other clinical variables to suggest the best treatment for prostate cancer (Knezevic et al. 2013).

A whole category of tests is still under development, but shows good promise: profiling microRNA (miRNA) expression. MiRNAs are very much involved in gene regulation and expression, and show differential expression in cancer cells (Lorio and Croce 2009). They are being developed as diagnostic and prognostic tools, and it seems that small sets of miRNAs (assessed by PCR methods) have good predictive value for several types of cancer affecting lung, ovaries or pancreas (Schultz et al. 2014). Corresponding tests should soon be available for clinical purposes.

In all cases the expression tests performed by RT-PCR can be performed on FFPE samples, and benefit both from the recognition of RT-PCR as the gold standard and from the availability of many fixed samples for retrospective studies. This has certainly played an important part in the success of Oncotype Dx Breast, together with its "first-player" status and with the avoidance of lengthy FDA approval procedures thanks to its performance as a laboratory-developed test (LDT).

17.3   An Array-Based Expression Test: Mammaprint (and Some Others)

17.3.1   An Early, High-Impact Study

The aim of the Mammaprint test is similar to that of Oncotype: obtaining prognostic information in the case of early breast cancer, with the objective of avoiding unnecessary chemotherapy. It is based on some of the most significant early papers on expression profiling in oncology, published in 2002 (van 't Veer et al. 2002; van de Vijver et al. 2002). This was one of the first studies using oligonucleotide arrays representing the whole set of human genes (manufactured at the time by Rosetta Inpharmatics, Kirkland, USA), done at the Netherlands Cancer Institute on an extensive set of fresh-frozen samples from patient tumours. It delineated a set of 70 genes that gave good discrimination between "good prognosis" and "bad prognosis" patients (Fig. 17.2). The test was later validated in an independent set of

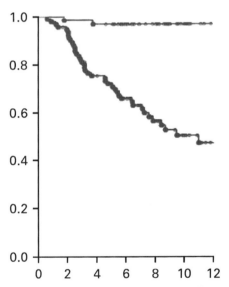

Fig. 17.2 Performance of the Mammaprint test developed by Agendia: Overall survival (ordinate) for lymph-node negative breast cancer patients as a function of years after treatment (abscissa). *Top* curve, patients with a "good prognosis" expression signature; *bottom* curve, patients with a "poor prognosis" expression signature (60 and 91 patients initially). (Data from Fig. 2D of van de Vijver et al. 2002)

patients, and a company called Agendia (Amsterdam, the Netherlands), was spun off the academic centre in 2003 to commercialize the test.

17.3.2 A Long Path to Wide Commercialization

The test, using a custom microarray manufactured by Agilent, was launched in Europe in 2004, and it was later incorporated into a long-term prospective study called MINDACT (Microarray In Node-negative and 1–3 positive lymph node Disease may Avoid ChemoTherapy), that aims to find out whether or not the test has clinical utility—this clinical trial is still ongoing (Cardoso et al. 2007). Meanwhile, Agendia began commercialisation of the test in the USA and was soon ordered by the FDA to obtain clearance from that agency (although the test was technically an LDT, being performed at the Agendia laboratories on frozen samples sent to them). The FDA developed a new formalism for such tests, called IVDMIA (*In Vitro* Diagnostic Multivariate Index Assay), and meant to apply to complex tests whose interpretation depends on non-transparent, proprietary bioinformatics algorithms that cannot be independently derived or verified by the end user. Mammaprint was finally cleared by FDA in 2007 and marketed in the USA, at a price similar to that of Oncotype (4,200 USD). The test has been recognized as providing prognostic (Goldhirsch et al. 2011) and possibly predictive value (Goldhirsch et al. 2013). Since Agendia is a privately held company (in contrast to Genomic Health, listed on Nasdaq and therefore providing financial reports), sales figures are not available, but the uptake seems to have been slow, and the number of tests sold per year is probably in the low thousands, much less than for Oncotype. From 2011 on, Agendia has been able to accept fixed (FFPE) samples (Mittempergher et al. 2011), with a significant impact on sales that have more than doubled year over year. In the US, a number of insurance companies have agreed to reimburse the Mammaprint test, with a coverage claimed by Agendia to represent most of the potential patients.

17.3.3 Other Microarray-Based Expression Tests

Other companies have developed expression tests with clinical validity and in some cases have obtained FDA approval. The TOO test developed by Pathwork Diagnostics (Redwood City, USA) is an interesting example. This aims to define the tissue of origin for cancer that is discovered only once it has metastasized to several locations: knowing the organ in which the original tumour appeared can have important implications for therapy. Accordingly, the test examines a 1500-gene expression profile using an Affymetrix gene chip, and determines its similarity with expression profiles for 15 tissues or organs (Dumur et al. 2008). Results are fairly clear-cut, as the original expression profiles are largely conserved in metastatic cells (Fig. 17.3). The test was commercially launched in 2006, as a laboratory-developed test using fresh-frozen samples. It received FDA approval in 2008, and a version using fixed

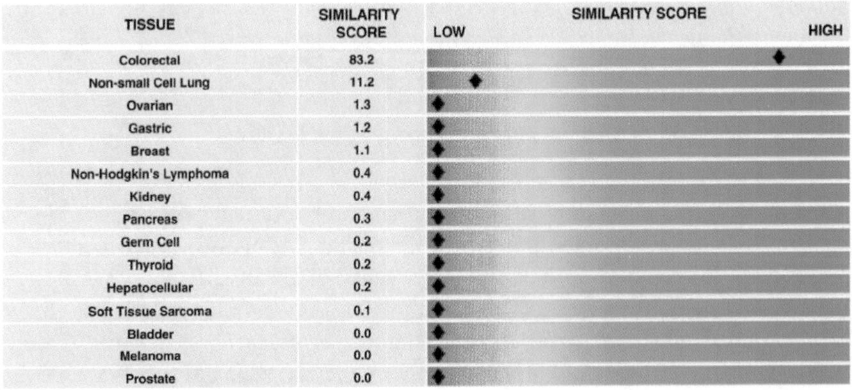

TISSUE	SIMILARITY SCORE	SIMILARITY SCORE LOW	HIGH
Colorectal	83.2		◆
Non-small Cell Lung	11.2	◆	
Ovarian	1.3	◆	
Gastric	1.2	◆	
Breast	1.1	◆	
Non-Hodgkin's Lymphoma	0.4	◆	
Kidney	0.4	◆	
Pancreas	0.3	◆	
Germ Cell	0.2	◆	
Thyroid	0.2	◆	
Hepatocellular	0.2	◆	
Soft Tissue Sarcoma	0.1	◆	
Bladder	0.0	◆	
Melanoma	0.0	◆	
Prostate	0.0	◆	

Fig. 17.3 Performance of the Tumor Of Origin (TOO) test developed by Pathwork diagnostics. The *diamonds* show the similarity score between the patient's (secondary) tumour and 15 tumour types, and clearly point to colorectal cancer as the tumour of origin in the case presented. Adapted from the company's Web site, accessed May 2012 (no longer active as the company has failed)

samples was subsequently approved in 2010. A large validation study including 1100 samples was conducted, and coverage by Medicare was obtained in 2012. However, after spending more that 50 million USD on test development over almost a decade, the company went out of business in April 2013. This illustrates the "staying power" needed to achieve successful marketing of a complex test; it may also reflect the fact that the origin of a tumour is now considered less important than was believed 10 years ago[2].

Another clinical test based on RNA profiles is marketed by the Dutch company Skyline Dx (Rotterdam, the Netherlands) and is aimed at characterising subtypes of acute myeloid leukaemia. Using an Affymetrix array and analysing RNA from a bone marrow sample, it assays for the presence of several specific translocations (through presence or absence of mRNA from the corresponding fusion genes) while measuring expression levels for several other genes. Called the AML profiler, this test is CE-marked in Europe and awaiting FDA approval for the US.

Additional array-based expression test worth mentioning are Coloprint (from Agendia), assessing an 18-gene expression profile to indicate the prognosis of colon cancer, myPRS from Signal Genetics (New York, USA) that assesses 700 genes on an Affymetrix array to provide risk stratification of multiple myeloma patients… and quite a few others (Raman et al. 2013), none of which however have achieved the visibility and commercial success of Oncotype or Mammaprint.

---

[2] The traditional, organ-based classification of cancers tends to be replaced by characterization in terms of mutations present in the tumor cells.

17.4 The (Limited?) Promise of RNA-Seq

17.4.1 Profiles by RNA-Seq in Research and in the Clinic

For research applications, sequence-based approaches have essentially displaced microarrays: they provide digital information, can reach any required sensitivity by adjusting the sequencing depth, identify novel transcripts, and their cost is now competitive with a microarray experiment (Wang et al. 2009; McGettigan 2013). In terms of clinical applications, however, these advantages are less evident, and indeed no expression test based on RNA-seq technology has yet been commercially offered for clinical use. The sheer amount of data produced by NGS approaches, a definite advantage in research, can be detrimental in the clinical situation; regulatory issues are still somewhat unclear for NGS data, and the need for extensive bioinformatics processing of the results is a serious issue, both in terms of infrastructure and for regulatory purposes.

17.4.2 Tumour Profiling at the DNA (and RNA?) Level

New-Generation Sequencing (NGS) is indeed penetrating in the clinical space, for the purpose of characterising the mutations present in tumours using either sets of relevant genes ("panels" comprising from 50 to several hundred genes), whole exomes (a complete set of coding sequences) or in some cases whole-genome sequencing (Desai and Jere 2012). The aim here is to characterise the tumour by its mutation pattern and to find "actionable mutations", that is, mutations that are associated with an approved or experimental drug targeted at the corresponding protein. Such clinical applications are actively pursued and the various issues raised by their introduction into routine clinical practice are being dealt with—for example, the Illumina MiSeq sequencing system has recently been validated by the FDA for clinical use. In a few cases, the schemes being implemented also include characterization of the tumour by its expression pattern, naturally obtained in this context by RNA-seq (Balko et al. 2013). This may develop in the future, but the complexity of the data and the regulatory issues involved will probably limit the extent to which RNA-seq will become clinically significant.

17.5 Some Important Points Applying to All Expression-Based Clinical Tests

As can be gathered from this chapter, the road to actual clinical implementation of expression profiles (mostly in oncology) has been rather bumpy, with a limited number of successes and many failures. Now that we have described some of the actual implementations, let us review the issues involved.

17.5.1 Validity of Signatures

In the initial period of expression profiling for clinical purposes (done at that time essentially with microarrays), there have been some serious issues with the design of the studies. In experiments profiling a hundred samples on arrays representing 25,000 genes, with relatively noisy data, the opportunity for false discovery is high. This is called the "curse of dimensionality": with many more features than samples, there is a high risk of over fitting the data, i.e. finding classifications that have no foundation and will not hold on a new set of samples. Avoiding this requires rigorous statistical methods and, for example, a complete separation between the "training set" of samples (used to derive a clinically meaningful expression signature) and the "test set" on which validation is performed. This was not always the case in the first publications, where, for example, the test set was sometimes used to choose between alternative models derived using the training set—thus resulting in circular reasoning (Simon et al. 2003; Campbell 2004).

In addition, and even when there is no issue with the statistical methods, the set of genes included in the signature is not unique: alternative choices are possible with essentially equivalent results, as demonstrated in particular for the 70-gene signature used in the Agendia test (Ein-Dor et al. 2005). In fact, signatures based on randomly chosen sets of genes often turn out to have good predictive value for, e.g. breast cancer outcome! (Venet et al. 2011; Jordan 2012). Such profiles are still medically useful, if they conclusively identify groups of patients with clinically different prognosis; however, they cannot be used to draw conclusions on the involvement of particular genes in the properties of the cancer cells. In other words, these profiles may be clinically valid without having any particular scientific significance, and their biological interpretation should be extremely cautious.

17.5.2 Clinical Utility: Essential, but Hard to Prove

Even when the test has shown analytical validity (good experimental performance) and clinical validity (proven correlation between the test result and important clinical features such as duration of metastasis-free survival), its clinical utility remains to be demonstrated (Simon 2008; Koscielny 2010). Fig. 17.4 shows a simple example in which a test appears to have clinical utility according to the classification achieved with the training sample, but loses this potential when the test sample is used (Chen et al. 2007). A real assessment of clinical utility requires a prospective clinical trial, in which the outcome of cancer management with and without the test is ascertained over a period of 5–10 years, in order to determine if the use of the test results in significant improvement for the patient, in terms of disease-free survival and/or reduction of toxic side effects. Economic factors, such as savings due to avoidance of costly chemotherapies are also considered. These studies are very long and expensive, and in fact the first two aimed at Oncotype (TAILORx, Zujewski et al. 2008) and at Mammaprint (MINDACT, Cardoso et al. 2007) are still ongoing at this time.

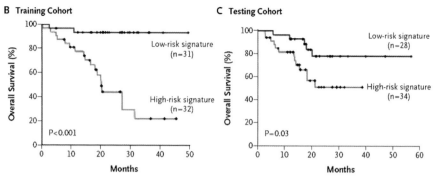

Fig. 17.4 An example of vanishing clinical utility (data from Chen et al. 2007). A five-gene expression signature separates non-small cell lung cancer patients into two groups with different prognosis (overall survival (%) versus time in months). As seen on the training cohort (*left*), the survival for patients with the *low-risk signature* is so good that further treatment could be omitted for these individuals, suggesting clinical utility for the test. However the results obtained on the testing cohort (that was not used to derive the signature, and provides the only valid test of the profile) do not confirm this hope: while the signature still has some clinical validity (predicting a differential prognosis), the survival curve for the low-risk patients is not good enough to forego treatment, thus the test cannot be expected to have clinical utility

17.5.3 Assay Technology: The Comeback of RT-PCR

The major application of DNA microarrays, as introduced in the mid-1990s, was the determination of expression profiles. Extensive clinical applications, notably in oncology, were expected, and sophisticated diagnostic tests were predicted to represent a major field of application for this technology in the 2000s. As outlined above, this has not happened and expression profiling for clinical purposes is currently dominated by quantitative RT-PCR approaches. In fact, tests are quite often developed using microarrays but the final assay is adapted for PCR. This is due to a number of factors: increased sophistication of PCR assay systems, through the introduction of microfluidic devices that make multiplex measurements practical (Ballester et al. 2013), growing reliance on FFPE samples (that are very difficult to profile on microarrays), and familiarity of the RT-PCR method that is recognized as the gold standard for expression and is "FDA-friendly", that is, well recognised by the agency that has defined procedures to deal with tests using this approach. In addition, it is now clear that the value added by including many genes (hundreds or thousands) in the signature is very limited, and that a small set of genes provides essentially equivalent performance (Haibe-Kains et al. 2012)—so that there is no incentive to perform the test in high multiplex mode. Given this, it also seems unlikely that RNA-seq approaches will have a great impact in this field, except possibly as part of extensive, integrated genomic profiling exercises. Expression profiling diagnostics will be dominated by RT-PCR approaches performed on a few tens of genes. Current efforts at performing diagnostic tests in a non-invasive fashion, using blood or urine samples rather than biopsies (Bidard et al. 2013), may eventually apply to expression tests. In fact some of the work currently performed on miRNA profiling is already done on blood samples (Schultz et al. 2014). Another possible avenue is the use of circulating tumour cells found in blood samples of patients (Pierga et al. 2012).

17.5.4 Moving Closer to Functional Information

For all its value, expression profiling is still steps away from actual biological function. It has been widely used, with some significant success, as a proxy for protein abundance and protein activities that could only be measured with difficulty and at low multiplex factors. The situation has however changed, and protein abundance can now be determined at high sensitivity and in multiplex fashion, using either aptamer or antibody arrays, or mass spectrometry (see for example Li et al. 2013). Effective multiplex methods are also now available for some types of enzymatic activities, for example for kinase activities that are extremely important in the context of oncology (Arsenault et al. 2011; Hilhorst et al. 2013). Such approaches can provide invaluable information, for example allowing testing of a drug on patient samples prior to treatment to find out if that particular drug is likely to be effective on that particular individual—a step toward real personalised medicine. Tests based on these methodologies can be expected in the coming years, and they are quite likely to be successful.

17.6 Conclusion

Expression profiling has not proven as medically useful as anticipated in the late 1990s—and its major implementation today involves quantitative RT-PCR, not arrays as initially anticipated. This is due to under-appreciated difficulties in developing tests that have proven clinical utility—part of the general problem with biomarkers, very few of which have actually resulted in useful diagnostics tools (Hayes et al. 2013). It is compounded, in the case of cancer, by the newly recognised genomic complexity of the disease, both from patient to patient and also within a given patient with the realisation of the common genetic heterogeneity of cancer cells that is probably to most serious hindrance to effective targeted treatment. However, the trend towards using more and more genomic information to evaluate prognosis and guide treatment is bound to continue, and gene expression will undoubtedly be an important facet of this "precision medicine" in the future.

References

Arango BA, Rivera CL, Glück S (2013) Gene expression profiling in breast cancer. Am J Transl Res 5:132–138.

Arsenault R, Griebel P, Napper S (2011) Peptide arrays for kinome analysis: new opportunities and remaining challenges. Proteomics 11:4595–4609.

Ballester M, Cordon R, Folch JM (2013) DAG expression: high-throughput gene expression analysis of real-time PCR data using standard curves for relative quantification. PLoS ONE 8:e80385

Balko JM, Giltnane J, Wang K et al (2013) Molecular profiling of the residual disease of triple-negative breast cancers after neoadjuvant chemotherapy identifies actionable therapeutic targets. Cancer Discov 4:232–245

Bidard FC, Weigelt B, Reis-Filho JS (2013) Going with the flow: from circulating tumor cells to DNA. Sci Transl Med 5:207ps14

Campbell G (2004) Some issues in the statistical evaluation of genetic and genomic tests. J Biopharm Stat 14:539–552

Cardoso F, Piccart-Gebhart M, Van 't Veer L et al (2007) The MINDACT trial: the first prospective clinical validation of a genomic tool. Mol Oncol 1:246–251

Chen HY, Yu SL, Chen CH et al (2007) A five-gene signature and clinical outcome in non-small-cell lung cancer. N Engl J Med 356:11–20

Desai AN, Jere A (2012) Next-generation sequencing: ready for the clinics? Clin Genet 81:503–510

Dumur CI, Lyons-Weiler M, Sciulli C et al (2008) Interlaboratory performance of a microarray-based gene expression test to determine tissue of origin in poorly differentiated and undifferentiated cancers. J Mol Diagn 10:67–77

Ein-Dor L, Kela I, Getz G et al (2005) Outcome signature genes in breast cancer: is there a unique set? Bioinformatics 21:171–178

Goldhirsch A, Winer EP, Coates AS et al (2011) Strategies for subtypes—dealing with the diversity of breast cancer: highlights of the St Gallen International Expert Consensus on the Primary Therapy of Early Breast Cancer 2011. Ann Oncol 22:1736–1747

Goldhirsch A, Winer EP, Coates AS et al (2013) Personalizing the treatment of women with early breast cancer: highlights of the St Gallen International Expert Consensus on the Primary Therapy of Early Breast Cancer 2013. Ann Oncol 24:2206–2223

Haibe-Kains B, Desmedt C, Loi S et al (2012). A three-gene model to robustly identify breast cancer molecular subtypes. J Natl Cancer Inst 104:311–325

Hayes DF, Allen J, Compton C et al (2013) Breaking a vicious cycle. Sci Transl Med 5:196cm6

Hilhorst R, Houkes L, Mommersteeg M et al (2013) Peptide microarrays for profiling of serine/threonine kinase activity of recombinant kinases and lysates of cells and tissue samples. Methods Mol Biol 977:259–257

Jordan B (2010) Is there a niche for DNA microarrays in molecular diagnostics? Expert Rev Mol Diagn 10:875–882

Jordan B (2012) Are expression profiles meaningless for cancer studies? Bioessays 34:730–733

Kittaneh M, Montero AJ, Glück S (2013) Molecular profiling for breast cancer: a comprehensive review. Biomark Cancer 5:61–70

Knezevic D, Goddard AD, Natraj N et al (2013) Analytical validation of the Oncotype DX prostate cancer assay—a clinical RT-PCR assay optimized for prostate needle biopsies. BMC Genomics 14:690

Koscielny S (2010) Why most gene expression signatures of tumors have not been useful in the clinic. Sci Transl Med 2:14ps2

Li XJ, Hayward C, Fong PY, Dominguez M et al (2013) A blood-based proteomic classifier for the molecular characterization of pulmonary nodules. Sci Transl Med 5:207ra142

Lorio MV, Croce CM (2009) MicroRNAs in cancer: small molecules with a huge impact. J Clin Oncol 27:5848–5856

McGettigan PA (2013) Transcriptomics in the RNA-seq era. Curr Opin Chem Biol 17:4–11

Mittempergher L, de Ronde JJ, Nieuwland M et al (2011) Gene expression profiles from formalin fixed paraffin embedded breast cancer tissue are largely comparable to fresh frozen matched tissue. PLoS One 6:e17163

Paik S, Shak S, Tang G et al (2004) A multigene assay to predict recurrence of tamoxifen-treated, node-negative breast cancer. N Engl J Med 351:2817–2826

Parker JS, Mullins M, Cheang MC et al (2009) Supervised risk predictor of breast cancer based on intrinsic subtypes. J Clin Oncol 27:1160–1167

Pierga JY, Hajage D, Bachelot T, Delaloge S et al (2012) High independent prognostic and predictive value of circulating tumor cells compared with serum tumor markers in a large prospective trial in first-line chemotherapy for metastatic breast cancer patients. Ann Oncol 23:618–624

Raman G, Avendano EE, Chen M (2013) Update on Emerging Genetic Tests Currently Available for Clinical Use in Common Cancers. Evidence Report/Technology Assessment. No. < # > . (Prepared by the Tufts Evidence-based Practice Center under Contract No. 290-2007-10055-I.) Agency for Healthcare Research and Quality, Rockville. http://www.cms.gov/Medicare/Coverage/DeterminationProcess/Downloads/id92TA.pdf. Accessed 22 Oct 2014

Schultz NA, Dehlendorff C, Jensen BV et al (2014) MicroRNA biomarkers in whole blood for detection of pancreatic cancer. JAMA 311:392–404

Simon R (2008) Lost in translation: problems and pitfalls in translating laboratory observations to clinical utility. Eur J Cancer 44:2707–2713

Simon R, Radmacher MD, Dobbin K et al (2003) Pitfalls in the use of DNA microarray data for diagnostic and prognostic classification. J Natl Cancer Inst 95:14–18

van de Vijver MJ, He YD, van't Veer LJ et al (2002) A gene-expression signature as a predictor of survival in breast cancer. N Engl J Med 347:1999–2009

van 't Veer LJ, Dai H, van de Vijver MJ et al (2002) Gene expression profiling predicts clinical outcome of breast cancer. Nature 415:530–536

Venet D, Dumont JE, Detours V (2011) Most random gene expression signatures are significantly associated with breast cancer outcome. PLoS Comput Biol 7:e1002240

Wang Z, Gerstein M, Snyder M (2009) RNA-Seq: a revolutionary tool for transcriptomics. Nat Rev Genet 10:57–63

Zujewski JA, Kamin L (2008) Trial assessing individualized options for treatment for breast cancer: the TAILORx trial. Future Oncol 4:603–610

Concluding Remarks and Perspectives

I hope this book has served as an overview of transcriptomics, from the fundamental concepts and methodology, its use in health and human disease, to the interpretation of the results. Following the completion of the human genome, transcriptomics has entered a new realm of research, including research that was traditionally conducted via reductionist approaches, in particular, immunology (Chaussabel and Baldwin 2014). Because the transcriptome of a cell, tissue or organs change according to the strict conditions set forth at any given moment, the study of the transcriptome holds tremendous promise for health and disease research. Even disciplines that rarely adopt genetic approaches, such as physiology or pharmacology, are now examining their model systems from a transcriptomics perspective. What was once an exclusive task for geneticists and molecular biologists, i.e., sequencing the genome, has been largely engaged by transcriptomics in the post-genome era and has opened doors for the entire biomedical research community, including mathematicians, biostatisticians and computer scientists. These fields have contributed to the construction of algorithms, programs for data analysis and improvement of bioinformatics pipelines, without which, we would be unable to interpret the enormous quantities of data being generated by these experiments. And of course, clinicians themselves have seen the potential of transcriptomics in diagnosis and prognosis. Unraveling the code of life no longer involves deciphering three-letter codons (as developed by scientists in the 1960s) or sequencing all 3 billion bp of the human genome (mid-1980–2000), but deciphering the human transcriptome in response to normal physiological conditions as well as different disease states. The mouse (*Mus musculus*) is often used as a model system to answer questions of human interest, which must then be validated in humans. This is another challenge of comparative transcriptomics, which although not explicitly discussed in this book, is currently making its mark in the literature. In fact, the core concept of the central dogma of molecular biology has not changed over the last several decades. Instead, what has happened is a reinterpretation of the data, such that the "dogma" can now become genome → transcriptome → proteome.

Geraldo A. Passos, Ribeirão Preto, August 2014

Chaussabel D, Baldwin N (2014) Democratizing systems immunology with modular transcriptional repertoire analysis. Nature Rev Immunol 14:271–280

© Springer International Publishing Switzerland 2014
G. A. Passos (ed.), *Transcriptomics in Health and Disease*,
DOI 10.1007/978-3-319-11985-4

Index

A

Actionable mutation, 335
Adult stem cells, 99–101, 104, 110
Analytical validity, 328, 336
Animal model, 35, 140, 147, 213, 214, 291,
  296, 317
Antifungal resistance, 252, 257
Arthritis, 144, 145, 212–220
Autoimmune disease, 138, 139, 148, 196,
  211, 223

B

Bioinformatics, 10, 11, 15, 16, 36, 51–56, 60,
  140, 146
Biomarkers, 70–75, 90, 139, 140, 148,
  168, 185, 201, 204, 291–298,
  300, 303

C

Cancer, 23, 74, 184, 329, 331, 332
Carlos chagas, 311
cDNA microarray, 196, 197, 198, 204, 205,
  320
Central dogma, 22
Chagas disease, 311–314, 316, 317, 321
Chagas disease cardiomyopathy, 316
Chemotherapy, 73, 328, 329, 331, 332, 333
Chromosomal location, 216
Circulating biomarkers, 184, 186
Clinical trial, 291, 298, 328, 331, 333, 336
Clinical tuberculosis, 289, 291
Clinical utility, 74, 328, 333, 336, 338
Clinical validity, 328, 329, 333, 336, 337
Complex network measurements, 87,
  89, 128
Computational methods, 11, 50, 58, 88

D

Data mining, 297, 302
Diagnostic biomarkers, 138, 151, 212, 291,
  328
Differential gene expression analysis, 19, 61,
  64
DNA
  damage, 163, 164, 166, 168, 169, 175,
    183, 184
  microarray, 75, 79, 80, 81, 89, 90, 99, 204,
    337
Drug response, 246, 253

E

Experimental tuberculosis, 291
Expressed sequence tags, 22
Expression
  profiling, 6, 7, 10, 74, 96, 204, 212, 338
  signatures, 73, 74, 75, 146, 183, 217, 328,
    329

F

Fungal invasion, 233, 257

G

Gene coexpression networks, 80, 125, 130
Gene expression
  regulation, 96, 141, 257
Genome, 96, 216
  project, 4, 23, 28

H

Heart, 125, 315–323
Heat map, 21
Hierarchical clustering, 21, 199, 204

© Springer International Publishing Switzerland 2014
G. A. Passos (ed.), *Transcriptomics in Health and Disease,*
DOI 10.1007/978-3-319-11985-4

Host-pathogen interaction, 230, 236, 257, 282, 299
Human leucocyte antigen complex, 139
Hybridization signatures, 71, 75, 96

I
Ifn-γ, 146, 199, 320
Immune response, 62, 130, 140, 144, 147, 149, 173, 185, 196, 198, 224, 230, 231, 232, 235–237, 271, 274, 280, 281, 293, 294, 319
Inflammation, 137, 139, 141, 144, 148, 149, 173, 174, 197, 213, 218, 233–236, 313, 317–319

K
Kissing bug, 313

L
Laboratory-developed test, 330–333

M
Major histocompatibility complex, 149, 212, 214, 232
Mesenchymal stem cells, 110, 111
Microarray, 10, 11, 31, 74, 75, 82, 90, 140, 145
MicroRNA, 141, 145, 187, 188
Molecular biology, 22, 50, 111, 196, 230, 317
Multiple sclerosis, 138, 148, 150, 198
Mycobacterium tuberculosis, 289, 302
Myocarditis, 317, 319, 321

N
Network visualization analysis, 85
Next generation sequencing, 15, 49, 51, 72, 96, 187, 224, 230

O
Organ gene expression, 109
Osteoblast differentiation, 115, 117
Oxidative stress, 100, 124, 131, 162, 164, 166-169, 171–173, 175, 184, 238, 239, 243, 245, 269

P
Paracoccidioides, 238, 266-272, 282
Paracoccidioidomycosis, 233, 266, 272
Peripheral blood mononuclear cells, 140, 145, 167, 197, 272, 297

Polysome profile, 99, 100, 102, 104
Posttranscriptional regulation, 97, 100, 104, 182
Predictive, 139, 205, 327-331, 333, 336
Profiling, 6-8, 10, 23, 73, 74, 96, 99, 100, 102–104, 114, 167, 174, 200–202, 204, 205, 212, 230, 240–243, 245, 257, 299, 318–320, 322, 327, 331, 332, 336–338
Prognostic, 73, 327, 328-333
Prophylactic biomarkers , 291, 292
Proteome, 4, 143
Protozoan infection, 266

R
Rheumatoid arthritis, 145, 146, 196, 211, 213, 214, 217, 223
Ribosome profile, 102, 104
RNA-sequencing, 60, 99

S
Selected mice , 214, 217
Sequencing, 15, 61, 335
Signature, 3, 73, 145, 146, 197, 198, 200, 205, 232, 292, 294, 297, 298, 328, 336, 337
Spondyloarthritis, 196, 203
Systemic lupus erythematosus, 138, 144, 198
Systems biology, 212, 236, 302

T
Target prediction, 169, 187, 188
Thymus gene coexpression networks , 123
Network community detection, 125, 128
Transcription, 25, 28, 32, 33, 97, 110, 111, 146, 147, 170, 174, 184, 188, 204, 217, 243, 253, 272, 296, 300
Transcription profiling, 167, 230, 240–243, 245, 257
Transcriptome, 4, 22, 28, 54, 61, 216, 230, 245, 250
Translation, 97, 103
Trisomy 21 gene dysregulation, 123
Trypanosoma cruzi, 312
Tumour, 328, 329, 332–335, 338
Type 1 diabetes mellitus, 138–141, 164, 165
Type 2 diabetes mellitus, 164, 169, 170

Printed by Printforce, the Netherlands